Supervision in the Hospitality Industry

Eighth Edition

Supervision in the Hospitality Industry

John R. Walker, DBA, FMP, CHA

McKibbon Professor of Hotel and
Restaurant Management and Fulbright Senior Specialist

University of South Florida, Sarasota-Manatee

Jack E. Miller

Library of Congress Cataloging-in-Publication Data

Names: Miller, Jack E., author. | Walker, John R., author.
Title: Supervision in the hospitality industry / John R. Walker, DBA, FMP,
    CHA, McKibbon Professor of Hotel and Restaurant Management and Fulbright
    Senior Specialist, University of South Florida, Sarasota-Manatee, Jack E. Miller.
Description: Eighth edition. | Hoboken, New Jersey : John Wiley & Sons, Inc.,
    [2017] | Earlier editions by Jack E. Miller and various other authors. |
    Includes bibliographical references and index.
Identifiers: LCCN 2015043071 (print) | LCCN 2015043639 (ebook) | ISBN
    9781119148463 (cloth : acid-free paper) | ISBN 9781119191995 (pdf) | ISBN
    9781119191971 (epub)
Subjects: LCSH: Hospitality industry—Personnel management.
Classification: LCC TX911.3.P4 M55 2017 (print) | LCC TX911.3.P4 (ebook) |
    DDC 647.9068/3—dc23
LC record available at http://lccn.loc.gov/2015043071

ISBN: 978-1-119-14846-3

Printed in the United States of America

10   9   8   7   6   5   4   3   2   1

Dedication

To the late Jack and Anita Miller,

A couple who loved and made a major contribution to hospitality education.

Jack was a scholar, a superb educator, and a gentleman.

and

To you, the professors and students who are dedicating yourselves to the future of hospitality and hospitality management.

Contents

Preface

Ask any hospitality manager what his or her greatest challenge is and the response will likely be "finding and keeping good employees." For many recent college graduates, supervision, whether on the providing or receiving end, is an added challenge because they have little or no experience with it. Likewise, how associates are supervised and lead is critical to the success of any organization and *Supervision in the Hospitality Industry, Eighth Edition* will help you prepare for a supervisory leadership role with associates in the hospitality industry.

A primer for the leadership and management of people in the hospitality industry, *Supervision* is about supervising and leading the people who cook, serve, tend bar, check guests in and out, carry bags, clean rooms, mop floors—the people on whom success or failure of every hospitality enterprise depends. It is a book about first-line supervision, written especially for the beginning leader, for the new supervisor promoted from an hourly job, and for students planning a career in the hospitality industry. *Supervision* is unique in that it does not solely rely on the supervisor's point of view; instead, it considers the viewpoints of all levels of associates to create an informed picture of management and supervision in the hospitality industry.

Hospitality is an industry heavily dependent on its human resources but plagued with people problems—its demands, its people, its pace, its long hours, the typical attitudes and habits of managers and workers, and the special problems of time pressure, of the unpredictable, of everything happening at once. *Supervision* gives you the tools you need to recognize and solve these problems.

Supervision is unique in focusing directly on leading human resources, especially front-line associates, and applying the wisdom of leadership theory and experience to the hard realities of the hospitality industry in down-to-earth terms. It is practical, concrete, and results-oriented. Real settings and real challenges are used to present the principles of good leadership in supervision. The primary objective remains to provide the reader with a basic yet comprehensive knowledge about the different elements of the supervisor's job, and a basic awareness and appreciation of the skills, attitudes, and abilities needed to lead associates successfully. A firm grasp of these basics can provide a solid foundation for increasing skills and knowledge on the job and ultimately for achieving success through leading associates at any level.

Yet, basics are not necessarily simple. There are no sets of exhaustive rules; rather, the concepts, theories, principles, and real-world applications behind good supervisory practices are presented in order to give depth to understanding. Terms are defined clearly and explained fully; then are shown how they apply, using examples and incidents from industry. In sum, *Supervision* has been written to be read, understood, absorbed, and put to work—a how-to book that provides the understanding necessary to adapt and use in one's own circumstances in one's own way.

For the Student

If you are a student, you will find in *Supervision* what you need to approach the realities you will meet in your first supervisory position. Even more important, you can gain the knowledge and insight that will help you to grow as a supervisor, to develop the skills and personal qualities you need, and to work out your own supervision style. We suggest you begin by assuming that you are a supervisor. Use the incidents, discussions, and your own creative ideas for solving problems and getting results and you will find yourself well prepared when you find yourself in a supervisory position

For the Instructor

Instructors will find *Supervision* not only satisfying in content but also easy to use and appealing to your students. It assumes no specific knowledge other than a general familiarity with hospitality, foodservice, or lodging operations. It can be used at any course level in a hospitality or culinary arts program after the first semester or the first year. It is also suitable for seminars and continuing

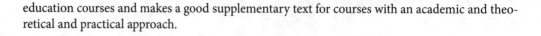

education courses and makes a good supplementary text for courses with an academic and theoretical and practical approach.

For the Supervisor

If you are in management, you will find this book useful in helping to develop your supervisory associates, especially those who have been promoted from hourly positions or are first-time supervisors from outside the industry. The material is solid, the scene familiar, and the presentation clear and easy to follow. It will help your supervisors to understand and develop the skills and abilities to work effectively with hourly employees. It will help you to bring supervisors to the level of productivity and the manner of performance that you want in your organization. In developing supervisors as the key people in your operations, you will help your enterprise serve its guests well and become more profitable.

Organization of the Text

This **Eighth Edition** of **Supervision** covers four key areas of hospitality supervision:

1. **Supervision:** An overview of the role of the supervisor and the importance of supervisor communication.
2. **Equal Opportunity, Diversity, Recruitment, and Performance Standards:** An in-depth explanation of equal opportunity laws and the importance of diversity in the workplace, a detailed section on recruitment best practices, and helpful information on maximizing performance effectiveness.
3. **Creating a Positive Work Environment:** Contains valuable information on how to motivate employees, develop teams, and successfully manage conflict in the workplace.
4. **Maintaining High Performance:** Describes how to excel as a supervisor through the use of discipline, decision-making tactics, and delegation.

New to This Edition

Supervision has been revised and updated to help supervisors and leaders of hospitality human resources meet the challenges and demands of the world's largest industry grouping, to be leaders, to possess excellent supervisory skills, to be highly productive and flexible. New and revised to this edition includes:

- **Learning objectives** have been inserted directly into the chapters to facilitate and monitor the learning process.
- An improved organizational structure, with special attention paid to **chunking large sections of content** into more manageable, easier-to-remember sections.
- Updated coverage on **delegation**, which includes sections on supervisors delegating and common mistakes in delegating.
- New and updated **case studies** have been added to the end of each chapter.
- Coverage of **diversity in the workplace** has been revised and expanded in Chapter 5.
- The opportunities and challenges of **supervising a restaurant shift** have been added to Chapter 7.
- Chapter 6 has the addition of brand new **social media recruiting and hiring** suggestions, detailing the use of LinkedIn and Instagram in the workplace.
- Chapter 7 contains new techniques for **evaluating on-the-job performance** and one-on-one performance management.

- Chapter 8 has contains new information on the **leadership behaviors**.
- Updated **team-building** techniques have been added to Chapter 9.
- New sections on teaching methods for **training and developing a job-training program** have been added to Chapter 10.
- Chapter 11 has updated information for resolving **conflict in the workplace**.
- New **industry profiles** have been added throughout the chapters.

Aids to Facilitate Learning

The writing is in a clear engaging conversational style with numerous industry examples for ease of understanding topics and concepts.

Following are the pedagogical features found within each chapter:

- The **chapter openings** help to structure assignments and set learning goals by describing a supervisory situation and listing the *chapter objectives*.
- **New photographs** enliven the text, and updated diagrams, flow charts, and sample materials provide focal points for discussion.
- **Industry profiles** allow supervisors and leaders in the hospitality industry to give their perspective on supervision and leadership issues.
- **Case studies** allow students to analyze real-life scenarios from various segments of the hospitality industry
- **Key points** summarize the important concepts in the chapter.
- **Key terms** are highlighted in the text, reemphasized in the end-of-chapter list *Key Terms*, and assembled in the *Glossary* for reference and review.
- **Review questions** enable students to reinforce mastery of the materials presented in the text and likely improve their test scores.
- **Activities** allow an opportunity to practice human resources leadership.
- **Applications** can be used to test knowledge, spark interest, bring out opposing views and different approaches, and involve students in typical supervisory problems and situations.

Additional Resources

To aid students in retaining and mastering hospitality human resources, there is a **Study Guide** (ISBN 978-1-119-14847-0), which includes learning objective reviews, study notes and chapter outlines, key terms and concept reviews, and quizzing exercises.

An **Instructor's Manual** and a set of **PowerPoint slides** to accompany the textbook are available to qualified adopters upon request from the publisher, and are also available for download at www.wiley.com/college/walker.

The **Test Bank** has been specifically formatted for **Respondus**, an easy-to-use software program for creating and managing exams that can be printed to paper or published directly to Blackboard, WebCT, Desire2Learn, eCollege, ANGEL, and other eLearning systems. Instructors who adopt *Supervision* can download the Test Bank for free. Additional Wiley resources also can be uploaded into your LMS coursed at no charge.

A **Book Companion Website** (www.wiley.com/college/walker) provides readers with additional resources as well as enabling instructors to download the electronic files for the **Instructor's Manual**, **PowerPoint slides**, and **Test Bank**.

Interactive Case Studies allow readers to view the cases presented in each chapter in virtual scenarios and answer questions pertaining to the cases.

WileyPLUS Learning Space

A place where students can define their strengths and nurture their skills, WileyPLUS Learning Space transforms course content into an online learning community. WileyPLUS Learning Space invites students to experience learning activities, work through self-assessment, ask questions, and share insights. As students interact with the course content, each other, and their instructor, WileyPLUS Learning Space creates a personalized study guide for each student. Through collaboration, students make deeper connections to the subject matter and feel part of a community.

Through a flexible course design, instructors can quickly organize learning activities, manage student collaboration, and customize your course—having full control over content as well as the amount of interactivity between students.

WileyPLUS Learning Space lets the instructor:

- Assign activities and add your own materials
- Guide your students through what's important in the interactive e-textbook by easily assigning specific content
- Set up and monitor group learning
- Assess student engagement
- Gain immediate insights to help inform teaching

Defining a clear path to action, the visual reports in WileyPLUS Learning Space help both you and your students gauge problem areas and act on what's most important.

Acknowledgments

The authors greatly acknowledge the following reviewers of this and previous editions: Susan Annen, Johnson County Community College; Leslie Bilderback and Diana Altieri, Southern California School of Culinary Arts; Pat J. Bottiglieri, The Culinary Institute of America, Kathleen Fervan, Augusta State Technical College; Alisa Gaylon, The Art Institute of Pittsburg Online Division; Ava Gritzuk, Central Piedmont Community College; Chad Gruhl, Metropolitan State University of Denver; G. Michael Harris, Jr., Mohave Community College; Brenda Hodgins, Red Deer College; Anthony McPhee and Theresa A. White, Cooking and Hospitality Institute of Chicago; Robert A. Palmer and William B. Martin, California State Polytechnic University, Pomona; Greg Quintard, Nash Community College; Kyle Richardson, Joliet Junior College; Andrew Rosen, Johnson & Wales University—Miami; Gary Schwartz, Asheville Buncombe Technical Community College; Diane Watson, The Art Institute of Atlanta; Matt Williams, Texas Culinary Academy; and James Zielinski, College of DuPage.

We are very grateful to Holly Loftus for her invaluable assistance in various aspects of manuscript preparation and for her contributions to Chapters 7, 11, and 12. To my colleague Dr. Chad Gruhl, a sincere thank you for all your excellent comments, observations, and contributions that helped shape the chapter on equal opportunity laws and diversity. Thanks also to the industry reviewers: Carol Newberry, V.P of People at First Watch Restaurants; Bob Haber, Director of Human Resources at the Grand Hyatt Tampa Bay; Laurie Bennett, Executive Director of Human Resources at Sarasota Memorial Hospital; Debbie Rios, Director of Human Resources at the Colony Resort, Sarasota; Rachana Dinkar, Inclusion Programs Director, OSI Restaurant Partners; Bonnie Smith and Carolyn Dyson, University of South Florida; Shirley Ruckl, Director of Human Resources at the Longboat Key Club & Resort, Sarasota; Jo Askren, Director of the Culinary Innovation Laboratory and Chef Gary Colpitts, CEC, University of South Florida Sarasota-Manatee Culinary Innovation Laboratory and Suzette Marquette, CCC of Manatee Technical Institute, Manatee County, Florida; and particularly Gerard Violette of Sarasota Memorial Hospital; Michelle Tarullo, Director of Human Resources at the Hyatt Sarasota; Charlotte Jordan, former Director of Human Resources at the Ritz-Carlton Sarasota; and industry professionals who allowed us to profile them. A special thanks to the Hilton Garden Inn, the Longboat Key Club & Resort, Sodexho, and Sarasota Memorial Hospital for kindly allowing photos to be taken.

Special thanks to my wonderful editor Gabrielle Corrado for all her hard work on the manuscript and to JoAnna Turtletaub for overseeing the project. You were both a pleasure to work with.

WileyPLUS Learning Space

An easy way to help your students learn, collaborate, and grow.

Diagnose Early

Educators assess the real-time proficiency of each student to inform teaching decisions. Students always know what they need to work on.

Facilitate Engagement

Educators can quickly organize learning activities, manage student collaboration, and customize their course. Students can collaborate and have meaningful discussions on concepts they are learning.

Measure Outcomes

With visual reports, it's easy for both educators and students to gauge problem areas and act on what's most important.

Instructor Benefits

- Assign activities and add your own materials
- Guide students through what's important in the interactive e-textbook by easily assigning specific content
- Set up and monitor collaborative learning groups
- Assess learner engagement
- Gain immediate insights to help inform teaching

Student Benefits

- Instantly know what you need to work on
- Create a personal study plan
- Assess progress along the way
- Participate in class discussions
- Remember what you have learned because you have made deeper connections to the content

Chapter 1

The Supervisor as Manager

Y ou are now a boss, or soon will be. Being a new supervisor is exciting; there will be challenges, opportunities, and rewards. Your company has invested its trust in you and has expectations of your performance. But how do you feel?

Well, you wouldn't be alone if you felt some apprehension because you are responsible not only for your own work but also for the work of others. The team members that you will be supervising will probably take a wait-and-see attitude until they get to know you. A good approach for a new supervisor is to talk to the previous supervisor and your boss, since they best know the details of the job and the people you will be supervising. Of course there is a caution: What if the previous supervisor was incompetent or had biases? That's why you check carefully with your new boss to get his or her perspective.

Another wise move is to review the files of the employees you will be supervising. By looking over their files and evaluations, you should be able to gain a better understanding of your new employees. The best way to start your first day as a new supervisor is to have your boss introduce you in a formal capacity, followed by a chance for informal interaction.

Ever wonder about the impact that supervisors can have on the success of a hospitality company? Here is an example: On Restaurant Row in one city, one family restaurant has had 12 different busboys in two months. In the restaurant next door, the food is superb one week and terrible the next.

Across town, students at the local community college would rather eat at the local fast-food restaurants than in the school's cafeteria, where the pizza and burgers leave much to be desired. On the outskirts of the city, students at the state university rave about the quality of the food and the tremendous choices they have. Many of the students look for jobs working for the university foodservice.

In many of the city's hotels, the employee turnover rate is high. Every seven days, they turn over thousands of employees in the industry. They don't have a "labor" crisis. They have a turnover crisis.[1] Service is poor and guests complain—but then, that's just part of the game, isn't it? Yet several hotels in town have few staffing problems and happy guests.

Throughout the city, a common cry in the hospitality industry is that you just can't get good people these days. People don't work hard the way they used to, they don't do what you expect them to, they come late and leave early or don't show up at all, they are sullen and rude—the complaints go on and on.

Is this true? If it is true, what about those establishments where things run smoothly? Can it be that the way in which the workers are managed has something to do with the presence or absence of problems? You bet it does! And that is what this book is all about. In this chapter, we explore the management aspect of the supervisor's job. After completion of this chapter you should be able to:

- Explain the supervisor's role in decision making, problem solving, and delegation of duties.
- Identify the obligations and responsibilities of a supervisor or executive chef.
- Describe the functions of management.
- Compare and contrast the major theories of people management as they relate to hospitality employees.
- List examples of technical, human, and conceptual skills used by hospitality supervisors.
- List three to five best practices for new supervisors.

The Supervisor's Role

LEARNING OBJECTIVE: Explain the supervisor's role in decision-making, problem solving, and delegation of duties.

In the hospitality industry, almost everything depends on the physical labor of many hourly (or nonmanagerial) workers: people who cook, serve tables, mix drinks, wash dishes, check guests in and out, clean rooms, carry bags, mop floors. Few industries are as dependent for success on the performance of **hourly workers** as the hospitality industry. These employees make the products and they please the guests—or drive them away!

How well these employees produce and serve depends largely on how well they are supervised. If they are not supervised well, the product or the service suffers and the operation is in trouble. It is the people who supervise these employees who hold the keys to the success of the operation.

A **supervisor** is any person who manages employees who make products and/or perform services. A supervisor is responsible for the output of the people supervised—for the quality and quantity of the products and services. A supervisor is also responsible for meeting the needs of employees and the production of goods and services only by motivating and stimulating employees to do their jobs properly. Today's employees are different than they were 10 to 20 years ago; they no longer give their allegiance to the supervisor automatically in exchange for a paycheck. Instead, they give their supervisor the right to lead them.

Usually, a supervisor is the manager of a unit or department of an enterprise and is responsible for the work of that unit or department. In large enterprises, there are many levels of supervision, with the people at the top responsible for the work of the managers who report to them, who, in turn, are responsible for the performance of those they supervise, and so on, down to the frontline supervisor who manages the hourly workers. The **first-line supervisor** and unit manager are the primary focus of this book. Figure 1.1 shows the levels of employees in a large company.

Organizational charts for a large hotel and a large restaurant are shown in Figures 1.2 and 1.3. An **organizational chart** shows the relationship among and within departments. **Line functions** (associates directly involved in producing goods and services) and staff functions (the advisers) are spelled out. The human resource and training departments are examples of staff who advise line departments, such as the food and beverage department, on matters including hiring, disciplining, and training employees.

Using the organization chart, you can also see the various levels of management, with authority and responsibility handed down from the top, level by level. **Authority** can be defined as the right and power to make the necessary decisions and take the necessary actions to get the job done. **Responsibility** refers to the obligation that a person has to carry out certain duties and activities. First-line supervisors represent the lowest level of authority and responsibility, and hourly workers report to them.

Many supervisors—station cooks, for example—also do some of the work of their departments alongside the workers they supervise. Thus, they are typically in close daily contact with the people they supervise and might even at times be working at the same tasks. They are seldom isolated in a remote office but are right in the middle of the action. They are known as **working supervisors** (but as we see later in the text, these supervisors may not qualify as exempt employees).

Each supervisor's job is described in terms of a job title and the scope of the work required, rather than in terms of the people to be supervised. An executive chef, for example, is responsible for all kitchen food production. An assistant executive housekeeper in a hotel is responsible for getting the guests' rooms cleaned and made up. A food and nutrition supervisor in a hospital might be responsible for overseeing the service of patient meals. A restaurant manager or a unit manager in a food chain is responsible for the entire operation. Thus, the focus is placed on the work rather than on the employees. But because the work is done by people, *supervision is the major part of the job.*

---

hourly workers
Employees paid on an hourly basis who are covered by federal and state wage and hour laws and are therefore guaranteed a minimum wage.

supervisor
A person who leads and manages employees who are performing services or making products.

first-line supervisor
A supervisor who leads and manages hourly paid employees.

organizational chart
A diagram of a company's organization showing levels of management and lines by which authority and responsibility are transmitted.

line functions
The employees directly involved in producing goods and services.

authority
Possessing the rights and powers to make the decisions and take the requisite actions to get the job done.

responsibility
The duties and activities assigned to a given job or person, along with an obligation to carry them out.

working supervisors
Supervisors who take part in the work itself in addition to supervising.

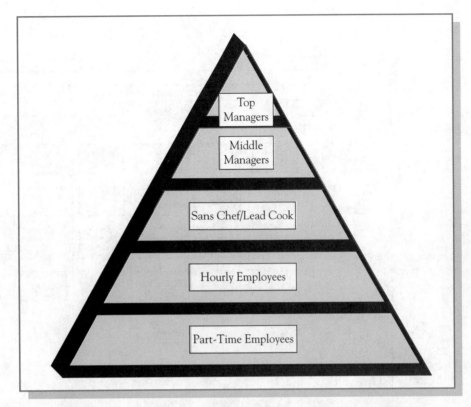

FIGURE 1.1: The levels of employees in a large company.

exempt employees
Employees, typically managerial, who are not covered by the wage and hour laws and therefore do not earn overtime pay.

nonexempt employees
Employees who are paid by the hour and are not exempt from federal and state wage and hour laws. Also called *hourly employees*.

As a supervisor, you depend for your own success on the work of others, and you will be measured by their output and their performance. *You will be successful in your own job only to the degree that your team members allow you to be,* and this will depend on how you manage them. This will become clearer as you explore this book.

Other organizational terms with which you need to become familiar include **exempt employees** and **nonexempt employees**. Hourly employees are considered nonexempt employees because they are not exempt from federal and state wage and hour laws. In other words, they are covered by these laws and are therefore guaranteed a minimum wage and overtime pay after working 40 hours in a workweek. Supervisors are considered exempt employees; they are not covered by the wage and hour laws and therefore do not earn overtime pay when certain conditions are met: when they spend 50 percent or more of their time managing, when they supervise two or more employees, and under federal law when they are paid $455 or more per 40-hour week (or more if the state imposes a higher standard). More information on exempt and nonexempt employees can be found on the Department of Labor's website at www.dol.gov/elaws/esa/flsa/screen75.asp.

Obligations and Responsibilities of a Supervisor/Executive Chef

LEARNING OBJECTIVE: Identify the obligations and responsibilities of a supervisor or executive chef.

When you begin to supervise the work of other people, you cross a line that separates you from the hourly workers—you step over to the management side. In any work situation, there are two points of view: the hourly workers' point of view and management's point of view.

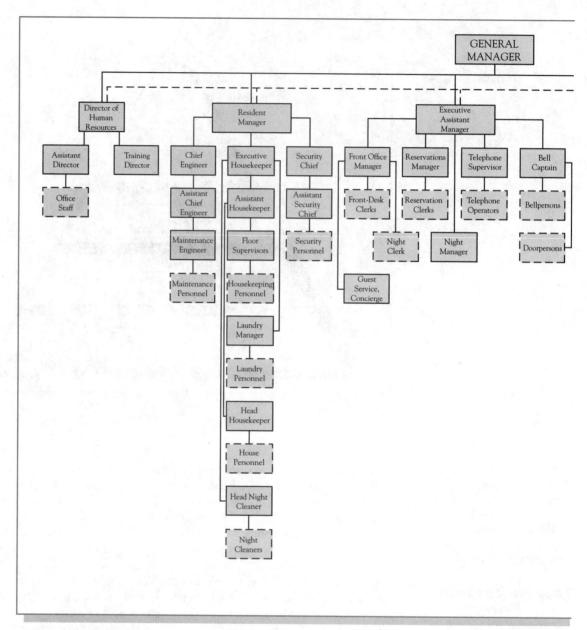

FIGURE 1.2: Organization chart for a large hotel. Boxes with dashed lines indicate hourly workers. Dashed reporting lines indicate staff (advisory) positions.

The line between them is clear-cut; there are no fuzzy edges, no shades of gray. When you become a supervisor, your responsibilities are management responsibilities, and you cannot carry them out successfully unless you maintain a manager's point of view. Now, you will be a part of setting the standards rather than seeking to attain performance goals set by others. You will be held accountable for achieving department goals and keeping your team motivated and productive.

In order to maintain a reputation of excellence, you should realize the importance of being responsible for 10 things:

1. Achieve or exceed the expected results, on time and on budget: planning = determining priorities; organizing = scheduling; motivating = creating a positive work environment; controlling = monitoring and taking corrective action if deviations are outside acceptable limits.

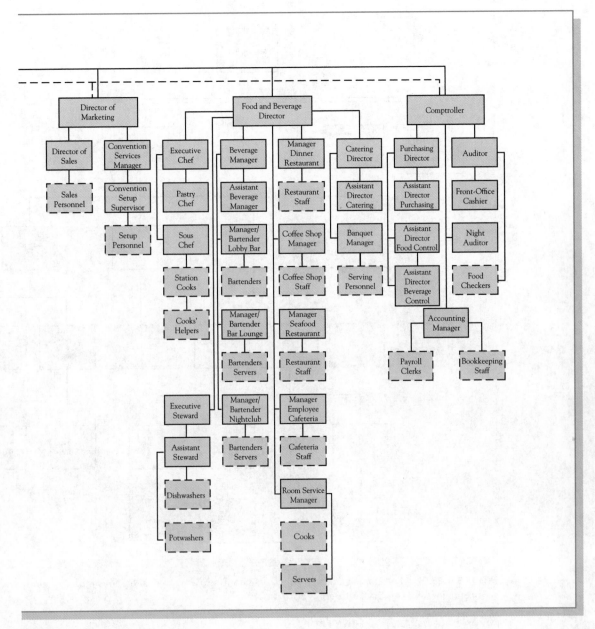

FIGURE 1.2: (continued)

2. Communicate effectively.

3. Build a winning team.

4. "Walk your talk" as a leader, setting a good example. Plus, you should be able to do the work of those you supervise.

5. Create a positive work environment.

6. Motivate your team.

7. Work efficiently and effectively with your manager and peers.

8. Coach and mentor your team.

9. Get the resources necessary for your team to do the job.

10. Treat all team members fairly and equally.

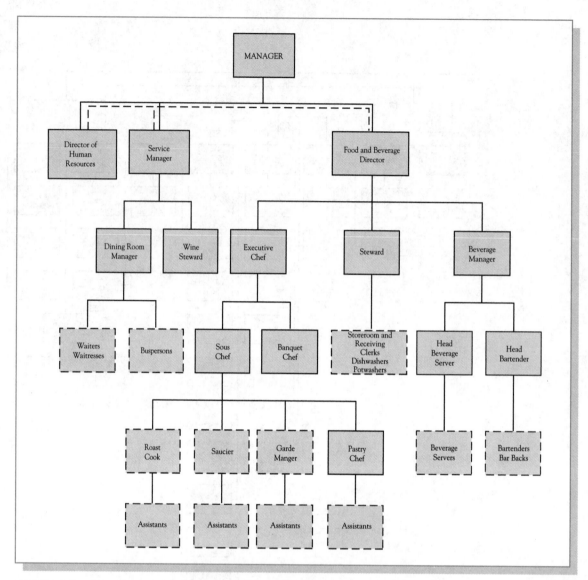

FIGURE 1.3:　Organization chart for a large restaurant. Boxes with dashed lines indicate hourly workers. Dashed reporting lines indicate staff (advisory) positions.

Now is a good time to reflect on your past supervisors and see how you would describe them. What made them good or bad supervisors? You could take this supervisor assessment to check on your supervisory skills. Figure 1.4 illustrates a supervisor's assessment.

No one is more responsible for a business's success than the manager. It takes a lot of savvy to manage a bar or restaurant well, and not just anyone can pull it off. There are some qualities that make up the all-pro manager. For example, all-pro managers not only have a working knowledge of the business they operate but also possess a sound grasp of business in general. A good manager knows his or her market, knows the competition and what it is doing, and responds accordingly. Other all-pro manager traits include a desire to lead, maturity/stability, good money sense, possession of street smarts, and legal knowledge.

Supervisor's Assessment

	1 Great Need for Improvement	2 Need for Improvement	3 Acceptable	4 Good	5 Excellent
1. You know the company's mission and department goals.					
2. You know the tactics to meet goals.					
3. You would include associate input into formulating goals.					
4. You know the policies and procedures.					
5. You are fair, consistent, and treat everyone the same.					
6. You give praise for job well done.					
7. Your associates clearly understand what is expected of them.					
8. You are a good example.					
9. You have enthusiasm for your work.					
10. You are a good listener.					
11. You are a good coach / trainer.					
12. You respect your associates.					
13. You are a good communicator.					
14. You are a good decision maker.					
15. You are passionate about the success of your team / department.					
16. You are a good delegator.					
17. Your team gets the job done on time and on budget.					
18. You are a good motivator.					
19. You give positive reinforcement.					
20. You represent your associates well with management.					

FIGURE 1.4: For best results, compare the supervisor's score with one from the boss and an established score from the associates who are being supervised.

❋ THE SUPERVISOR IN THE MIDDLE

Now that you are a supervisor yourself, you've got bosses all around you. Your philosophy of engagement requires you to take the position that you work for your employees—always making sure they have what they need to do their best work and caring for them personally.[2]

As a hospitality supervisor, you have obligations to the owners, the guests, and the employees you supervise, which puts you right in the middle of the action (Figure 1.5). To your employees, you represent management: authority, direction, discipline, time off, more money, and advancement. As

Supervisors have obligations and responsibilities to guests, employees, and employers.
Robert Kneschke/Shutterstock

a supervisor, you have the day-to-day responsibility for what goes on in the workplace and for ensuring that the work is carried out in such a way that no one's security, safety, or health is jeopardized. You may also play a critical role in supporting a drug-free workplace program and enforcing the policy. However, you are not expected to perform the role of police officer or counselor.[3]

To the owners and your superiors in management, you are the link with the workers and the work to be done; you represent productivity, food cost, labor cost, quality control, and guest service; you also represent your people and their needs and desires.

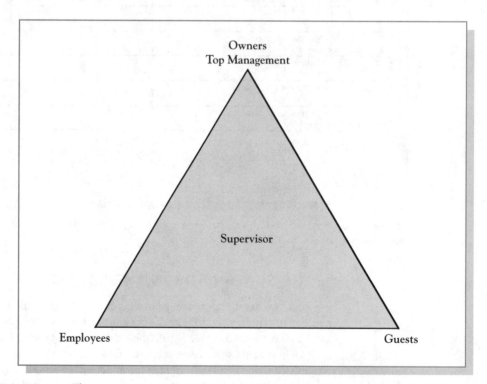

FIGURE 1.5: The supervisor is right in the middle of everything and everybody—the link between top management and employees and guests.

CORPORATE HIERARCHY

When you become a supervisor, your responsibilities are management responsibilities, and you cannot carry them out successfully unless you maintain a manager's point of view.
maximummmmum/Shutterstock

To the guests, your output and your employees represent the enterprise: If the food is good, it's a good restaurant; if the doorperson and the desk clerk and the server in the hotel restaurant provide high-quality service, it's a good hotel. No matter how modest your area of responsibility, it is a tough assignment.

Many new supervisors are promoted from hourly jobs and suddenly find themselves supervising people with whom they have worked side by side for years. You worked and socialized together, and you now find yourself on the other side of that line between management and workers. Now you might be carrying out policies you used to complain about. You might have to work your entire crew on Christmas Day. You might have to discipline your best friend.

It is lonely on that side of the line, and the temptation is great to slip back to your buddies, to the old attitudes, the old point of view. We call this **boomerang management**—going back to where you came from. It doesn't work. You've got to maintain a manager's point of view; you've got to stay in charge.

There is no compromise. You can empathize with your workers; you can listen and understand. But your decisions must be made from management's point of view, meaning there is a code of conduct—ethics, policies, and procedures—to abide by when making decisions. Your employer expects it, and your people expect it. If you try to manage from your workers' point of view, they will take advantage of you. They really want you to manage.

> **boomerang management**
> Reverting from management's point of view to the employees' point of view.

❋ OBLIGATION TO OWNERS

Your primary obligation to the owners is to make their enterprise profitable. They have taken the risk of investing their money, and they expect a reasonable return on that investment. The major part of your responsibility to them is to run your part of their business to produce that return. That is what they are interested in; that is what they have hired you to do. Remember, you as a supervisor are to run the business as if it were your own.

They also want you to run things *their way*. If they tell you how they want it done, you have an obligation to do it that way, even if you see better ways to do it. They are paying you to do it their way, and you have an obligation to do anything they require that is morally and legally correct.

Suppose that the owners have a system for everything. They don't want you to change anything; they want you to oversee their system. Suppose you don't agree with it. Suppose you think they don't put enough Asian slaw on each plate; you think the guest should have more, and so does the guest. Here, portion control equals cost management using the proper-sized ladle, spoon, or portion scoop. Suppose you see a better way of doing something. You don't take it upon yourself to make a change. You go back to the owner or to your supervisor and you explain your idea and why it would be better. The two of you must agree on what changes, if any, you are going to make.

Suppose the people who hire you don't tell you how they want things done. This happens all too often in the restaurant business; they hire you and put you to work without telling you what to do. They may have definite expectations of you, but they do not verbalize at least 50 percent of what they ought to; they expect you to know what they want. You must find these things out for yourself—ask questions, get things straight.

What are the guidelines and procedures? What authority do you have? Where, if anywhere, do you have a free hand? Since you must manage as they want you to manage, it is your obligation to find out what they want and be sensitive to their expectations. Because of your daily interactions with guests and employees, you also have an obligation to communicate their desires to the owners.

Suppose that you are hired to run a hospital kitchen. Here the emphasis is not on profit; your obligation to follow the system is absolute because the health of the patients is at stake. Patient health is the purpose of the hospital, and food is a basic element in patient care. Every recipe must be followed to the letter and to the quarter ounce. Every grain of salt is important. Every sanitation procedure is critical. You must not change a thing without having the authority to do so.

❋ OBLIGATIONS TO GUESTS

Your second obligation as a supervisor is to the guests. They are the reason a hotel or restaurant exists, and they are the source of its profits. They come to your enterprise by choice. If they are treated well, they might continue to come. If they are not, you will probably never see them again. The importance of customer service seems obvious, yet poor service is all too common, and it is one of the big reasons for failure in the hospitality industry. Most of the people who never come back are responding to poor service or to the fact that an hourly employee was insensitive to their needs.

Consider the following scenario. You are a guest arriving at a hotel after a long and tiring trip, and you tell the desk clerk you have a reservation. She runs your name through her computer and says, "No, you don't have a reservation." How do you feel? You are frustrated and angry because you know you made a reservation. "Well," you ask, "do you have any rooms?" "Yes," she says, "we have rooms, but you don't have a reservation."

Now not only are you frustrated and angry but also you are beginning to feel rejected. "Well," you say, "may I have a room?" She lets you (*lets you*) have a room. As you head to the elevator with your bags in a huff, she says, as though to a bad child, "But you didn't have a reservation!" Will you stay at that hotel again? Not if you can help it.

That desk clerk obviously had not been trained by her supervisor in guest relations. Furthermore, chances are good that she picked up her attitude and behavior from the supervisor. It is very easy, when your mind is on a million other things, to blame the guests for being demanding and unreasonable, and you often feel that if it weren't for the guests, you could get twice as much work done in half the time. You forget that if it weren't for the guests, you would have no work to do. As a manager you must fulfill their needs and desires, and that also means training your people to assume this obligation. Never forget that your employee is a direct reflection of *you*.

In a hotel or restaurant, guests usually encounter only hourly employees. Hotel guests see the desk clerk, the bellperson, the server in the restaurant, and the guest services agent. Restaurant patrons see the host or hostess, servers, and perhaps a bartender or cashier. These hourly employees represent you, they represent the management, and they convey the image of the entire operation.

As a supervisor, you have an obligation to guests to see that your employees are delivering on the promises of service and product that you offer—giving the guests what they came for. And you should be visible in person—guests like to feel that the supervisor or manager cares, and your employees work better when you are present and involved in the action.

In a hospital or nursing facility kitchen, you have an obligation to the patients to see that they get the kind of food the doctor ordered and that it is not only nourishing and germ-free using safety and sanitation procedures (such as HACCP [Hazard Analysis of Critical Control Point] procedures) in place. But the food must also look visually appealing and taste good. The food cannot help the patients recover if they don't eat it.

For many people in hospitals, food is the most important part of their day. They lie in bed with nothing to do, and breakfast, lunch, and dinner become major events. You have an obligation to those patients to speed their recovery by giving your best effort to making their meals a pleasure.

As a supervisor in a school cafeteria, you have an obligation to the students. As a supervisor in a U.S. Army or Navy kitchen, you have an obligation to your country. Wherever you work, you have an obligation to the consumer of the product your workers prepare and to the user of the services your people provide.

❉ OBLIGATIONS TO EMPLOYEES

Your third obligation as a supervisor is to the people you supervise. It is up to you to provide these employees with an environment in which they can be productive for you. This is something you need because you are directly dependent on them to make you successful. You certainly can't do all the work yourself.

The most important value for most employees is the way the boss treats them. They want to be recognized as individuals, listened to, told clearly what the boss expects of them, and why. If they are going to be really productive for you, they want a climate of acceptance, approval, open communication, fairness, and belonging. With most employees today, the old hard-line authoritarian approach simply does not work. You owe it to your employees and to yourself to create a work climate that makes them willing to give you their best.

A poor **work climate** can cause high labor turnover, low productivity, and poor quality control and can ultimately result in fewer customers—problems that are all too common in restaurants, hotels, and hospitals. It is easy for employers to blame these problems on "the kind of employees we have today" and to look at these employees as a cross they have to bear.

There is an element of truth here: Hotels and restaurants are dependent on large numbers of people to fill low-wage, entry-level jobs that have little interest and no perceived future. Washing pots, busing tables, dishing out the same food every day from the same steam table, lifting heavy bags, mopping dirty floors, cleaning restrooms, straightening up messy rooms left by unheeding customers every single day can become very tiresome. Workers take these jobs either because no special skill, ability, or experience is required, or because nothing else is available.

Some of these people consider the work demeaning. Even though they are doing demanding work that is absolutely essential to the operation, management often looks down on them. They are frequently taken for granted, ignored, or spoken to only when scolded. Given the nature of the work and the attitudes of management and sometimes of other employees, it is no wonder that turnover is high.

Another level of hourly employee is the skilled or semiskilled: the front desk clerk, the cashier, the bartender, the cook, the waiter and waitress. These jobs are more appealing—the money is better, and there is sometimes a chance for advancement. Yet here, too, you often find temporary employees—students, moonlighters, people who cannot find anything in their own fields. They are working there *until* they have enough money for college, they get married and move away, they find a better-paying job—that is, *until something better comes along*. Many employers assume that their employees will not stay long, and most of them do not. Even the best supervisors cannot keep such workers indefinitely. But good supervisors can keep them longer than they might otherwise stay.

According to a 2010 National Restaurant Association Restaurant Industry Operations Report, the turnover rate for hourly workers in full-service operations is 71 percent when the average check is under $15.[4] That means that your typical full-service restaurant will lose every one of its hourly employees within a year's time and have to fill every position. If we were to ask employees to explain why they left their jobs, the most frequently cited reasons would likely include bad supervision and management, better opportunities, a better work schedule, and more enjoyable working conditions.

The fact is, though, that two operations hiring from the same labor pool can get radically different results according to the work climate their supervisors create. You can see this in multiunit

work climate
The level of morale in the workplace.

operations: One unit might be consistently better than the others—one that is cleaner, has a better food cost, has a better labor cost, and has more satisfied guests. It is the supervisor or manager and good employees who make this difference, and usually this supervisor has created a climate in which the employees will give their best.

Hospitality Employees by Generation

There is no valid stereotype of today's hospitality employee. The industry employs people of all ages and backgrounds. You will most likely be supervising people from one of these *generations*:[5]

> **Generation X**, those born between 1966 and 1976 = 41 million
>
> **Generation Y**, those born between 1977 and 1994 = 71 million
>
> **Mellennials**, those born between 1980 and 2000 = 81 million
>
> **Generation Z**, those born between 1995 and 2012 = 25 million

Many of the foodservice workforce, as well as a big presence in hotels, are employees from 18 years old to their mid-30s, a group referred to as Generation Y (those born between 1977 and 1994). There are 71 million Generation Yers. Xers and Yers will work hard, but they will also make some demands. They want to do work that they consider worthwhile as well as work they enjoy doing. The employees want their supervisors to let them be more involved by listening to them and by allowing them to participate in decision making. Not surprisingly, employees do not want supervisors to bark orders in a militant fashion; they want training and expect management to invest time and money in their training and development. Generation Z, the 25 million and rapidly growing group born between 1995 and 2012, will grow up with highly sophisticated expectations of media and computers.[6]

Growing Diversity in Hospitality Employees

An already-diverse workplace is becoming more diverse than ever. The National Restaurant Association declares that the restaurant industry employees 13.5 million in the United States. That is 10 percent of the overall U.S. workforce. To illustrate the diversity in the industry of restaurant managers, 35 percent speak a language other than English at home and 38 percent are minorities. Of supervisors, 22 percent speak a language other than English at home and minorities make up 37 percent. Of chefs, 45 percent speak a language other than English at home and minorities make up 56 percent. Of cooks, 42 percent speak a language other than English at home and minorities make up 58 percent. Diversity of the hospitality workforce is discussed in more detail in Chapter 5.

Many of today's employees tend to have a higher expectation level and a lower frustration tolerance than employees of past generations. They expect more out of a job than just a paycheck. But even needing that paycheck does not guarantee that a person will work well on the job. That is why it is necessary to have supervisors and managers.

> A lack of leadership is a problem in our industry. Part of the problem is that we're not talking about leadership in our meetings at a unit level. We, as managers, need to realize that people don't want to be managed; they want to be led.[7]

✳ SO WHO'S NUMBER ONE?

There's an expression in the hospitality industry that goes like this: "If you [the supervisor or manager] take care of the employees, the employees will take care of the guests, and the profits will take care of themselves." As a supervisor, your number one concern is your employees. You need to be committed to serving the employees who serve the guest because the way you treat your employees will be reflected in how they treat guests.

When you treat employees the way you want them to treat guests (with consideration, respect, and so on), employees then tend to provide high-quality service. This keeps guests happy and

---

Generation X
A group of Americans born between 1966 and 1976.

Generation Y, or Millennials
A group of Americans born between 1977 and 2000.
   Mellennials are "connected" and tech savvy, with gadgets that enable them to multitask. They are used to working in teams and like to make friends at work. Millennial employees work well with a diverse group of coworkers and have a positive attitude about work and prefer frequent and immediate feedback on how they are doing.

Generation Z
A group of Americans born between 1995 and 2012.

The work environment is one of the most important aspects of supervision.
Pressmaster/Shutterstock

increases the chances of return visits, meaning more business and more money spent. This, in turn, helps build profits and keeps the owners happy. Studies involving various companies show that those who value their employees highly (by providing training, rewards, and so on) have higher guest satisfaction and profitability.

Michael Klauber, owner and general manager of the restaurant Michael's on East, says that with 110 employees, ranging in age from 18 to around 40, his biggest challenge is creating an atmosphere in which people want to work. He places his employees before the guests because if the employees are happy, the guests will be happy.[8]

Functions of Management

LEARNING OBJECTIVE: Describe the functions of management.

Are first-line supervisors really managers? Yes, indeed. A **manager** is a person who directs and controls an assigned segment of the work in an enterprise. Although supervisors often do not have the title *manager* and midlevel managers and top executives in large enterprises may not regard them as part of the management team, supervisors have crossed that line from the employees' side to the management side, and they perform the functions of management in their area of control. So we need to examine the functions of management.

Theoretically, there are four main functions of management:

Planning: Looking ahead to chart goals and the best courses of future action. Planning involves, for example, determining who, what, why, when, where, and how work will be done.

Organizing: Putting together the money, staff, equipment, materials, and methods for maximum efficiency to meet an enterprise's goals. Organizing affects such aspects as operating a restaurant kitchen or hotel room division.

Leading: Interacting and guiding employees about getting certain goals and plans accomplished. Leading involves many skills, such as communicating, motivating, delegating, instructing, supporting, developing, and mentoring employees.

Controlling and evaluating: Monitoring and evaluating results in terms of goals and standards previously agreed upon. These include such things as performance and quality standards, and taking corrective action when necessary to stay on course.

manager
One who directs and controls an assigned segment of work in an operation.

planning
Looking ahead to chart the best course of future action.

organizing
Putting together the money, personnel, equipment, materials, and methods for maximum efficiency to meet an enterprise's goals.

leading
Guiding and interacting with employees regarding getting certain goals and plans accomplished; involves many skills, such as communicating, motivating, delegating, and instructing.

controlling
Measuring and evaluating results in terms of goals and standards previously agreed upon, such as performance and quality standards, and taking corrective action when necessary to stay on course.

Does this list seem remote and unreal to you? It does to many people who run hotels and foodservice operations, and there is nothing like management experience to upset management theory. A busboy quits, the dishwashing machine breaks down, two servers are fighting in the dining room, the health inspector walks in, an official from the liquor control board is coming at 2:00, and someone rushes up and tells you there's a fire in the kitchen. How does management theory help you at a moment like this?

✳ THE REALITY

There is nothing wrong with management theory; it can be useful, even in a crisis. The challenge is how to apply it. In the circus we call the hospitality industry, nothing comes in neat and tidy packages. Supervisors seldom have control over the shape of their day. The situation changes every few seconds; you blink and the unexpected usually happens.

In a foodservice operation you are manufacturing, selling, and delivering a product, all within minutes. In a hotel you may have 5,000 guests one day and 500 the next. You deal with your superiors, you deal with your subordinates, and you deal with your guests, all coming at you from different directions. Salespeople, deliveries, inspectors, guest complaints, and applicants for jobs interrupt you. You are likely to have only a few seconds available when you make many important decisions. Figure 1.6 shows the interactions of a supervisor.

PROFILE Jim Sullivan

Courtesy of Jim Sullivan

I like what I do. Every year, I arrange dozens and dozens of service and sales-building seminars for successful companies around the world. I also help overhaul and redesign manager and server training manuals and programs for a variety of successful chains and independent restaurants. And in doing so, I get to assimilate a wide variety of best practices relating to customer service, employee retention, same-store sales-building, cost controls, and creative management. I also see subtle patterns, trends, and evolutions occurring in hospitality management theory and practice. In case you hadn't noticed, a sea change of behavior is in full swing right now. I'd like to outline and possibly debunk nine customer service myths that used to hold water in our industry and now are losing value as operating principles. Do you agree or disagree with the following points and counterpoints? The way you think about each one may provide a road map for your operation's success in the looming new century.

No. 1: "The customer comes first." Really? Today, you need good employees more than they need you. As Wally Doolin, former chief executive of restaurant chain TGI Friday's parent company, Carlson Restaurants Worldwide, pointed out at the recent Multi-Unit Foodservice Operators confab: "Our employees are our first market." Amen. So, instead of ranking relationships between customers and employees, we should focus on establishing equity instead. In other words, never treat a customer better than you do an employee. Service, like charity, begins at home, and if you're not investing in serving your team as well as you serve your customers, you're headed for trouble, pure and simple.

No. 2: "A satisfied customer comes back." Customer "satisfaction" is meaningless. Customer loyalty is priceless. People don't want to be "satisfied" as customers. Heck, Kmart can "satisfy" customers. They want fun, flair, and memorable experiences. A satisfied customer doesn't necessarily ever come back. As the noted New York restaurateur Danny Meyer says, "Give your guests what they remember and give them something new each time they visit."

No. 3: "We've got to focus on the competition." That's right. But what you may not realize is that your competition is the customer, not other restaurants. So stop looking across the street and focus on the face above the tabletop or at the counter.

No. 4: "Comment cards, social media, Trip Advisor and 'secret' shoppers are popular methods used to accurately measure service." Measuring customer satisfaction in your restaurant merely by tallying mystery shopper scores and comment cards is like judging chili by counting the beans. Measure what matters: Same-store sales increases, higher customer traffic, and lower employee turnover are just as important—if not more so. Mystery shopping is effective, but only if it measures the good as well as the bad and if the "shoppers" are people with hospitality experience who know the subtleties. Quality for your staff first, and they will build a happy customer. A happy customer buys more.

No. 5: "People are our most important asset." That old adage is wrong. The right people are your most important asset. The right people are not "warm bodies." The right people are those servers, cooks, hostesses, or managers who exhibit the desired team and customer service behavior you want, as a natural extension of their character and attitude,

(continued)

regardless of any control or incentive system. Hire the personality; train the skills. Where do you find them? See No. 6.

No. 6: "There's a labor crisis." According to the National Restaurant Association and the Bureau of Labor Statistics, staff turnover was 62.6 percent.[9] Yikes. But where do they go? Is it to other industries or other restaurants? Get straight on this: We don't have a "labor" crisis. We've got a turnover crisis. So the tough question you have to ask yourself about your operation is not, "Are there enough people available to work?" but rather, "Are there enough people available to work who want to work for us?" Make your operation a fun, reputable, and caring place to work.

No. 7: "Invest first in building the brand." Sorry, I disagree. Invest first in people, second in brand, third in bricks and mortar. Mike Snyder, then president of Red Robin International Inc., summed it up this way: "Give me a Weber [barbecue] and a tent in a parking lot along with the best service-oriented people who take care of the customer and each other, and I'll beat the _____ off the restaurant with the multimillion-dollar physical plant every shift."

No. 8: "Information is power." Know the difference between *information* and *communication*. Those two words often are used interchangeably, but in fact mean two different things. Information is "giving out"; communication is "getting through." Training is your secret weapon, but I suspect that much of your training informs more than it communicates. Besides, the belief that information is power leads managers to hoard it, not share it, and that's backward thinking. Sharing information not only enlightens but also shares the burden of leadership and engages the creativity and solutions of the entire team.

No. 9: "We need new ideas to progress." Why do companies always want new ideas? I'll tell you why: Because "new ideas" are easy. That's right. The hard part is letting go of ideas that worked for you two years ago and are now out of date. So before you and your team brainstorm dozens of new ideas that get listed on flip charts, give everyone a warm fuzzy feeling, and are never implemented, allow me to suggest a different angle. The newest and most innovative thing you can do for your business may be to master the "basics" that everyone knows and no one executes consistently. I'm referring to caring behavior, service with flair, and employee appreciation. Because, unlike Nehru jackets and the Backstreet Boys, the basics of great service never go out of style.

In summary, remember that there is no silver bullet for guaranteeing great service and a great team. Maybe Darrell Rolph, chief executive of Carlos O'Kelly's, says it best: "Keep it fresh, keep it focused, and remember to say thank you."

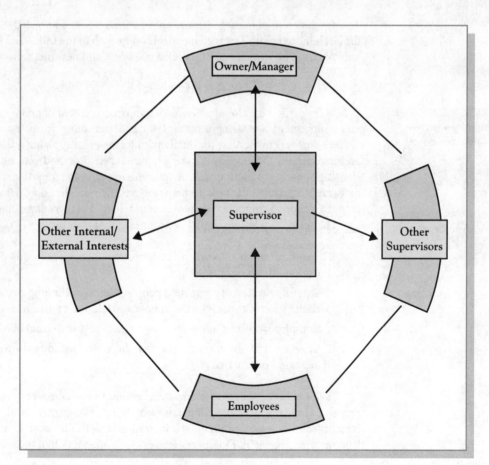

FIGURE 1.6: The interactions of a supervisor.

In such circumstances, supervisors usually react to situations rather than act on them according to a preconceived plan plotted out in the quiet of an office. Managing becomes the ability to adjust actions and decisions to given situations according to the demands of those situations. It is a *flex style of management*, calling on theory, experience, and talent. It is a skill that cannot be taught but has to be developed in supervised experience on the job. It means doing what will be most effective in terms of the three elements involved: *the situation*, *your workers*, and *yourself*. It means developing techniques and applying principles of management in ways that work for you.

According to the Center for Creative Leadership, a research firm in Greensboro, North Carolina, poor interpersonal skills are one of the leading reasons that managers fail.[10] Managers who fail are often poor listeners, can't stimulate their employees, don't give and take criticism well, and avoid conflict.

Managers need to learn how to lead people, just as you do, through supervised experience on the job. They must learn how to convert classroom theories into practical applications that are accepted by the people they supervise. No one can teach you; it is theory, then practice, then experience.

As a flex-style manager reacting to constantly changing situations, your on-the-spot decisions and actions are going to be far better if you too can draw on sound principles of management theory and the accumulated experience of successful managers. In this book, we introduce you to those principles and theories that can help you to work out your own answers as a supervisor in a hotel or foodservice setting. They can provide a background of knowledge, thoughts, and ideas that will give you confidence and a sense of direction as you meet and solve the same problems that other managers before you have met and solved.

Theories of People Management

LEARNING OBJECTIVE: Compare and contrast the major theories of people management as they relate to hospitality employees.

The development of management as an organized body of knowledge and theory is a product of the last hundred years. It was an inevitable outgrowth of the Industrial Revolution and the appearance of large enterprises needing skilled managers and new methods of running a business.

✳ SCIENTIFIC MANAGEMENT

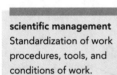

scientific management
Standardization of work procedures, tools, and conditions of work.

One of the earliest developments affecting the management of people was the scientific management movement appearing around 1900 and stemming from the work of Frederick Taylor. Taylor's goal was to increase productivity in factories by applying a scientific approach to human performance on the job. Using carefully developed time-and-motion studies, he analyzed each element of each production task. By eliminating all wasted motions, Taylor arrived at the "one best way" to perform the task. In the same way, he established a "fair day's work," which was the amount of work a competent employee could do in one workday using the one best way.

The system Taylor developed had four essential features:[11]

1. Standardization of work procedures, tools, and conditions of work through design of work methods by specialists

2. Careful selection of competent people, thorough training in the prescribed methods, and elimination of those who could not or would not conform

3. Complete and constant overseeing of the work, with total obedience from the employees

4. Incentive pay for meeting the fair day's work standard—the employee's share of the increased productivity

Taylor believed that his system would revolutionize labor–management relations and would produce "intimate, friendly cooperation between management and men" because both would benefit from increased productivity. Instead, his methods caused a great deal of strife between labor and management. Employees who once planned much of their own work and carried it out with a craftsman's pride were now forced into monotonous and repetitive tasks performed in complete obedience to others.

Taylor believed that higher wages—"what they want most"—would make up for having to produce more and losing their say about how they did their work. But his own workers, with whom he was very friendly, did everything they could to make his system fail, including breaking their machines. The craft unions of that day fought Taylor's system bitterly, and relations between management and labor deteriorated. Productivity, however, increased by leaps and bounds, since fewer employees could do more work.

Another innovator, Frank Gilbreth, carried forward the search for the one best way of performing tasks, or **work simplification**. Using ingenious time-and-motion study techniques, he developed ways of simplifying tasks that often doubled or tripled what a worker could do.[12] His methods and principles had a great impact in foodservice kitchens, where work simplification techniques have been explored extensively.

Taylor's innovations began a revolution in management's approach to production. His theories and methods were widely adopted (although the idea that the employee should share in the benefits of increased productivity seldom went along with the rest of the system). A whole new field of industrial engineering developed in which efficiency experts took over the planning of the work. In this process, the employees came to be regarded as just another element of the production process, often an adjunct to the machine. Their job was to follow the rules, and the supervisors saw to it that they did. This became the prevailing philosophy of people management.

Everything is systematized, and the employee is simply taught to run the machines, follow the rules, and speak given phrases. When the bell rings, the worker turns the hamburgers on the grill. To make a pancake, the worker hits the batter dispenser one time. There is no room for deviation.

Such standardization has many benefits to the enterprise. It maintains product consistency from one unit to the next. It allows the use of unskilled labor and makes training quick, simple, and inexpensive. It is well suited to short-term employees on their first jobs. But such complete standardization does not work in every setting or for all employees. When there is no room for deviation, there is no opportunity for originality, no relief from monotony. In many organizations, this would lead to high turnover with problems in training and morale.

Probably the very nature of the hotel and food businesses makes them unsuitable for totally scientific management. Still, there are important elements of the method that can be used to

work simplification
The reduction in repetitive tasks to the fewest possible motions, requiring the least expenditure of time and energy.

In the foodservice industry you can see the influence of scientific management in some fast-food operations.

Kondor83/Shutterstock

increase productivity, achieve consistent results, make customers happy and patients well, increase profit, and make a manager's life much easier, all without making workers into human machines.

For example, you can see scientific management at work in the standardized recipe, the standardized greeting, the standardized hotel registration procedures, and the standardized making of a hospital bed. Scientific management as a whole is practiced in restaurants and hotels far less than it could be. We have the methods and techniques, but we seldom use them. We may have standardized recipes, but except in baking, many cooks never look at them. We may standardize procedures, but we seldom enforce them. We hire in panic and in crisis; we take the first warm body that presents itself and put it to work. We use the magic apron training method: We give the new employee an apron and say, "Go." We assume that anyone knows how to do some of our entry-level jobs. We hang on to inefficient workers because we are afraid that the next ones we hire will be even worse. As for overseeing their work, who has time to do that? Standardization could help alleviate some of the tension that arises when time is tight and human resources are limited.

✳ HUMAN RELATIONS THEORY

In the 1930s and 1940s, another theory of people management appeared—that of the human relations school. This was an outgrowth of studies made at the Hawthorne plant of Western Electric Company.

Researchers testing the effects on productivity of changes in working conditions came up with a baffling series of results that could not be explained in the old scientific management terms. During a prolonged series of experiments with rest periods, for example, the productivity of the small test group rose steadily whether the rest time was moved up or down or was eliminated altogether. Furthermore, employees from the test group were out sick far less often than the large group of regular employees, and the test group worked without supervision. It became obvious that the rise in productivity was the result of something new, not the economic factor of a paycheck or the scientific factors of working conditions or close supervision.[13]

Elton Mayo, the Harvard professor who conducted the experiments, concluded that a social factor, the sense of belonging to a work group, was responsible. Other people had other theories to explain the increased productivity: the interested attention of the researchers, the absence of authoritarian supervision, participation in the planning and analysis of the experiments. People are still theorizing about what **human relations theory** can do. We explore this in later chapters.

Participative management was the real meaning of the Hawthorne experiments, but everyone agrees that the experiments shifted the focus of human resources to the people being managed. Now, enter the human relations theorists, who stressed the importance of concern for employees as individuals and as members of the work group or team. "Make your employees happy and you will have good workers," they said. "Listen to your people, call them by name, remember their birthdays, help them with their problems."

This was the era of the company picnic, the company newspaper, the company bowling team, the company Boy Scout troop. But happiness, it turns out, does not necessarily make people productive. You can have happy employees who are not productive. There is more to productivity than that. Yet we do need nearly everything the human relations theorist emphasized. It isn't happiness that will make your employees produce; it is your own ability to lead your employees. Some human relations techniques, such as listening and communicating and treating people as individuals, can make you a better leader, and this is the biggest thing that human relations theory can do for you.

✳ PARTICIPATIVE MANAGEMENT

Building on the new interest in the worker, a trend toward **participative management** developed in the 1960s and 1970s. In a participative system, employees participate in the decisions that concern them. They do not necessarily make the decisions; this is not democratic management by majority vote. The manager still leads and usually makes the final decision, but he or she discusses plans and procedures and policies with the work groups who must carry them out, and considers their input in making final decisions. In taking part in such discussions, employees come to share the concerns and objectives of management and are more likely to feel committed to the action and to being responsible for the results.

human relations theory
A theory that states that satisfying the needs of employees is the key to productivity.

participative management
A system that includes employees in making decisions that concern them.

Participative management as a total system is probably not suited to the typical foodservice or lodging enterprise. Nevertheless, certain of its elements can work very well. Discussing the work with your people, getting their ideas, and exchanging information can establish a work climate and group processes in which everybody shares responsibility to get results. You might call it *management by communication*.

✻ TOTAL QUALITY MANAGEMENT

total quality management (TQM)
A process of total organizational involvement in improving all aspects of the quality of a product or service.

Total quality management (TQM) is a participative process that empowers all levels of employees to work in groups to establish guest service expectations and determine the best way to meet or exceed those expectations. TQM is a continuous improvement process that works best when supervisors and managers are good leaders. A successful company will employ leader-managers who create a stimulating work environment in which guests and employees become an integral part of the mission by participating in goal and strategy setting.

Installing TQM is exciting because once everyone becomes involved, there is no way of stopping the creative ways in which employees solve guest-related problems and improve service. Other benefits include cost reductions, increased guest and employee satisfaction, and ultimately, profit. TQM is discussed at length in Chapter 7.

✻ HUMANISTIC MANAGEMENT

What is likely to work best in the hotel and foodservice industries is selective borrowing from all three systems of management: scientific, human relations, and participative. We need to apply many of the principles of scientific management. We need standardized recipes, we need to train employees in the best ways to perform tasks, and we need systems for controlling quality, quantity, and cost. But one thing we do not need from scientific management is its view of the employee as no more than a production tool.

Here we can adapt many features of the human relations approach. If we treat employees as individuals with their own needs and desires and motivations, we can do a much better job of leading them and we are far more likely to increase productivity overall. From participative management we can reap the advantages of open communication and commitment to common goals, so that we are all working together. The successful manager will blend all three systems, deliberately or instinctively, according to the needs of the situation, the workers, and his or her personal style of leadership. We call this humanistic management.

humanistic management
A blend of scientific, human relations, and participative management practices adapted to the needs of the situation, the employees, and the supervisor's leadership style.

Like Frederick Taylor and all the theorists since, today's supervisor is concerned with productivity: getting people to do their jobs in the best way, getting the work done on time and done well. This is an age-old problem: When Pope John XXIII was asked how many people worked for him, he answered, "About half of them."[14] It is sad but human that many people will do as little work as possible unless they see some reason to do better. Often, they see no reason.

This is where leadership comes in: the supervisor interacting with the employees. Look at it as a new form of ROI, not *return on investment* but *return on individuals*. As a supervisor you will succeed only to the degree that each person under you produces; you are judged on the performance, the productivity, and the efficiency of others. The only means for your success is a return on each person who works for you. As a leader you can give them reasons to do better. Use your *Is*: imagination, ideas, initiative, improvement, interaction, innovation, and—why not?—inspiration. It is the personal interaction between supervisor and worker that will do the trick.

Managerial Skills

LEARNING OBJECTIVE: List examples of technical, human, and conceptual skills used by hospitality supervisors.

Management at any level is an art, not a science providing exact answers to problems. It is an art that can be learned, although no one can really teach you. You do not have to be born with certain talents or personality traits. In fact, studies of outstanding top executives have failed to identify a

FIGURE 1.7: Levels of technical, human, and conceptual skills required by top, middle, and supervisory managers.

managerial skills
The three sets of skills that a manager needs: technical, human, and conceptual.

technical skill
The ability to perform the tasks of the people supervised.

human skill
The ability to manage people through respect for them as individuals, sensitivity to their needs and feelings, self-awareness, and good person-to-person relationships.

conceptual skill
The ability to see the whole picture and the relationship of each part to the whole.

common set of traits that add up to successful leadership, and experts have concluded that successful leadership is a matter of individual style.

There are, however, certain managerial skills essential to success at any level of management: technical skill, human skill, and conceptual skill. At lower levels of management, technical and human skills are most important because managers here are concerned with the products and the people making them. Conceptual skill is necessary, too, but not to the same degree. In top-management positions in large corporations, conceptual skill is all-important, and the other skills come into play less often (see Figure 1.7).

All these skills can be developed through exercise, through study and practice, and through observation and awareness of one's self and others. We look at them as they apply to supervisors in foodservice and hospitality enterprises. Then we add to the list of skills a fourth essential for managerial success: some personal qualities that will enable you to survive and prosper.

❋ TECHNICAL SKILLS

The kind of technical skill useful to you as a supervisor is the ability to do the tasks of the people you supervise. You need such knowledge to select and train people, plan and schedule the work in your department, and take action in an emergency. Most important, your technical skills give you credibility with your workers. They will be more ready to accept and respect you when they know that you have competence in the work you supervise. If you have been an hourly employee, you may already have the technical skills you need. In large organizations, some supervisors are required to go through the same skills training as the employees.

❋ HUMAN SKILLS

The skill of handling people successfully is really the core of the supervisor's job. Such skill has several ingredients and is not achieved overnight.

First in importance is *your attitude toward the people who work for you.* You must be able to perceive and accept them as human beings. If you don't—if you think of them as cogs in the wheels of production, or if you look down on them because you are the boss and they scrub floors for a living—they will not work well for you or they will simply leave. *They will not let you succeed.*

A second ingredient of human skills is sensitivity, the ability to perceive each person's needs, perceptions, values, and personal quirks so that you can work with each one in the most productive way. You need to be aware that José still has trouble with English, that Rita will cry for days if you speak sharply to her, and that Charlie won't do anything at all unless you almost yell. You need to realize that when Jim comes in looking like thunder and not saying a word, he's mad about something and you had better find a way to defuse him before you turn him loose among the guests. You need to be able to sense when a problem is building by noticing subtle differences in employee behavior.

A third ingredient is self-awareness. Have you any idea how you come across to your employees? You need to be aware of your own behavior as it appears to others. For example, in your concern for quality, you may always be pointing out to people things they are doing wrong. They

You need to establish person-to-person relations with individual employees. This will allow you to form a better working relationship with them and allow them to see you as an individual as well as a boss.

Monkey Business Images/Shutterstock

probably experience this as criticism and see you as a negative person who is always finding fault. If you become aware of your habits and their reactions, you can change your manner of correcting them and balance it out with praise for things well done.

You also need to be aware of your own perceptions, needs, values, and personal quirks and how they affect your dealings with your employees. When you and they perceive things differently, you will have trouble communicating. When you and they have different needs and values in a work situation, you may be working at cross-purposes. Human skills come with practice. You have to practice treating people as individuals, sharpening your awareness of others and of yourself, figuring out what human qualities and behaviors are causing problems and how these problems can be solved. This is another instance in which the flex style of management figures: responding to your people, yourself, and the situation. It is a continual challenge because no two situations are ever exactly alike. The ultimate human skill is putting it all together to create an atmosphere in which your employees feel secure, free, and open with you and are willing to give you their best work.

❋ CONCEPTUAL SKILLS

Conceptual skills require the ability to see the whole picture and the relationship of each part to the whole. The skill comes in using that ability on the job. You might need to arrange the work of each part of your operation so that it runs smoothly with the other parts—so that the kitchen and the dining room run in harmony, for example. Or you may need to coordinate the work of your department with what goes on in another part of the enterprise.

For example, in a hotel the desk clerk must originate a daily report to the housekeeper showing what rooms must be cleaned, so that the housekeeper can tell the cleaning associates to draw their supplies and clean and make up the rooms. When they have made up the rooms, they have to report back to the housekeeping department so that the rooms can be inspected and okayed, and then the housekeeping department has to issue a report back to the desk clerk that the rooms are ready for occupancy.

If you are the front office manager, you must be able to see this process as a whole even though the front desk cares only about the end of it—are those rooms ready? You must understand how the front desk fits into a revolving process that affects not only housekeeping and cleaning personnel but laundry, supplies, storage, and so on, and how important that first routine report of the desk clerk is to the whole process, to everyone involved, to customer service, and to the success of the enterprise.

In departments where everyone is doing the same tasks, conceptual skill is seldom called on, but the more complicated the supervisor's responsibilities, the greater the need for conceptual skill. A restaurant manager, for example, has a great deal of use for conceptual ability since he or she is responsible for both the front and the back of the house as well as the business end of the operation.

Consider what happened to this new restaurant manager. Things got very busy one night, and a server came up to him and said, "We can't get the tables bused." "Don't worry about it," he said, and he began to bus tables. Another server came to him and said, "We can't get the food out of the kitchen quick enough to serve the people." "Don't worry about it," he said, "I'll go back there and help them cook the food." While he was cooking, another server came in and said, "We can't get the dishes washed to reset the tables." "Don't worry about it, I'll wash the dishes," he said. While he was washing the dishes, another server came in and said, "We can't get the tables bused."

And so it went: the tables, the food, the dishes, the tables, the food. At this point, the owner came in and said, "What the hell are you doing?" "I'm washing the dishes, I'm busing the tables, I'm cooking the food, and before I leave tonight I'm gonna empty the garbage," said the manager with pride. "Look at this place!" the owner shouted. "I hired you to manage, not to bus tables and cook and wash dishes!"

That manager had not been able to see the situation as a whole, to move people about, to balance them out where they were needed most: to manage. He had boomeranged back to doing the work himself because it was easier than managing, more familiar than dealing with the whole picture. Supervisors who are promoted from hourly positions often have this problem. When they finally learn to look at the whole picture and deal with it, they have truly attained the management point of view.

❋ PERSONAL SKILLS AND QUALITIES

In addition to managing others, supervisors must be able to manage themselves. This, too, is a skill that can be developed through awareness and practice. It means doing your best no matter what you have to cope with, putting your best foot forward and your best side out, keeping your cool. It means setting a good example; it means self-discipline. You cannot direct others effectively if you cannot handle yourself. It also means having self-control and supporting your own supervisor even when you personally disagree with a decision or action.

Managing yourself also means thinking positively. According to Manz, two different patterns of thinking are opportunity thinking and obstacle thinking.[15] When faced with a challenging situation, **opportunity thinkers** concentrate on constructive ways of dealing with the circumstances, whereas **obstacle thinkers** focus on why the situation is impossible and retreat. Let's say that you are a supervisor and your evening dishwasher calls in sick. If you are an opportunity thinker, you'll look at the work schedule to see who else may be able to fill in and take steps to contact them.

If you are an obstacle thinker, you'll think there's nothing you can do except ask the staff to switch to disposable dishes, even though you know it costs a lot more and the food doesn't look as good. Work on being an opportunity thinker, and if you make mistakes, learn from them and don't make yourself miserable. Everyone makes mistakes, even the people at the top of the organization chart. Guilt and worry will wear you down; self-acceptance and self-confidence will increase your energy.

Your own moods will affect your employees, too; they can run right through your whole department. Your employees watch you more carefully than you think; they can tell if your day is going fine or if you just had a frustrating meeting with your boss. When you get right down to it, your employees need a boss with a consistently positive outlook and attitude on the job.

You need to build a good, strong self-image. You have obligations to yourself as well as to others. Give yourself credit when you are right; face your mistakes when you are wrong and correct them for the future. You need to know yourself well, including your strengths and weaknesses, to work out your personal goals and values as they apply to your job, to know where you stand and where you are going.

In addition to having faith in your own ability to reach goals, you need to believe that employees will perform effectively when given a reasonable chance. You need to realize that you are also responsible for developing your employees through techniques such as coaching and counseling.

---

opportunity thinkers
Those who concentrate on constructive ways to deal with a challenging situation.

obstacle thinkers
Those who focus on why a situation is impossible and will retreat from it.

Another pair of useful personal qualities are flexibility and creativity. No hospitality manager can survive for long without flexibility—the ability to respond effectively to constantly changing situations and problems, to adapt theory to the reality of the moment, to think creatively because there are no pat answers. You must be able to respond to changes in the industry, too; yesterday's solutions will not solve tomorrow's problems. These again are skills that you can learn and practice; you do not have to be born with them.

Finally, being a supervisor requires high energy levels and the ability to work under great pressure. The time pressure in the hospitality field is unlike that in many other businesses; the meals must be served in a timely fashion; the rooms have to be ready in time for the next guest; a hospital patient who has diabetes needs his snacks at 2:00 P.M. and 8:00 P.M. Much stamina is needed to deal with these pressures.

You need to make a conscious and deliberate decision to be a manager. Here are three questions you must answer:

1. **Do you really want it?** Is there something about being a manager in the hospitality business (or wherever you are) that provides the responsibility, the challenge, and the fulfillment you want from the work you do?

2. **What is the cost?** Without tips or overtime pay, you'll probably make less money than some of your workers do. The hours are long, you'll work on weekends when everyone else is playing, the responsibility is unremitting, and the frustration level is high. You are squarely in the middle of all the hassle: Your employer is telling you, "I want a lower food cost, I want a lower labor cost, and I want this place cleaned up." Your employees are saying, "I can't be here Friday night, that's not my job, get somebody else to do it, and I want more money." The guests are saying, "The food is cold, your service is slow, and your prices are too high." And your family is saying, "You're never home, we never get to go out together, you don't have time to help us with our homework, and what do you mean, *fix you a cheese sandwich*, after you've been down there with all that food all day?" You work with people all day long, and yet it is a lonely job.

3. **Is it worth the cost?** Is the work itself satisfying and fulfilling? Will you learn and grow as a professional and as a person? Are you on the path you want to be on? Do you want to be a manager enough to pay the price?

If your answer is yes, then pay the price—pay it willingly and without complaint; pay it gladly. This may be the most important quality of all—to have the maturity to decide what you want and accept the tough parts with grace and humor; or to see it clearly, weigh it carefully, and decide you are not going that way after all. One successful manager is Tim Stanton, a joint venture partner with restaurant chain Outback Steakhouse, who is described by a colleague as being incredibly driven and extremely motivated, detail-oriented, passionate about food quality, and demanding but fair.[16]

Tips for New Supervisors

LEARNING OBJECTIVE: List three to five best practices for new supervisors.

As we are beginning to realize, being a new supervisor can be a daunting task, but it can also be a stimulating and rewarding time of personal accomplishment. New supervisors need to be prepared, particularly in the hospitality industry. They need to hit the ground running.

Here are some tips:

- Start as you mean to continue. Set your standards and keep them.

- Develop a game plan with your boss of what you and your employees are to achieve and the best ways to go about it.

- Be you—don't try to be someone else now that you have some authority. Be objective; treat everyone the way you would like to be treated.

- Praise the good work that you and your team have done in the past.
- Your employees have needs—ask how you can help them do a better job.
- Begin getting a feel for the workplace by listening and asking questions. You might see changes that should be made, but don't rush into making hasty decisions from the get-go. It's much better to solicit the ideas and questions from your team.
- Be positive, upbeat, and ready to share the knowledge of your team and encourage everyone's participation.
- Outline the team's strengths, accomplishments, and the challenges ahead. Explain that this will be a *we-and-us* not an *I-and-you* situation because as a team more can be accomplished and obstacles can be overcome.
- Know the company's vision, mission, goals, and strategies.
- Know the company's philosophy and culture.
- Check the organization chart for reporting relationships.
- Check the budget: What are your budgeting responsibilities? What percentage is discretionary and what can be moved from one item to another?
- Know the policies and procedures but don't be afraid to say, "I need to check on that and get back to you." Then do so.
- Set a good example, arrive early, dress appropriately, and do not do personal business on company time. Remember that your behavior, attitude, and work habits will influence your team.

Before you became a supervisor, you were in a position where you were aware of the department's goals and your own responsibilities but your main concern was contributing your part toward the success of the department. Now, however, you are responsible for the work of others and a productive team. Now is the time to determine your priorities and plan the work to be done.

As we progress through this book, we learn more about these other types of successful supervisor behavior. The next chapter examines supervisory leadership.

 ## CASE STUDY: The Good-Guy Supervisor

Three weeks ago, José was promoted from assistant manager to manager of the employee cafeteria at City Hospital. It was a big move up for him, but he had always been a conscientious, hardworking, and loyal employee, and he felt that his promotion was well deserved. He knows the operation backward and forward, having worked both in the kitchen and on the line, and he has always been well liked by his fellow employees, so he figured he would have no problems in the new job. When he took the job, José promised himself that he would never forget how it felt to be an employee and be ordered around, always being told what you're doing wrong, like Debra, the manager before him, used to do. Everyone hated her. He was determined to do things differently and not be such a drill sergeant. He'd be friendly, relaxed, and helpful, and people would do their best for him in return.

But things are not going quite the way he thought they would. Several of the line servers have become very careless about clean uniforms and wearing hairnets, and one of them, Maria, has come in late several times, and he has had to fill in for her on the line. He finally spoke to her about it, but when she told him her problems with her husband, who brings her to work, he could see how it kept happening, and he felt sorry for her. But it is worrying him.

A couple of other things are worrying him, too. Erma, who makes the salads, does not always have them ready on time, and yesterday the lettuce was gritty and several people complained. This morning José helped her wash it, which made her mad, and he couldn't understand that. Dan, the head cook and a good friend, saw the whole thing and laughed at his reaction. "Wise up, José," he said. "You're never gonna make it this way." And he lit a cigarette right in front of José after he took the chicken out of the oven.

Case Study Questions

1. What did Dan mean by his remark?
2. Why was Erma mad?
3. What should José do about Maria? About Erma? About Dan and the smoking, which is strictly against the rules?
4. Why isn't José able to maintain performance standards?
5. If you were José's supervisor, what advice would you have given him before he started on the job? What would you say to him now?
6. Do you think that José's supervisor is in any way to blame for José's problems?
7. What is José doing when he takes Maria's place on the line and helps Erma wash the lettuce?
8. What is the fundamental principle of supervision involved in José's case?

KEY POINTS

1. A supervisor is any person who manages people making products or performing services. A supervisor is responsible for the quality and quantity of the services and products for meeting the needs of employees. Only by motivating and stimulating the employees to do their jobs properly will supervisors produce high-quality products and services.
2. Using an organization chart, you can see line and staff functions, as well as how authority and responsibility are handed down from the top level of management to the first-line supervisors.
3. As a supervisor, you depend for your own success on the work of others, and you will be measured by their output and their performance. You will be successful in your own job only to the degree that your workers allow you to be, and this will depend on how you manage them.
4. As a hospitality supervisor, you have obligations to the owners, guests, and employees. To your employees, you represent management. To the owners and your bosses, you are the link with the employees and the work to be done. You represent productivity, cost control, quality control, and customer service. You also represent your people and their needs and desires. To the guests, your output and your employees represent the company.
5. As a supervisor, you've got to maintain the management point of view. You can't go back to where you came from (boomerang management).
6. As a supervisor, if you take care of the employees, the employees will take care of the customers, and the profits will take care of themselves. Your principal concern is your employees.
7. Some of the most important management activities you will be involved in are planning, organizing, leading, and controlling.
8. Managing is the ability to adjust actions and decisions to given situations. A flex style of management calls on theory, experience, and talent. It is a skill that cannot be taught in a classroom but has to be developed on the job.
9. The successful supervisor will blend principles of scientific management, human relations, and participative management, according to the needs of the situation and the employees, into a style referred to as *humanistic management*.
10. For success, managers need technical skills, human skills, conceptual skills, and certain personal skills.

KEY TERMS

authority
boomerang management
conceptual skill
controlling
exempt employees
first-line supervisor
Generation X

Generation Y
Generation Z
hourly workers
human relations theory
human skill
humanistic management
leading

line functions	planning
manager	responsibility
managerial skills	scientific management
nonexempt employees	supervisor
obstacle thinkers	technical skill
opportunity thinkers	total quality management (TQM)
organizational chart	work climate
organizing	work simplification
participative management	working supervisors

REVIEW QUESTIONS

Answer each question in complete sentences. Read each question carefully and make sure that you answer all parts of the question. Organize your answer using more than one paragraph when appropriate.

1. In one paragraph, describe what a supervisor does. Is a supervisor a manager? Why or why not?
2. Are supervisors exempt or nonexempt employees? What is the difference between the two categories?
3. Describe and explain the necessary change in point of view when an hourly employee becomes a supervisor.
4. Briefly describe a supervisor's responsibilities to owners, guests, and employees. Who should be the most important, and why?
5. Give an on-the-job example of each of the management functions discussed.
6. Compare and contrast the principles of scientific management, human relations theory, and participative management. What elements of each school of thought are appropriate in the hospitality industry?
7. What types of human skills does a supervisor need? Why does a supervisor need human skills?
8. Give three examples of personal skills and qualities that supervisors need.

ACTIVITIES AND APPLICATIONS

1. Discussion Questions
- Describe the process of management through effective communication skills.
- Do you think it is true that "you just can't get good employees these days"? If so, why can't you? If it isn't true, why do so many managers believe that it is?
- Why can't a supervisor manage employees successfully from an employee's point of view?
- Which three human skills do you consider most important in supervising employees at work? Why?
- Can you think of a situation in which a supervisor is given responsibility for seeing that a certain job gets done but has no authority to see it through?
- Why is the supervisor called a linking pin?
- What problems does boomerang management cause?
- Do you agree that a supervisor's principal priority should be his or her employees? Why or why not?
- What should be the priority of the hourly employees? Why?

2. Organizational Charts
Bring in organizational charts from various hospitality operations. Working in small groups, identify the number of management levels, the titles of all first-line supervisors, the titles of hourly employees, and any staff functions.

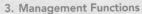

3. Management Functions

Mrs. R. is a manager in a white-tablecloth restaurant. Each of her activities described here are examples of management functions. Identify each management function or functions.

A. Mrs. R. writes out the work schedule for her employees for the next two-week rotation.

B. It is brought to Mrs. R.'s attention that a server has been late enough times to possibly be disciplined. Mrs. R. reviews the records and sits down with the server for a talk.

C. Every day, Mrs. R. observes the dining room to see how her employees are performing.

D. Every day, Mrs. R. greets and chats with many of the restaurant's guests.

E. Mrs. R. works out the next year's budget for her area.

F. Mrs. R. talks with the chef about improving communications between the kitchen and the dining room.

G. Mrs. R. holds a training session for all servers.

H. At the end of the shift, Mrs. R. totals up sales and compares the figure to projections.

I. Two servers want to switch shifts. Mrs. R. decides that they can do so.

J. Mrs. R. is made aware of a problem between the busers and the servers. Mrs. R. starts to gather information about it.

4. Brainstorming

Brainstorm examples of technical, human, and conceptual skills, and also the personal skills and qualities needed by hospitality supervisors.

 # ENDNOTES

1. Jim Sullivan, personal communication, July 16, 2004.
2. Martha I. Finney, *The Truth About Getting the Best From People* (Upper Saddle River, NJ: Financial Times Press, 2008), p. 150.
3. www.dol.gov/elaws/asp/drugfree/drugs/supervisor/screen50.asp. Retrieved October 22, 2014.
4. www.restaurant.org/News-Research/News/Economist-s-Notebook-Hospitality-employee-turnover. Retrieved December 15, 2014.
5. http://factfinder2.census.gov/faces/tableservices/jsf/pages/productview.xhtml?src=bkmk. Retrieved October 21, 2014.
6. www.socialmarketing.org/newsletter/features/generation3.htm. Retrieved May 18, 2010.
7. Garry Colpitts, Chef, University of South Florida Culinary Innovation Laboratory, personal communication, November 13, 2014.
8. Michael Klauber, owner of Michael's on East Restaurant, personal communication, November 13, 2014.
9. www.restaurant.org/Search?searchtext=turnover&searchmode=anyword. Retrieved November 13, 2014.
10. www.skillsyouneed.com/interpersonal-skills.html. Retrieved on November 13, 2014.
11. *The Principles of Scientific Management*, Copyright © 1911 by Frederick W. Taylor. Published in Norton Library, 1967.
12. Frank and Lillian Gilbreth Papers, Archives and Special Collections, Purdue University Libraries.
13. Jeffrey A. Sonnenfeld, "Shedding Light on the Hawthorne Studies," *Journal of Organizational Behavior*, vol. 6, no. 2 (April 1985), pp. 111–130.
14. Gerald Twomey, "Anniversary Thoughts," *America: The National Catholic Weekly* (October 7, 2002). Available at americamagazine.org/content/article.cfm?article_id=2516.
15. Manz, C. C., *Mastering Self-Leadership* (Englewood Cliffs, NJ: Prentice-Hall, 1992).
16. Duecy, "The NRN 50 General Managers," pp. 134–135.

Chapter 2

The Supervisor as Leader

If you were to ask any hospitality leader what his or her greatest challenge is, the likely answer would be finding and keeping great employees motivated. Given the high turnover in the hospitality industry and the resultant cost, we begin to understand some of the leadership challenges that supervisors face.

The idea that a supervisor must be a leader comes as a surprise to people who have never thought about it before. The term *leader* is often associated with politics or religious movements or guerrilla-warfare situations in which people voluntarily become followers of the person who achieves command. Although it is not necessarily true, it is generally assumed that the one who is followed is a "born leader" whose influence is based at least partly on charisma or personal magnetism. In a work situation, the supervisor is in command by virtue of being placed there by the company and its superiors. In the hospitality industry, the term *supervisor* refers to a manager at a lower organizational level who supervises entry-level or other employees who themselves do not have supervisory responsibilities. The employees are expected to do what the boss tells them to do—that's just part of the job, right?

But if employees simply do what they are told, why is labor turnover so high, productivity so low, and absenteeism so prevalent? Why is there conflict between labor and management? The truth of the matter is that the boss is in charge of the employees, but that does not guarantee that the employees will put all of their efforts into the job. This is where leadership comes in.

In this chapter, we explore the kinds of interactions between a supervisor and the employees that relate to the building of leadership in work situations. After completion of this chapter, you should be able to:

- Describe the characteristics of leadership.
- Explain the foundations of leadership development.
- Compare and contrast the different leadership styles.
- Name the necessary steps in creating your own leadership style.
- Identify the ethical considerations of a true leader.
- Apply leadership skills to become an employee mentor.

Characteristics of Leaders

LEARNING OBJECTIVE: Describe the characteristics of leadership.

If we were to examine great leaders of the past, we would likely come up with a list similar to the 14 leadership traits from the *Guidebook for Marines*: justice, judgment, dependability, initiative, decisiveness, tact, integrity, enthusiasm, bearing, unselfishness, courage, knowledge, loyalty, and endurance (known by the acronym JJ DIDTIEBUCKLE). Of these, a Marine would likely say that integrity is the most important. Integrity to a Marine means to do something right even if no one is aware of it.[1]

Experience has shown that effective leaders have six traits that distinguish them from non-leaders: drive, the desire to influence others, honesty and moral character, self-confidence, intelligence, and relevant knowledge (see Figure 2.1).

A person's *drive* shows that he or she is willing and able to exert exceptional effort to achieve a goal. This high-energy person is likely to take the initiative and be persistent.

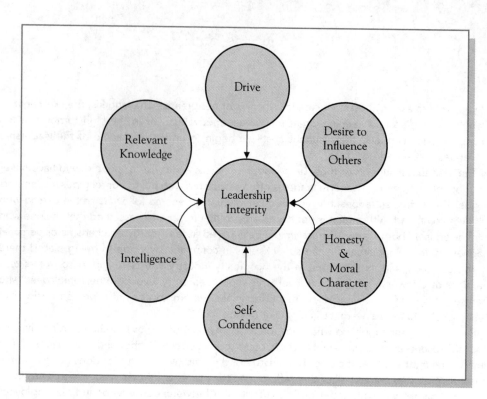

FIGURE 2.1: Characteristics and traits of effective leaders.

Leaders need *integrity*, which is doing the right thing even if no one is looking.

Leaders have a *desire to influence others*. This desire is frequently seen as a willingness to accept authority. A leader also builds trusting relationships with those supervised by being truthful. By showing consistency between their words and actions, leaders display *honesty and moral character*.

Leaders have *self-confidence* to influence others to pursue the goals of the organization. Employees tend to prefer a leader who has strong beliefs and is decisive over one who seems unsure of which decision to make.

Influencing others takes a *level of intelligence*. A leader needs to gather, synthesize, and interpret a lot of information. Leaders create a **vision**, develop goals, communicate and motivate, problem-solve, and make decisions. A leader needs a high level of *relevant knowledge*—technical, theoretical, and conceptual. Knowledge of the company, its policies and procedures, the department, and the employees are all necessary to make informed decisions.[2] Chefs need the ability to have empathy toward staff and take the time to listen.

John C. Maxwell in his *21 Indispensible Qualities of a Leader* rightly suggests that leadership truly develops from the inside out. If you can become the leader you *ought* to be on the *inside*, you will be able to become the leader you *want* to be on the *outside*. People will want to follow you.[3] Followers love leaders with charisma, and John C. Maxwell has the following suggestions for improving your charisma:[4]

- *Change your focus.* Observe your interaction with people during the next few days. As you talk to others, determine how much of the conversation is focused on yourself. Determine to tip the balance in favor of others.

- *Play the first-impression game.* Try an experiment. The next time you meet someone for the first time, try your best to make a good impression. Learn the person's name. Focus on his or her interests. Be positive. And most important, treat them as a "10." If you can do this for a day, you can do it every day. And that will increase your charisma overnight.

vision
The articulation of the mission of the organization in such an appealing way that it vividly conveys what it can be like in the future.

Approximately half of the foodservice workforce, as well as a big presence in hotels, are employees from 17 to 34 years old, referred to as Generation Y.

Dmitry Kalinovsky/Shutterstock

■ *Share yourself.* Make it your long-term goal to share your resources with others. Think about how you can add value to five people in your life this year. Provide resources to help them grow personally and professionally, and share your personal journey with them.

Reflecting on John C. Maxwell's *21 Indispensible Qualities of Leadership*, there are several that stand out, and there are hospitality leaders to match them. You can explore more about each hospitality leader if you like—just Google them.

Effective leaders are able to influence others to behave in a particular way. This is called **power**. There are four primary sources of power:[1-3]

power
The capacity to influence the behavior of others.

1. *Legitimate power* is derived from an individual's position in an organization.
2. *Reward power* is derived from an individual's control over rewards.
3. *Coercive power* is derived from an individual's ability to threaten negative outcomes.
4. *Expert power* is derived from an individual's personal charisma and the respect and/or admiration the individual inspires.

Many leaders use a combination of these sources of power to influence others to goal achievement.[5]

The Nature and Foundations of Leadership

LEARNING OBJECTIVE: Explain the foundations of leadership development.

leader
A person in command whom people follow voluntarily.

You are going to be a **leader**. Now, you may wonder, "What is a leader, and how is it any different from being a manager?" These are good questions. As a part of the management staff, one is expected to produce goods and services by working with people and using resources such as equipment and employees. That is what being a manager or supervisor is all about. As discussed in Chapter 1, an important managerial function is to be a leader. A good leader allows staff to give their input on vision and marketing opportunities to the chef or dining room or restaurant manager.

A *leader* can be defined as someone who guides or influences the actions of his or her employees to reach certain goals. A leader is a person whom people follow voluntarily. What you, as a supervisor, must do is to direct the work of your people in a way that causes them to do it voluntarily. You don't have to be a born leader, you don't have to be magnetic or charismatic; you have to get people to work for you willingly and to the best of their ability. That is what leadership is all about.

Although it is true that many leadership skills are innate and that not all managers make great leaders, it is also true that most managers will benefit from leadership training. Moreover, natural leaders will flourish in an environment that supports their growth and development.

There are seven steps to establishing a foundation for leadership development:

1. Commit to investing the time, resources, and money needed to create a culture that supports leadership development.

2. Identify and communicate the differences between management skills and leadership abilities within the organization.

3. Develop quantifiable measurables that support leadership skills. These include percentage of retention, percentage of promotables, and percentage of cross-trained team members.

4. Make leadership skills a focus of management training. These include communication skills (written, verbal, nonverbal, and listening), team-building skills (teamwork, coaching, and feedback), proactive planning skills (transitioning from managing shifts to managing businesses), and interpersonal skills (motivation, delegation, decision making, and problem solving).

5. Implement ongoing programs that focus on leadership skills, such as managing multiple priorities, creating change, and improving presentation skills.

6. Know that in the right culture, leaders can be found at entry level.

7. Recognize, reward, and celebrate leaders for their passion, dedication, and results.

In theory, you have authority over your employees because you have formal authority, or the right to command, given to you by the organization. You are the boss and you have the *power*, the ability to command. You control the hiring, firing, raises, rewards, discipline, and punishment. In all reality, your authority is anything but absolute. Real authority is conferred on your subordinates, and you have to earn the right to lead them. It is possible for you to be the formal leader of your work group as well as have someone else who is the informal leader actually calling the shots.

The relationship between you and your people is a fluid one, subject to many subtle currents and cross-currents between them and you. If they do not willingly accept your authority, they have many ways of withholding success. They can stay home from work, come in late, drag out the work into overtime, produce inferior products, drive your guests away with rudeness and poor service, break the rules, refuse to do what you tell them to, create crises, and punish you by walking off the job and leaving you in the lurch. Laying down the law, the typical method of control in hospitality operations, does not necessarily maintain authority; on the contrary, it usually creates a negative, nonproductive environment.

What this all adds up to is that your job as a supervisor is to direct and oversee a group of employees who are often untrained, all of whom are different from each other, and some of whom would rather be working somewhere else. You are dependent on your employees to do the work for which you are responsible. You will succeed only to the degree that they permit you to succeed. It is your job to get the workers to do their best for the enterprise, for the guests, and for you. How can one do this?

---

leadership
Direction and control of the work of others through the ability to elicit voluntary compliance.

formal authority
The authority granted by virtue of a person's position within an organization.

real authority
The authority that employees grant a supervisor to make the necessary decisions and carry them out.

formal leader
The person in charge based on the organization chart.

informal leader
The person who, by virtue of having the support of the employees, is in charge.

do the right things right
To be a leader and manager; to be both effective and efficient.

management by walking around (MBWA)
Spending a significant part of the day talking to employees, guests, and peers while listening, coaching, and troubleshooting.

Peter Drucker, a distinguished leadership expert, noted, "Managers are people who do things right, and leaders are people who do the right things." Think about that for a moment. In other words, managers are involved in being efficient and in mastering routines, whereas leaders are involved in being effective and turning goals into reality. As a supervisor and leader, your job is to **do the right things right**, to be both efficient and effective. An effective supervisor in the hospitality industry is one who, first, knows and understands basic principles of management, and second, applies them to managing all the resource operations.

In the hospitality industry we use a technique referred to as **management by walking around (MBWA)**—spending a significant part of your day talking to your employees, your guests, and your peers. As you are walking around and talking to these various people, you should be performing three vital roles discussed in this book: listening, coaching, and troubleshooting.

leadership style
A pattern of interaction that a supervisor or manager uses in directing subordinates.

Choosing a Leadership Style

LEARNING OBJECTIVE: Compare and contrast the different leadership styles.

The term **leadership style** refers to your pattern of interacting with your subordinates: how you direct and control the work of others, and how you get them to produce the goods and services for which you are responsible. It includes not only your manner of giving instructions but also the methods and techniques you use to motivate your workers and to ensure that your instructions are carried out.

There are several different forms of leadership style—autocratic, bureaucratic, democratic, and laissez-faire being the most popular styles today (Figure 2.2). Before choosing a style of leadership, one must identify the pros and cons of each and then decide if it will be the most effective style in the hospitality industry.

Autocratic leadership style can be identified with the early, classical approach to management. A supervisor practicing an autocratic style is likely to make decisions without input from staff, to give orders without explanation or defense, and to expect the orders to be obeyed. When this style of leadership is used, employees become dependent on supervisors for instructions. The wants and needs of employees come second to those of the organization and the supervisor.

In *bureaucratic* leadership style, a supervisor manages "by the book." The leader relies on the property's rules, regulations, and procedures for decisions that he makes. To the employees, their leader appears to be a "police officer." This style is appropriate when employees can be permitted no discretion in the decisions to be made.

Democratic (also called *participative*) leadership style is almost the reverse of the autocratic style discussed previously. Democratic supervisors want to share decision-making responsibility. They want to consult with group members and to solicit their participation in making decisions and resolving problems that affect employees. Employers strongly consider the opinions of employees and seek their thoughts and suggestions. All employees are informed about all matters that concern them. One could compare a democratic supervisor to a coach who is leading his or her team. Chefs should share responsibility with team members within the restaurant/organization, but holding them accountable for decisions and inactions.

Laissez-faire (also called *free-reign*) leadership style refers to a hands-off approach in which the supervisor actually does as little leading as possible. In effect, the laissez-faire supervisor delegates all authority and power to the employees. The supervisor relies on the employees to establish goals, make decisions, and solve problems. At best, the laissez-faire style has limited application to the hospitality industry.

Forms of Leadership Style

Autocratic	Bureaucratic	Democratic	Laissez-Faire
Sees themselves as sole decision maker	Strictly by the book	Almost a reversal of autocratic	Hands-off approach
Shows little concern about others' opinion	Relies on rules and regulations	Wants to share responsibilities	Turns over control; delegates authority
Focuses on completing goals	Act like they are a police officer	Collaborates opinions when decision making	Works well when employees are self-motivated
Dictates tasks to be accomplished	Appropriate when employees are permitted no discretion	Is a concerned *coach* of the team	Little application in the hospitality industry

FIGURE 2.2: The pros and cons of each leadership style.

Choosing your leadership style will be one of your most important career decisions.
EDHAR/Shutterstock

✳ THE OLD-STYLE BOSS

reward and punishment
A method of motivating performance by giving rewards for good performance and by punishing poor performance.

In the hospitality industry, the traditional method of dealing with hourly employees has generally been some variation of the command–obey method combined with carrot-and-stick motivation of **reward and punishment**. The motivators relied on to produce the work are money (the carrot) and fear (the stick)—fear of punishment, fear of losing money by being fired. All too often, the manner of direction is to lay down the law in definite terms, such as cursing, shouting, and threatening as necessary to arouse the proper degree of fear to motivate the employee.

autocratic method
Behaving in an authoritarian or domineering manner.

People who practice this **autocratic method** of managing employees believe that it's the only method that employees will understand. Perhaps that is the way the supervisor was raised, or perhaps it is the only method the supervisor has ever seen in action. In any case, it expresses the view of the people involved that "employees these days are no good."

Some employees are simply bad workers. However, cursing, shouting, and threatening seldom helps them improve. Many employees do respond to a command–obey style of direction, but those employees often come from authoritarian backgrounds and have never known anything else. This style is traditional and military—perhaps the style of dictatorship in countries from which some immigrants come. However, for your average American employee, it does not work. It might be enough to keep people on the job, but they are not working to their full capacity.

When coupled with a negative view of the employee, this style of direction and control is far more likely to increase problems than to lessen them, and to backfire by breeding resentment, low morale, and adversary relationships. In extreme cases, the boss and the company become the bad guys, the enemy, and workers give as little as possible and take as much as they can. In response, close supervision and tight control are required to see that nobody gets away with anything. In this type of atmosphere, guest service suffers and patrons go somewhere else.

We are also learning more about what causes employees to work productively, including many of the things we have been talking about, such as positive work climate, person-to-person relations, and other people-oriented methods and techniques. At this point, let us look at some current theories of leadership and see how—or whether—they can be applied in hotel and food-service settings. These theories emerged in the 1950s and 1960s, following the discovery that making employees happy does not necessarily make them productive. The theories are based on what behavioral scientists, psychologists, and sociologists tell us about human behavior. They explore what causes people to work productively and how this knowledge can be used in managing employees.

✳ THEORY X AND THEORY Y

Theory X
The managerial assumption that people dislike and avoid work, prefer to be led, avoid responsibility, lack ambition, want security, and must be coerced, controlled, directed, and threatened with punishment to get them to do their work.

In the late 1950s, Douglas McGregor of the MIT School of Industrial Management advanced the thesis that business organizations based their management of employees on assumptions about people that were wrong and were actually counterproductive. He described three faulty assumptions about the average human being as **Theory X**:

1. They have an inborn dislike of work and will avoid it as much as possible.
2. They must be "coerced, controlled, directed, and threatened with punishment" to get the work done.
3. They prefer to be led, avoid responsibility, lack ambition, and want security above all else.

Theory Y
The hypothesis that work is as natural as play or rest; people will work of their own accord toward objectives to which they feel committed.

McGregor argues: "These characteristics are not inborn." He believed people behaved this way on the job because they were treated as though these things were true. In fact, he stated, "This is a narrow and unproductive view of human beings," and he proposed **Theory Y**:

1. Work is as natural as play or rest; people do not dislike it inherently.
2. Control and the threat of punishment are not the only means of getting people to do their jobs. They will work of their own accord toward objectives to which they feel committed.
3. People become committed to objectives that will fulfill inner personal needs, such as self-respect, independence, achievement, recognition, status, and growth.

4. Under the right conditions, people learn not only to accept responsibility but also to seek it. Lack of ambition, avoidance of responsibility, and the desire for security are not innate human characteristics.

5. Capacity for applying imagination, ingenuity, and creativity to solving on-the-job problems is "widely, not narrowly, distributed in the population."

6. The modern industrial organization uses only a portion of the intellectual potential of the average human being.

Thus, if work could fulfill both the goals of the enterprise and the needs of the workers, they would be self-motivated to produce, and consequently, coercion and the threat of punishment would be unnecessary.

Theory X fits the old-style hospitality manager to a T, and it is safe to say that this pattern of thinking is still common in many other industries as well. However, behavioral science theory and management practice have both moved in the direction of Theory Y. Theory Y is a revised view of human nature with emphasis on using the full range of workers' talents, needs, and aspirations to meet the goals of the enterprise.

A popular way of moving toward a Theory Y style of people management is to involve one's employees in certain aspects of management, such as problem solving and decision making. Usually, such involvement is carried out in a group setting: meetings of the employees for the specific purpose of securing their input. The degree of involvement the boss allows or seeks can vary from merely keeping the employees informed of things that affect their work to delegating decision making entirely to the group.

The participative management style, mentioned in Chapter 1, results when employees have a high degree of involvement in such management concerns as planning and decision making. Enthusiasts of a participatory style of leadership believe that the greater the degree of employee participation, the better the decisions and the more likely they are to be carried out.

However, others point out that the degree of participation that is appropriate for a given work group will depend on the type of work, the people involved, the nature of the problem, the skill and sensitivity of the leader, and the pressures of time—the situational leadership approach, to be discussed shortly. The degree to which the boss involves the employees may also vary from time to time, depending on circumstances. You are not going to make a group decision when a person who is intoxicated is making a scene in the dining room or when a fire alarm is going off on the seventh floor.

❋ SITUATIONAL LEADERSHIP

situational leadership
Adaptation of leadership style to fit the situation.

In the **situational leadership** model developed by Kenneth Blanchard and Paul Hersey, leadership behaviors are sorted into two categories: directive behavior and supportive behavior.[6] *Directive behavior* means telling an employee exactly what you want done, as well as when, where, and how to do it. The focus is to get a job done, and it is best used when employees are learning a new aspect of their jobs. *Supportive behavior* is meant to show caring and support for your employees by praising, encouraging, listening to their ideas, involving them in decision making, and helping them reach their own solutions. This method is best used when an employee lacks commitment to do a job.

directing style
Leadership style that is high on directive and low on supportive behaviors.

By combining directive and supportive behaviors, Hersey and Blanchard came up with four possible leadership styles for different conditions. When an employee has much commitment or enthusiasm but little competence to do a job, a **directing style** is needed; this is high on directive and low on supportive behaviors. Suppose that you have a new employee full of enthusiasm who knows little about how to do the job. A directing style is appropriate: You train the new employee by giving multiple instructions, you make the decisions, you solve the problems, and you supervise closely. Enthusiastic beginners need this direction. A directing style is also appropriate when a decision has to be made quickly and there is some risk involved, such as when there is a fire and you need to get your employees out of danger.

As new employees get into their jobs, they often lose some of their initial excitement when they realize that the job is more difficult or not as interesting as they originally envisioned. This is

Transactional leaders motivate employees by appealing to their self-interest.

Monkey Business Images/Shutterstock

coaching style
Leadership style that includes a lot of directive behaviors to build skills and supportive behaviors to build commitment.

supporting style
Leadership style that is high on supportive behaviors and low on directive behaviors.

delegating style
Leadership style that is low on directive and supportive behaviors because you are turning over responsibility for day-to-day decision making to the employee.

transactional leaders
Leaders that motivate employees by appealing to their self-interest.

the time to use a coaching style, with a lot of directive behaviors to continue to build skills and supportive behaviors to build commitment. In addition to providing much direct supervision, you provide support. You listen, you encourage, you praise, you ask for input and ideas, and you consult with the employee. As employees become technically competent on the job, their commitment frequently wavers between enthusiasm and uncertainty. In a situation like this, the use of a supporting style that is high on supportive behaviors and low on directive behaviors is required. If an employee shows both commitment and competence, a delegating style is suitable. A delegating style of leadership is low on directive and supportive behaviors because you are turning over responsibility for day-to-day decision making to the employee doing the job. These employees don't need much direction, and they provide much of their own support.

Using this view of situational leadership, you need to assess the competence and commitment level of your employee in relation to the task at hand before choosing an appropriate leadership style (Figure 2.2). As a supervisor, your goal should be to build your employees' competence and commitment levels to the point where you are using less time-consuming styles, such as supporting and delegating, and getting quality results.

❀ TRANSACTIONAL LEADERSHIP

Transactional leaders motivate workers by appealing to their self-interest. In other words, workers do their jobs and give their compliance in return for rewards such as pay and status.

James MacGregor Burns, a prominent leadership researcher, wrote a significant book titled *Leadership*.[7] In the book, Burns describes leadership as falling within two broad categories: transactional and transformational. Transactional leadership seeks to motivate followers by appealing to their own self-interest. Transactional leaders stress communication of job assignments, work standards, goals, and so on in order to maintain the status quo. Its principles are to motivate by the exchange process. In other words, employees do their jobs and give their compliance in return for desired rewards such as pay and status.

Transactional leadership behavior is used to one degree or another by most leaders. However, as the old saying goes, "If the only tool in your toolbox is a hammer, every problem will look like a nail." Transactional leadership seeks to influence others by exchanging work for wages, but does

Courtesy of Laura Horetski

I was asked what leadership means to me as the front office manager of a major hotel chain. When you look at the definition of a leader, it states, "one who leads or guides." And we've all heard the phrase "lead by example." I don't think that is enough. There are at least seven qualities of leadership that I can think of that make a good leader.

A good leader is someone who is not afraid to get his or her hands dirty. Someone who will do the same job, duty, or task alongside subordinates, peers, and supervisors, while keeping a positive attitude. This helps build and gain respect. Besides, how else can you expect someone to do the job you ask him or her to do if you do not know how or are not willing to do it yourself?

A person who listens, not just hears. Pay complete attention to what people are saying. Look them in the eyes, acknowledge them, and don't interrupt. Ask questions of clarification, reiterate what they are saying, and ask the person if you understand correctly. But listening doesn't stop there. You need to follow through on the conversation and do what you said you would do. Build integrity and trust.

Make good business decisions but show compassion when needed. The bottom line is the bottom line. You don't have to be cruel to accomplish tough results. Be honest, state the facts, ask for suggestions, and make the best decision. A lot of times things look good on paper but don't really work in reality. Sometimes those who are on the front lines and performing the job everyday give the best answers. Not only do you get the answer you may be looking for, you also build confidence and develop future managers and supervisors.

Treat others fairly, including yourself. Favoritism has no place at work. Is it hard not to rely solely on those who are the strongest? Absolutely. But as a leader, it's your job to encourage and improve your super performers. Favoritism also provides an impartial playing field for everyone. Learn to delegate to improve teamwork and lighten the load for everybody.

Learning never stops. I try to learn something new every day, sometimes without even seeking it out. You also need to be open to learning from subordinates, peers, and supervisors. There is no one person who has all the answers. The workforce is always changing in every aspect, and you need to be able to adapt. It's important to stay fresh and current. Think outside the box; there's usually more than one way to accomplish a goal. If the way you tried doesn't work, you've learned, and it's what you take from the experience that's important.

Develop those under you. The fastest way to move up is to train someone to take your job. This is one of the best ways to show leadership. Too often, people are afraid of "losing their jobs" because someone else knows how to do their job. This is not the case. This frees up time for you to develop your skills in another position you are interested in, while developing your successor.

Finally, you need to be able to admit that you've made mistakes. As I said earlier, no one person has all the answers. You're going to stumble, trip, and even fall. But those who are honest and admit their failures will gain the respect of others and will learn the most. There's a saying, "No question is a dumb question." I say, "No mistake is a mistake."

I have had many teachers throughout my career, and I have taken pieces of their leadership styles along with me. You are never done learning how to lead. Each circumstance has its own manner in how to approach it. Above all else, a good leader is fair and ever changing. Have fun!

not build on the employee's need for meaningful work or tap into their creativity. The most effective and beneficial leadership behavior to achieve long-term success and improved performance is transformational leadership.[8] Rewards to consider in motivating and maintaining kitchen staff, for example, would be to pay for training for their professional development or professional organizations such as the Florida Restaurant & Lodging Association (FRLA), American Culinary Federation (ACF), and any industry certifications or restaurant and club associations. To "lead by example" as in action is often stronger and more effective than words. In a small kitchen environment people often indirectly analyze those their working alongside.

❋ TRANSFORMATIONAL LEADERSHIP

Transformational leadership is about finding ways of long-term higher-order changes in follower behavior. It is the process of gaining performance above expectations by inspiring employees to reach beyond themselves and do more than they originally thought possible. This is accomplished by raising their commitment to a shared vision of the future. As illustrated in Figure 2.3,

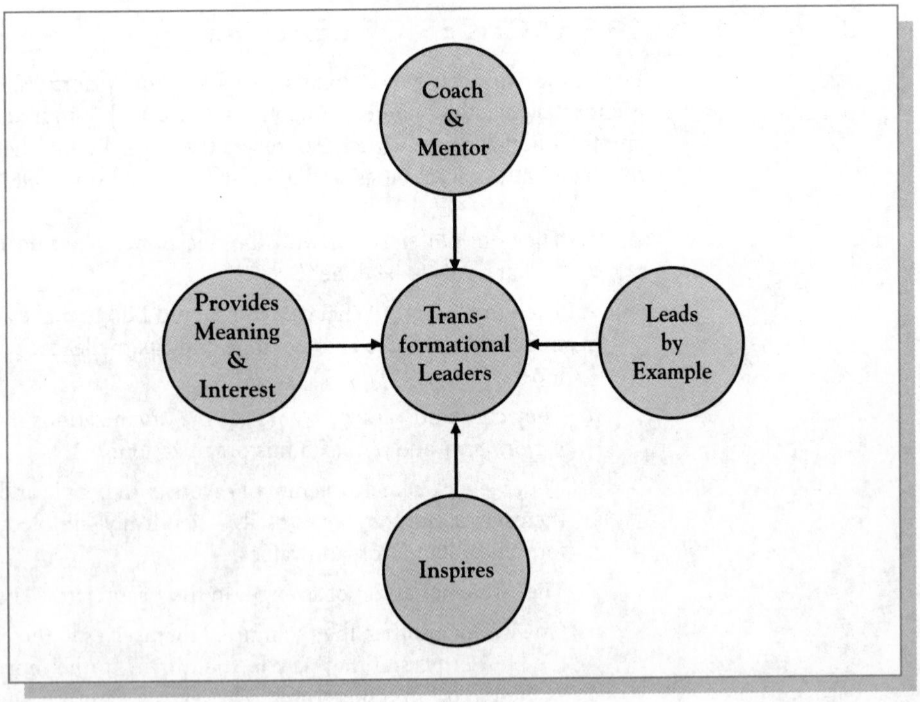

FIGURE 2.3: Transformational leaders.

transformational leaders
Leadership that motivates employees by appealing to their higher-order needs, such as providing employees with meaningful, interesting, and challenging jobs, and acting as a coach and mentor.

instead of using rewards and incentives to motivate employees, **transformational leaders** do the following:

1. Communicate with and inspire employees about the mission and objectives of the company.
2. Provide employees with meaningful, interesting, and challenging jobs.
3. Act as a coach and mentor to support, develop, and empower employees.
4. Lead by example.

By appealing to employees' higher-order needs, transformational leaders gain much loyalty that is especially useful in times of change. Transformational leaders generally have lots of charisma. One of the most transformational leaders was Dr. Martin Luther King, Jr. Dr. King dedicated his life to achieving rights for all citizens by nonviolent methods. In 1964, Dr. King won the Nobel Peace Prize and is perhaps best remembered for his "I Have a Dream" speech. Delivered in front of the Lincoln Memorial in the summer of 1963, Dr. King inspired his listeners to feel that history was being made in their very presence.

One hospitality example of a transformational leader was Horst Schultze, who developed Ritz-Carlton hotels and led the Ritz-Carlton Hotel Company (now part of Marriott International) to win the Malcolm Baldrige National Quality Award for service in 1992. A more recent example is the 2010 winner of the Malcolm Baldrige National Quality Award for health care, Advocate Good Samaritan Hospital. Its vision statement is to provide an exceptional patient experience marked by superior health outcomes and service.[9] Another example is Bob Basham's People Dedicated to Quality (PDQ). Basham set a vision and was able to communicate it so well to all employees that they went the extra mile to ensure the company's and their own success.

❋ SIX EMOTIONAL LEADERSHIP STYLES

Daniel Goleman, Richard Boyazits, and Annie McKee proposed Six Emotional Leadership Styles theory in their book *Primal Leadership*. The theory highlights the strengths and weaknesses of the six leadership styles: visionary, coaching, affiliative, democratic, pacesetting, and commanding.

✳ PRACTICES OF LEADERS

Leaders vary in their values, managerial styles, and priorities. Peter Drucker, renowned management scholar, author, and consultant, discussed with hundreds of leaders their roles, their goals, and their performance. Drucker observes that regardless of their enormous diversity with respect to personality, style, abilities, and interest, effective leaders all behave in much the same way:[10]

1. They did not start out with the question, "What do I want?" They started out asking, "What needs to be done?"

2. Then they asked, "What can and should I do to make a difference?" This has to be something that both needs to be done and fits the leader's strengths and the way she or he is most effective.

3. They constantly asked, "What are the organization's *mission* and *goals*? What constitutes *performance* and *results* in this organization?"

4. They were extremely tolerant of diversity in people and did not look for carbon copies of themselves. But they were totally—fiendishly—intolerant when it came to a person's performance, standards, and values.

5. They were not afraid of strength in their associates. They gloried in it.

6. One way or another, they submitted themselves to the *mirror test*—that is, they made sure the kind of person they saw in the mirror in the morning was the kind of person they wanted to be, to respect, and to believe in. This way they fortified themselves against the leader's greatest temptations—to do things that are popular rather than right and to do petty, mean, sleazy things. Finally, these leaders were not preachers, they were doers.[11]

✳ EMPOWERMENT

empowerment
To give employees additional responsibility and authority over their decisions, resources, and work.

Empowerment, which is also discussed in Chapter 7, is a technique used by participative leaders to share decision-making authority with team members. Empowerment means giving employees more control over their decisions, resources, and work. The relationship between employees and the company is more of a partnership, where the employees feel responsible for their jobs and have a share of ownership in the enterprise.[12] Empowered employees take responsibility and seek to solve problems; they see themselves as a network of professionals all working toward the same goals.

Empowered employees have the power to make decisions without a supervisor. They can even bend the rules if they think it will provide the guest with the service they are looking for. Empowered employees are more likely to take ownership of their jobs when they have the power to make a difference.

An example of empowered employees making a difference happened at Hampton Inns after it began a program of refunds to guests who were dissatisfied with their stays. The refund policy created far more additional business than it cost, but a surprise bonus was the increased morale when employees—everyone from front-desk associates to housekeepers—were empowered to give refunds. With greater participation and job satisfaction, employee turnover fell by more than half.[13] Empowerment has strong links to total quality management, which is discussed at length in Chapter 7.

Developing Your Own Style

LEARNING OBJECTIVE: Identify the necessary steps in creating your own leadership style.

Applying theory to reality is going to be something you work out for yourself. No one can teach you. Since even the theorists disagree among themselves, the choice is wide open. But don't throw it all out; a lot of what the behavioral scientists are saying can be very useful to you. There does seem to be general agreement, supported by research and experience, that the assumptions

Theory X makes about people are, at best, unproductive and, at worst, counterproductive, if not downright destructive.

> Success in life is measured by what we have overcome to become what we are and by what we have accomplished. Who we are is more important than what position we have.

However, an authoritarian style of leadership can be effective and even necessary in many situations, and there is actually no reason why it cannot be combined with a high concern for the workers and achieve good results.

As for Theory Y, probably two-thirds of the workforce has the potential for a Theory Y type of motivation—that is, working to satisfy such inner needs as self-respect, achievement, independence, responsibility, status, and growth. The problem with applying this theory in the hospitality industry is really not the workers. It is the nature of the work, the number of variables you have to deal with (including high worker turnover), the unpredictability of the situation, the tradition of authoritarian carrot-stick management, and the pressures of time.

The pace and pattern of the typical day do not leave much room for group activity or for planning and implementing changes in work patterns to provide such motivation. Furthermore, your own supervisor or your company's policies may not give you the freedom to make changes. In conclusion, Theory Y does not always work for everyone.

However, it is remarkable what is possible when an imaginative and determined manager sets out to utilize this type of motivation and develop this type of commitment. We talk more about motivation in Chapter 7.

The best style of leadership, for you, is whatever works best in terms of these three basics: your own personality, the employees you supervise, and the situations you face. It should be a situational type of leadership, just as your management style must be a flex style that reacts to situations as they arise.

You might give an order to Todd, but say "please" to Louis. You might stop a fight in the kitchen with a quick command when server Jenni and server Chris keep picking up each other's orders, and then later, you might spend a good hour with the two of them helping them reach an agreement to stop their running battle. You might see responsibilities you could delegate to Evelyn or Tyler. You might see opportunities to bring workers in on solving work problems, or you might solve them yourself because of time pressures or because the problems are not appropriate for group discussion.

You can borrow elements and techniques of Theory Y without erecting a whole system of participative management. If something does not work for all three of you—yourself, the workers, the situation—don't do it.

Much attention has been focused on corporate leadership and the associated scandals, including misuse of power, embezzlement, lack of moral and ethical behavior, lying, and other forms of improper behavior that have shaken the public's confidence in corporate leadership. Add to this the huge salaries many of these leaders are (and were) paid—even if their company did poorly, they still received a large salary increase. Events such as these have caused public opinion to turn against corporate leadership, with demands that corporate leaders become more ethical and moral in their behavior and make better decisions.

✻ VISION AND AWARENESS

As a leader, you will need to have a vision that is realistic and credible—a vision that everyone in the organization (or department) can rally around. Your vision—to be the best, or to be the most popular—needs to be complemented by the company purpose and **mission statement**. It needs to be ambitious and inspire enthusiasm. Leaders make things happen because they have developed the knowledge, skills, and attitude to positively motivate others to reach common goals.

What you need most in sculpting a vision and finding what works best is *awareness*: awareness of yourself and the feelings, desires, biases, abilities, power, and influence you bring to a situation; awareness of the special needs and traits of your various workers; and awareness of the situation, the big picture, so you can recognize what is needed as far as conceptual and human skills.

mission statement
Describes the purpose of the organization and outlines the kinds of activities performed for guests.

Engaged employees are more productive.
wavebreakmedia/Shutterstock

Leadership is also about change. As you develop awareness and vision, you will see an obvious need for change. Remember, there is a six-step method of making changes: (1) state the purpose; (2) involve others; (3) test the plan before you implement it companywide; (4) introduce the change; (5) maintain and reinforce the change; and (6) follow up.[14]

The best style of leadership is to be *yourself*. Trying to copy someone else's style usually does not work—the situation is different, you are different, the shoe does not fit.

Today, hospitality leaders are expected to have the communication skills to mobilize the energy and resources of a management team. Leaders are expected to be visionaries who see the future clearly and articulate the vision so that others can follow.

❋ EMPLOYEE ENGAGEMENT

As a people leader, your job is to inspire your employees to bring their personal greatness to work every day and to invest their best in your business—and that's a hard job. Getting the best is about building a culture of trust, connection, growth, and service. That culture is sustained and enlivened by supervisors and managers, one person at a time. That's employee engagement.[15]

Another aspect of employee engagement is whether the organization has a compelling reason for being and whether its employees believe in the reason. If leaders live this way, rather than just talk this way, and if compensation, bonuses, products, and services all subscribe to the reason for being, then success will result.

Companies with a highly engaged culture have shown consistent growth and profitability. One study has shown that companies with 60–100 percent engaged employees report an average shareholder return of 20.2 percent. But companies with less than 40 percent engagement show a 9.6 percent return.[16] So, the message is to show personal interest and concern to all employees, even when you are so busy you think you don't have the time to spare.

Ethics

LEARNING OBJECTIVE: Identify the ethical considerations of a true leader.

Although there are many definitions of **ethics**, ethics can generally be thought of as a set of moral principles or rules of conduct that provide guidelines for morally right behavior. To give you an idea of how ethics are involved in your job as a hospitality supervisor, let's look at three scenarios:

ethics
The study of standards of conduct and moral judgment; also, the standards of correct conduct.

1. You've completed interviewing a number of candidates for a security position. One of the top three candidates is a relative of a supervisor in another department with whom you are close friends. You've been getting pressure from your friend to hire this candidate and you don't want to alienate him, so you hire his relative, even though one of the other candidates is more suited for the job.

2. Business at the hotel could be better on weekends, so you advertise 25 percent discounts on rooms. To keep profitability high, you inflate the room rate before taking the discount.

3. As a purchasing manager, you know that the policy is not to accept free gifts from vendors. But one day when you are out to lunch with a vendor, he offers you free tickets to a Major League Baseball game and you accept them. You can't wait to take your son to the game.

As you can see from these examples, moral principles and standards of conduct are just as necessary in the workplace as they are in your personal life. There are ethical considerations in many of the decisions that you will make, from personnel management issues to money issues to purchasing and receiving practices. Unfortunately, the hospitality industry as a whole has not written its own code of ethics, but you will find that some operations have written their own.

Why is a code of ethics needed for hospitality operations? Just look at the temptations: stockrooms full of supplies that can be used at home and are often loosely inventoried, any kind of alcoholic beverage you want, empty hotel rooms, gambling, high-stress jobs, irregular hours, pressures to meet guests' needs. It can be easy to lose a sense of right and wrong in this field.

Stephen Hall suggests five questions that you can use to help decide how ethical a certain decision is:[17]

1. Is the decision legal?
2. Is the decision fair?
3. Does the decision hurt anyone?
4. Have I been honest with those affected?
5. Can I live with my decision?

These questions can provide much guidance.

Companies that create an excellent reputation built on mutual trust and those with ethical guidelines that help employees avoid risks and problems are more likely to be successful. Integrity, always doing the right thing, and dedication are both very important aspects relating to ethics in the workplace; so is aligning the work of individuals and teams with the organization's vision, mission, and goals.

The Supervisor as Mentor

LEARNING OBJECTIVE: Apply leadership skills to become an employee mentor.

This topic is a wonderful way to finish this chapter on supervision. As you become more experienced and proficient at being a hospitality supervisor, it is more likely that you will be a mentor to those who are less experienced and less skilled. A **mentor** is a leader, an excellent role model, and a teacher. A supervisor often functions as a mentor to a worker by providing guidance and knowledge on learning the operation and moving up the career ladder. The relationship often resembles that between a teacher and a student. At other times, the mentor simply provides an example of professional behavior with minimal or no interaction with the worker. Being a mentor can provide feelings of pride and satisfaction because you have contributed to someone else's career development.

mentor
An experienced and proficient person who acts as a leader, role model, and teacher to those less experienced and less skilled.

CASE STUDY: Firm, Fair, and Open?

Latisha has just been hired as the dining room supervisor on the noon shift in the coffee shop of a large hotel. She came from a similar job in a much smaller hotel, but she feels confident that she can handle the larger setting and the larger staff. Because she is eager to start things off right, she asks all the servers to stay for 10 minutes at the end of the shift so that she can say a few words to everyone. She begins by describing her background and experience and then proceeds to her philosophy of management. "I expect a lot of my people," she says. "I want your best work, and I hope you want it, too, for your own sake. You will not find me easy, but you will find me fair and open with you, and I hope you will feel free to come to me with suggestions or problems. I can't solve them all, but I will do my best for you." She smiles and looks at each one in turn.

"Now, the first thing I want to do," she continues, "is to introduce a system of rotating your stations so that everyone gets a turn at the busiest tables and the best tips and the shortest distance to the kitchen. I've posted the assignments on the bulletin board, and you will start off that way tomorrow and keep these stations for a week."

She says, "I will be making some other changes, too, but let's take things one at a time. Are there any questions or comments?" Latisha pauses for three seconds and then says, "I am very particular about being on time, about uniforms and grooming, and about prompt and courteous customer service. I advise you all to start off tomorrow on the right foot and we'll all be much happier during these hours we work together. See you tomorrow at 10:25."

Case Study Questions

1. What kind of impression do you think that Latisha is making on the employees?
2. What are the good points in her presentation?
3. What mistakes do you think she is making?
4. Why did nobody ask questions or make comments?
5. From this first impression, what would you say is her management style?
6. Do you think that people will feel free to come to her with suggestions and problems?
7. Do you think that she will set a good example?
8. Is she fair in her demands?
9. Do you think that her people will "start off on the right foot," as she suggests?
10. Do you think that she sees herself clearly? Is she aware of her impact on others?

KEY POINTS

1. Being a leader means guiding or influencing the actions of your employees to reach certain goals. A leader is a person whom people follow voluntarily.
2. As a supervisor, you have been given the formal authority to oversee your employees. Your subordinates confer real authority, and you have to earn the right to lead them.
3. As a supervisor and leader, your job is to do the right things right.
4. Leadership style refers to your pattern of interacting with your subordinates, how you direct and control the work of others, and how you get them to produce the goods and services for which you are responsible.
5. The old-style boss uses an autocratic method of managing employees that relies on the motivators of money or fear.
6. According to McGregor, the autocratic style is typical of Theory X bosses. Theory Y bosses believe that workers will work of their own accord toward objectives to which they feel committed.
7. In situational leadership, the leadership style is adapted to the uniqueness of each situation. The four primary styles of leading are directing, coaching, supporting, and delegating.
8. Transactional leaders appeal to employees' self-interest. Transformational leaders appeal to employees' higher-order needs.
9. Employers and employees must develop mutual respect for success.
10. Ethics can be thought of as a set of moral principles or rules of conduct that provide guidelines for morally correct behavior. The five questions presented in the chapter provide guidance for making ethical decisions.

11. A supervisor often functions as a mentor to a worker by providing guidance and knowledge on learning the operation and moving up the career ladder. The relationship resembles that between a teacher and a student.

KEY TERMS

autocratic method

coaching style

delegating style

directing style

do the right things right

empowerment

ethics

formal authority

formal leader

informal leader

leader

leadership

leadership style

management by walking
  around (MBWA)

mentor

mission statement

power

real authority

reward and punishment

situational leadership

supporting style

Theory X

Theory Y

transactional leader

transformational leader

vision

REVIEW QUESTIONS

Answer each question in complete sentences. Read each question carefully and make sure that you answer all parts of the question. Organize your answers using more than one paragraph when appropriate.

1. Identify typical hourly jobs in foodservice and lodging establishments. Include both skilled and unskilled jobs.
2. If a restaurant's turnover rate is 100 percent, what does that mean?
3. Define *leader* and *leadership*.
4. What is meant by "do the right things right"?
5. Compare and contrast the concepts of formal and real authority.
6. Why does a fear-and-punishment approach to supervision usually create a negative, nonproductive environment?
7. In two sentences, describe the essence of each of the following leadership styles: autocratic, Theory X, Theory Y, situational, transactional, and transformational.
8. Identify the six practices of successful managers.

ACTIVITIES AND APPLICATIONS

1. Discussion Questions

- Why do you think turnover is high in hotels and restaurants? If you resigned from a hospitality job, what were your reasons? What could be done by management to reduce turnover?
- Why might it be difficult to supervise employees in minimum-wage or low-wage hospitality jobs that require no special skills? What kinds of problems might arise? What can be done to solve these problems or avoid them?
- Which view of people is more accurate: Theory X or Y? Give examples from your own work experience to support your view.
- Under what circumstances might you need to be an autocratic leader?
- Describe situations in which each of the four styles of situational leadership would be appropriate.

2. Group Activity

Using the three situations described in the Ethics section, use the five questions from Hall that are listed in that section to examine how ethical each decision was. Discuss each question as a group and have one person record your ideas.

3. Leadership Assessment

Using Figure 2.4, assess your leadership abilities.

Using the letters in the word *leader*, think of a leader's qualities and actions that make him or her a good leader and fit as many as possible into L-E-A-D-E-R. For example, L—lends a hand; E—ethical; A—aware; D—. . . , and so on.

Directions: Answer each question realistically using the following scale.
               1—I do this seldom or never.
               2—I do this occasionally.
               3—I do this always or most of the time.

If your total points are:

  175–210   You are an excellent leader.

  120–175   You are probably new to leadership and trying hard. Work on the areas where you rated yourself a "1."

Below 120   You need to improve in many areas.

Personal Qualities

SELF-CONFIDENT
_____   1. I believe in myself.

CONSISTENT/COMMITTED
_____   2. I stay focused on the vision.
_____   3. I keep my word.

UPBEAT/POSITIVE
_____   4. I am a positive thinker.
_____   5. I am an optimist—my glass is half full.

HONEST/OPEN
_____   6. I am up-front and honest with others.
_____   7. I do not get defensive in conversation.

INTEGRITY
_____   8. I honor my commitments and promises.

FUNNY
_____   9. I use my sense of humor.
_____   10. I love to laugh at myself.

RISK-TAKING
_____   11. I take calculated risks when appropriate.
_____   12. I let myself and others make mistakes.

CREATIVE/DIVERGENT & ABSTRACT THINKER
_____   13. I encourage and try to look at things in new and different ways.

INTELLIGENT/COMPETENT
_____   14. I am knowledgeable and competent in my field.
_____   15. I can make the complex simple.
_____   16. I am a lifelong learner.

WIN/WIN ORIENTATION
_____   17. In interactions with others, I want everyone to be a winner.

ETHICAL
_____   18. I maintain ethical standards.

ORGANIZED
_____   19. My work and paperwork are well-organized.

FIGURE 2.4: Leadership assessment tool.

LOOKS TO FUTURE

_____ 20. I keep an eye and ear directed to trends in my industry.

_____ 21. I try to innovate.

CONGRUENT

_____ 22. I walk the talk.

FLEXIBLE

_____ 23. I keep an open mind.

_____ 24. I can change my mind and change my plans when appropriate.

Vision

VISION

_____ 25. I let my company's vision be my guide.

PERSONAL VISION

_____ 26. I write and revise my personal mission statement yearly.

Managing Relationships

SUPPORTING

_____ 27. I seek first to understand, then to be understood.

_____ 28. I genuinely show acceptance and positive regard toward staff.

_____ 29. I refrain from rudeness and treat others diplomatically and politely.

_____ 30. I maintain the self-respect of all individuals.

_____ 31. I have an open-door policy.

DEVELOPING/MENTORING

_____ 32. I believe developing and mentoring others is part and parcel of being a professional, and that this will enhance, not detract, from my career.

_____ 33. I actively develop and act as a mentor.

EMPOWERING

_____ 34. I actively empower staff to do their jobs in the manner they want as long as it supports our mission.

RECOGNIZING & REWARDING

_____ 35. I use a variety of techniques to recognize and reward staff for their achievements and contributions.

_____ 36. I provide fair, specific, and timely recognition and rewards.

_____ 37. I recognize and reward more people than just the top performers.

_____ 38. I use recognition and rewards that are desirable to the recipients.

MANAGING CONFLICT & CHANGE

_____ 39. I see conflict as an opportunity to grow.

_____ 40. I mediate conflicts and encourage constructive resolution of conflicts.

_____ 41. I work on building and maintaining cooperative staff relationships.

_____ 42. I realize that people generally don't resist change, but they do resist being changed.

TEAMBUILDING

_____ 43. I understand the teambuilding process.

_____ 44. I model teambuilding skills.

_____ 45. I help form and monitor teams.

FIGURE 2.4: _(continued)_

NETWORKING

——— 46. I actively network with people both within and outside of the industry.
——— 47. I keep in touch with members of my network.
——— 48. I am good at remembering names.

Managing the Work

PLANNING

——— 49. After considering input, I help establish clear priorities and goals for our unit/department.
——— 50. Unit/department policies and procedures are spelled out.
——— 51. Budgets are devised yearly and compared to monthly reports.

ORGANIZING

——— 52. The work of my unit/department runs efficiently.

DECISION MAKING

——— 53. I do much information gathering and get much input before making decisions.
——— 54. I build commitment for my decisions.
——— 55. I develop creative solutions.

PROBLEM SOLVING

——— 56. I identify problems and take responsibility for them.
——— 57. I use the problem-solving process including trying creative solutions.
——— 58. I don't ignore problem behaviors and I deal effectively and quickly with them.

CLARIFYING ROLES & OBJECTIVES

——— 59. My employees know what is expected of them.

INFORMING

——— 60. I interact with and inform my colleagues.
——— 61. I keep my supervisor informed about what I am doing.
——— 62. I prepare meeting agendas for all meetings I conduct.
——— 63. I keep staff informed about policies, procedures, and all changes.

MONITORING

——— 64. I monitor the performance of staff.
——— 65. I meet regularly with staff.
——— 66. I periodically walk around to talk with employees and guests.
——— 67. I attend monthly meetings of the local hospitality owners/operators.

MANAGING TIME

——— 68. I set daily priorities and do first things first.
——— 69. I set time aside every day for physical exercise.

DELEGATING

——— 70. I delegate appropriate tasks.

FIGURE 2.4:　(continued)

ENDNOTES

1. U.S. Marine Corps Association, *Guidebook for Marines*, 20th ed. (Quantico, VA: U.S. Marine Corps Association, 2014), pp. 43–49.

2. Larry J. Gitman and Carl McDaniel, *The Future of Business*, 5th ed. (Cincinnati, OH: South-Western Publishing, 2005), p. 209.

3. John C. Maxwell, *The 21 Indispensible Qualities of a Leader* (Nashville, TN: Thomas Nelson, 1999), p. x1.

4. Ibid.

5. Stephen P. Robins, David A. DeCenzo, and Robert M. Wolter, *Supervision Today*, 7th ed. (Upper Saddle River, NJ: Prentice Hall, 2012), pp. 235–236.

6. Paul H. Hersey, Kenneth H. Blanchard, and Dewey E. Johnson, *Management of Organizational Behavior*, 12th ed.— *Leading Human Resources* (Upper Saddle River, NJ: Prentice Hall, 2012).

7. Transactional and transformational leadership are discussed in depth in James MacGregor Burns, Leadership (New York: Harper & Row, 1978). Also see Bernard Bass, *Leadership and Performance Beyond Expectations* (New York: Free Press, 1985); and Joseph Selzer and Bernard Bass, "Transformational Leadership: Beyond Initiation and Consideration," *Journal of Management*, vol. 16, no. 4 (1990), pp. 693–703.

8. www.leadingtoday.org/onmag/jan03/transaction12003.html. Retrieved May 24, 2010.

9. Malcolm Baldrige National Quality Award recipients can be found at baldrige.nist.gov/Contacts_Profiles.htm.

10. Peter F. Drucker, "Foreword," in F. Hesselbein, M. Goldsmith, and R. Beckhard (eds.), *The Leader of the Future* (San Francisco: Jossey-Bass, 1966), pp. xii–xiv.

11. Ibid.

12. Larry J. Gitman and Carl McDaniel, *The Future of Business*, 5th ed. (Cincinnati, OH: South-Western Publishing, 2005), p. 211.

13. Ricky W. Griffen and Ronald J. Ebert, *Business*, 7th ed. (Upper Saddle River, NJ: Prentice Hall, 2004), p. 445.

14. John Horne, President, Anna Maria Oyster Bar, personal communication, December 17, 1014.

15. Martha I. Finney, "The Truth about Getting the Best from People," *Financial Times Press* (Upper Saddle River, NJ: Prentice Hall, 2008), p. 6.

16. Ibid.

17. Stephen S. J. Hall, Ed., *Ethics in Hospitality Management: A Book of Readings* (East Lansing, MI: Educational Institute, American Hotel & Lodging Association, 1992), p. 75.

Chapter 3

Planning, Organizing, and Goal Setting

As mentioned briefly in Chapter 1, there are various management functions, such as planning, organizing, leading, and controlling. As noted in Chapter 1, leading involves several aspects, including communicating, which is broken down here in a sequence of steps that you take to get your job done:

1. *Plan* what is to be done.
2. *Organize* how it is to be done (this includes coordinating and staffing).
3. *Lead* the work being done.
4. *Communicate* what needs to be done.
5. *Evaluate* whether the objectives of your original plan have been met.

During the final step, you monitor whether the objectives of your original plan have been met. You might end up revising your plan or leaving it intact. In this manner, you have gone full circle through the management process (Figure 3.1).

In this chapter we present planning, organizing, and goal setting as a means of solving some problems, avoiding others, maintaining better control over events, and giving the supervisor more time to manage. After completion of this chapter, you should be able to:

- Explain the planning process.
- Differentiate the various types of plans and planning.
- Describe how to plan for change, identifying the various employee responses.
- Discuss how hospitality supervisors can best plan their time on the job.
- Demonstrate how effective organization contributes to a department's maximum success.

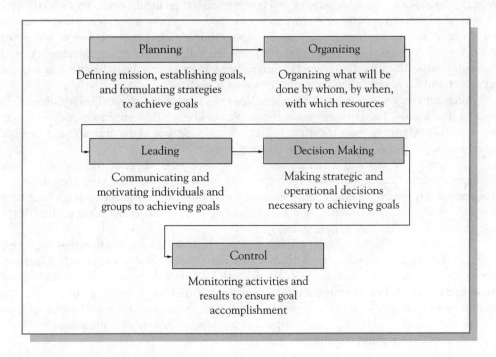

FIGURE 3.1: The management process.

The Nature of Planning

LEARNING OBJECTIVE: Explain the planning process.

planning
Looking ahead to chart the best course of action.

Planning means looking ahead to chart the best courses of future action. It is one of the basic functions of management (see Chapter 1), and it typically heads the list in management textbooks, whether the list names four functions and activities or 40. The reason it comes first is that it provides the framework for other functions and activities, such as organizing, communicating, staffing, motivating, coordinating, controlling, and so on. You have to have a plan in order to know where you are going and how to get there, how to organize, how to staff, how to communicate, how to motivate, what to direct and coordinate, how to control. Planning sets the goals and formulates the strategy for the future course of action to meet the company mission.

Policies are broad guidelines for managerial and supervisory action. For example: "We will pay the highest wages and have the best benefits plan; do whatever it takes to 'wow' the guest; or whenever possible, we promote from within the company." These statements are examples of policy decisions that are made at the senior level and used by supervisors as a guide. For instance: If the policy is to pay the highest wages, that does not mean paying $25.50 per hour for a job that is worth between $10.50 and $14.50. If the going rate is within the range of $10.50 to $14.50, the supervisor may give advice as to the best rate for a new employee. An important planning tool for chefs is a "battle plan" as a blueprint for planning and distributing/delegating what needs to be done. Additionally, add names and times of completion so it can easily be monitored throughout the production.

❋ LEVELS OF PLANNING

The future begins with the next few minutes and extends indefinitely. How far ahead does a manager plan? That depends.

Top-management planners in a large organization make long-range strategic plans. They should be looking at a three-, five-, or 10-year horizon and deciding where the organization should be heading over that time span and how it will get there. Such planning includes setting organizational goals, strategies, policies, and programs for achieving the goals. This level of planning is called **strategic planning**. It provides a common framework for the plans and decisions of all managers and supervisors throughout an organization. Leaders must now work within the framework of strategic goals that define the long-range direction of an organization. A solid strategic plan is critical to effective and high-quality decision making, provides an organizational direction, and aligns systems to reach organizational goals.[1]

strategic planning
Long-range planning to set organizational goals, strategies, tactics, and policies to meet or exceed goals.

Middle managers implement the organization's long-term goals on a smaller level and within shorter time frames. They typically make annual plans, although they sometimes plan for longer periods. These plans carry out the strategies, tactics, and programs of the organization's strategic plans within a manager's own function and area of responsibility.

As plans move down through channels to first-line supervisors at the operating level, management translates them into specific supervisory duties and responsibilities. Here, the planning period is typically one month, one week, one day, or one shift, and the plans deal with getting the daily work done. Yet whatever that work is, getting it done carries out some portion of the long-range plans made by planners at the top.

In the hospitality industry, and especially in foodservice, long-range planning has not been a key part of management philosophy. The traditional entrepreneurial style was a reactive, "seat-of-the-pants" approach. There is still a great deal of this philosophy in the industry, but the managerial style is changing. When enterprises that once were managed single-handedly by a successful entrepreneur on a day-by-day basis grow to a certain size, the "seat-of-the-pants" approach is no longer big enough. To survive, the company must develop professional management with a commitment to long-range planning.

Whether they work for an old-style enterprise or a large corporation with a commitment to planning, most supervisors look at a short time span and do not think of themselves as being part of a larger plan. To the degree that they plan ahead, they see planning as the most efficient way to get their work done. Many do very little planning and simply react to events as they happen. Often, it is impossible to plan very far ahead or to provide for everything that might happen. And even when we plan, things often do not turn out the way we thought they would.

A good plan deals with just this kind of uncertainty. Reasoning from the past and present to the future, it can establish probabilities, reduce risks, and chart a course of action that is likely to succeed. Let us see how this works.

❋ THE PLANNING PROCESS

Planning is a special form of decision making: It makes decisions about future courses of action. Therefore, the steps in making a good plan resemble those in making a good decision:

1. Define the goal, purpose, or problem, and set strategies for how to meet the goals.
2. Collect and evaluate data relevant to forecasting the future.
3. Develop alternative courses of action.
4. Decide on the best course of action.
5. Carry out the plan.
6. Control and evaluate results.

Much of a supervisor's planning is directed toward deciding, day by day, based on conditions that exist, the details of what is to be done. Since most of the work is a repetitive carrying-out of standard duties in accordance with prescribed procedures, daily planning is often limited to adjusting the what, who, and how much to the probable number of guests.

Yet even for a limited problem such as this, planning is essential to efficiency. Careful planning will provide the best product and the best guest service at the lowest cost with the least waste and confusion. It will reduce the risk of emergency: running out of supplies, running late, not enough employees, poor communications, duplicate or overlapping assignments, poor guest service, and all the other hazards of attempting to manage the work as it is happening.

Defining the goal, purpose, or problem, setting objectives, and developing and carrying out the best course of action are done in much the same way as in decision making. The biggest difference between the two processes comes in the second step. In decision making, you focus on the present and gather mostly current facts. In planning, you gather data from the past and the predictable future as well as the present, and on the basis of these data, you forecast future conditions and needs.

❋ FORECASTING

forecasting
Predicting what will happen in the future on the basis of data from the past and present.

Given a particular set of conditions, what has happened in the past and is happening in the present is likely to happen again in the future. To forecast, you find out what happened in the past and what is happening today. Then, *if none of the conditions change*, you can predict what can reasonably be expected to happen in the future. Forecasting is a very important function because it controls staffing, purchasing, and production decisions. Forecasting business volume in the hospitality industry can be tricky and should be based on intelligent analysis.

You might be involved in forecasting the number of guests in the restaurant tomorrow night, the house count for next weekend in a hotel, the number of patients who will choose sliced roast beef tomorrow at lunch. For example, suppose that you want to plan the number of portions of each menu item to prepare for lunch tomorrow, a Monday. You go to the sales records of the past 30–60 days to see how many of each item you have sold on Mondays. If conditions tomorrow will be the same as on Mondays in the past, you are safe in assuming that sales will follow pretty much the same pattern.

Forecasting is the first step in the planning process.
fGoodluz/Shutterstock

But suppose that conditions will not be the same: Suppose that tomorrow is Washington's Birthday. If you are in a business district, you can expect fewer total sales because offices will be closed. If you are in a shopping or recreation area, sales will probably increase. To gather relevant data, you will have to go back to previous holidays to see what happened then. If your menu or your prices have since changed, you have to consider that, too. Other things also affect the expected sales: the weather forecast, an ad you are running in the paper, special outside circumstances such as conventions and civic events, a new restaurant holding a grand opening across the street. All these things must be evaluated when you make your forecast.

Sometimes you have reliable data on the future. You might have a party scheduled with a definite number of reservations and a preplanned menu. You might have other reservations, too, giving you specific data on numbers but uncertain data on items. These are welcome additions to your projections from the past and your calculations about tomorrow's conditions.

In the end, your decision is the best estimate that you can make on the basis of available data. Often, this is simply a logical conclusion rather than the choice of alternatives specified in the standard planning procedures. The alternatives you have will be limited to various ways of handling the risks. Will you prepare more than "enough" and risk leftovers? Will you attempt to define an exact number with a runout time and plan to substitute a quickly prepared item? Will you plan to prepare additional portions of the original items if demand is high early in the serving period?

❋ THE RISK FACTOR

risk
A degree of uncertainty about what will happen in the future.

The future is always more or less uncertain. You reduce the degree of uncertainty, the risk, when you collect the relevant data and apply it to your forecast. If you have less than 1 percent of the relevant data, conditions are completely uncertain and the degree of risk is 99 percent. You might have anything from no guests to a full house—you simply don't know. At the opposite extreme, if data and conditions are completely known, there is complete certainty and no risk. The more relevant data you gather and apply to your forecast, the higher your percentage of certainty and the lower your risk.

In some foodservices, the degree of certainty about tomorrow is high. In a nursing home, for example, the population is stable and the menu is prescribed or preordered. In a hospital, the situation is similar, although occupancy is less predictable because patient turnover is higher. On a cruise ship, the number of guests is known, the guests' choices are predictable from the past, and the menu mix is fixed based on historical data.

Airline catering is preplanned according to the number of seats reserved and is updated for each plane as boarding passes are issued. Which of two choices guests will select is a forecast based on historical data, and it is usually slightly off—the last passengers served get the baked chicken. But seldom does anyone go without a meal, and seldom are meals left over. Planning in this case often goes right up to the cabin door.

Hotel occupancy is also fairly predictable, since most people make reservations ahead. There are also a predictable number of walk-ins based on past experience. Planning for the front desk, housekeeping, guest services, room service, restaurant and bar operations, and most other functions is based on the anticipated house count or occupancy rate. Usually, a two-weekly forecast meeting is held in which all the managers meet and are told the forecast for the next two weeks: "We're going to run at a 90 percent occupancy rate on Monday, we'll be up to 98 percent on Tuesday, we'll drop to 50 percent on Wednesday, then back up to 75 percent on Thursday, down to 27 percent on Friday," and so on.

This is the starting point of everyone's planning and scheduling for the week. Using the occupancy forecast, department managers can forecast their own needs. Restaurant counts are based on the occupancy forecast plus known group meal functions and historical data on walk-in customers. The housekeeping department can determine labor, laundry, and supply needs using the occupancy forecast. The front desk and the maintenance and security departments do the same. One forecast feeds all the others.

In planning repetitive work where most of the data are known or predictable and the risk factor is low, it is easy to decide on a final plan without generating and evaluating several alternatives. Usually, there is one obvious conclusion. But where many factors are unknown or questionable, it is wise to develop several alternatives and go through the entire decision-making process as time permits.

contingency plan
An alternative plan for use in case the original plan does not work out.

You can also reduce the risk by having an alternative plan in reserve, known as a **contingency plan**. In deciding the number of portions to prepare on a holiday, for example, you make a contingency plan for running out of something. In scheduling, you plan to have extra people on call or have an alternative plan for dividing the work.

You can also reduce the risk factor for repetitive situations by keeping records that add to your data for projecting the future. If, for example, you keep records of portions served in relation to portions prepared, they can tell you how successful your planning was and give you additional data for future forecasting. Over a period of time, if you have a steady clientele, you can develop standard numbers of portions for specific days of the week, which you then vary for special conditions.

Another way of reducing risk is to consult people who have more experience or expertise than you do—your boss, or perhaps certain of your coworkers, or, depending on the problem, an outside expert.

❋ QUALITIES OF A GOOD PLAN

The following are characteristics of a good plan. Not all of them apply in detail to every type of plan, but they all express general principles that apply to all plans and planning:

- *A good plan provides a workable solution to the original problem and meets the stated objectives.* (Will it solve the problem? Can you carry it out?) This principle applies whether it is a small, specific plan for one person doing one 5-minute task or a long-range plan involving an entire enterprise. A good plan concentrates on the problem and does not attempt to deal with side issues. It is sometimes difficult to maintain this sharp focus. As the saying goes, when you are up to your neck in alligators, it is easy to forget that your goal is to drain the swamp.

- *A good plan is comprehensive.* A good plan raises all relevant questions and answers them. Have you thought of everything?
- *A good plan minimizes the degree of risk necessary to meet the objectives.* Are these the best odds? What can go wrong?
- *A good plan is specific.* It includes the time, place, supplies, tools, and people (numbers, duties, and responsibilities) needed to carry it out. (Can somebody else follow your plan?)
- *A good plan is flexible.* It can be adapted if the situation changes, or it is backstopped by a contingency plan. (What will you do if . . .?)

Flexibility is especially important in the unpredictable restaurant or hotel setting, where anything can happen and often does. Remember from Chapter 1 that a flex style of management is more suitable to hospitality operations than rigid, carefully structured plans and planning, simply because circumstances are in constant flux and problems often change before they can be solved.

This does not do away with the need for planning, however. It simply requires flexible plans and managers who are able to adjust them according to the needs of the situation and the needs of the people involved. A flexible plan needs a flexible manager to make the most of it.

Types of Plans and Planning

LEARNING OBJECTIVE: Differentiate the various types of plans and planning.

Hospitality supervisors are likely to be so busy managing today's work that they seldom think of the future. But certain kinds of plans for the future can simplify daily management for all the days to come.

✳ STANDING PLANS

standing plans
An established routine or set of procedures used in a recurring situation.

One way to simplify future planning and managing is to develop plans that can be used over and over whenever the same situation occurs. Known as **standing plans**, these are established routines, formulas, blueprints, or a set of procedures designed to be used in a recurring or repetitive situation. A daily report is a standing plan for reporting house count, income, meals served, rooms made up, and so on. A procedures manual is a standing plan for performing particular duties and tasks. A menu is a standing plan for the food to be prepared for a given meal. A recipe is a standing plan for preparing a dish.

Any standing plan will greatly simplify a supervisor's task of planning and organizing the work. It does away with the need for a fresh plan each time the repeating situation comes up. All that remains to be planned in each instance is to provide for special circumstances. When such plans are put to use, they standardize the action so that everyone does things in the same way.

management by exception
Training employees so the supervisor needs only oversee that workers are meeting the standards and then deal with unexpected events that the standing plan doesn't cover.

If the situation recurs every day and people are trained in the procedures, the supervisor's need to manage is reduced to seeing that the workers meet the standards set and to dealing with the unexpected event that the plan does not cover. This is known as **management by exception**.

Most employees are happier with standing plans than they are being dependent on the supervisor to tell them what to do. Knowing what to do gives them confidence and makes them more independent. Gaps in communication are minimized. Work can go forward if the supervisor is tied up dealing with a problem somewhere else.

Large companies usually have standing plans for all kinds of repeating situations, especially if these situations recur throughout the operation. But often, small operations and individual departments of larger operations do not have predefined ways to meet recurring needs for

Standing plans help ensure that everyone knows what to do in given situations.

Blend Images/Shutterstock

planned action. Then you have crisis planning and firefighting and improvised solutions that might differ from day to day. You can avoid this by developing standing plans of your own. Such daily planning as scheduling, figuring out production details, ordering supplies, assigning rooms, or planning special events can be reduced to a form to be filled out, as in Figure 3.2. You can set up a whole series of standing plans for each job that you supervise by developing a performance standard system.

> When planning, it is always best to assume the worst and plan backward. Therefore, if the worst happens, you will be prepared and know exactly what to do. There will be less stress for all involved.

You can also develop standing plans for training new employees and for recruiting and hiring and evaluating performance. Although you do not use such plans daily, they are well worth the time and effort because they are ready when you need them. They take advantage of past experience, and they give you comprehensive solutions and consistent results.

Every hospitality operation must have standing plans and policies for dealing with matters affecting health and safety, such as sanitation, fires, and accidents. The law requires such plans. Usually, they consist of two parts: preventive routines and standard emergency procedures. It is a supervisor's responsibility to see that the department has plans that conform to company plans and to develop departmental plans if they do not exist. Part of all emergency planning is to train employees in all techniques and procedures and to hold periodic tests and drills to see that equipment is in order and people know what to do.

Company rules, procedures, and policies are another form of standing plan (see Figure 3.3). In this case, the supervisor carries out plans rather than makes them.

cc: Dining Room Supervisor   *John Doe*
       Floor Supervisor   *Jane Roe*
       Bulletin Board

Dining Room Order

Day   *Friday*   Date   *Nov. 30*   Time: Arrival   *12 noon*   Service   *12 noon*
Location   *Gourmet Room*   Guests: Confirmed   *36*   Set up   *40*

Room setup diagram (please show waiter stations)

Mary 1-2-3-4-5

Frank 6-7-8-9-10

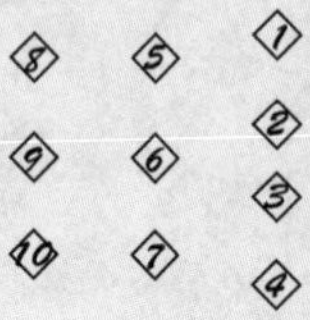

⟨8⟩   ⟨5⟩   ⟨1⟩
⟨9⟩   ⟨6⟩   ⟨2⟩
⟨10⟩   ⟨7⟩   ⟨3⟩
       ⟨4⟩

Servers   *Mary*
           Frank

Uniform
           White shirts & black slacks

Setup and serving duties

Place setting diagram

Menu   *Minestrone Soup*
        Vegetable Salad
        Banatha
        Italian Green Beans/Almonds
        Amaretto Cake

Equipment needed

Flatware		China		Glasses	
Knives	*40*	Plates		Water	*40*
Forks		Service		Other	
Dinner	*40*	Bread & butter	*40*	**Tablecloths**	*10, 54 × 54'*
Salad	*40*	Dinner	*40*		
Dessert	*40*	Salad	*40*	**Napkins**	*40*
Spoons		Dessert	*40*	**Miscellaneous**	(include rental equipment)
Bouillon	*40*	Cups & saucers	*40 & 40*		
Teaspoons	*40*	Soup cups	*40 & 40 underplates*		
Dessert		Fruit cups			*Coffee pots—5*
					Water pitchers—5
					Bud vases—10

Condiments

Salt & pepper	*10 sets*
Sugar & creamers	*10 sets*

Special arrangements

Head table:
Decorations:
Microphone:
Music:
Other

FIGURE 3.2: Standing plan for a dining room order for a specific event.

ABSENCES AND TARDINESS

POLICY: Employees are expected to be present during all scheduled hours of work. Unsatisfactory attendance* will be subject to disciplinary action.

PROCEDURE: Any employee who is unable to report to work as scheduled must notify his or her supervisor prior to reporting time. The cause of absence and expected duration must be reported.

Unsatisfactory attendance is defined as being absent from scheduled working hours more than five times in a period of eight months or late for scheduled working hours more than twice in one month.

FIGURE 3.3: Standing plan for dealing with absences and tardiness (page from a procedure manual).

You can also develop standing plans for special occasions such as customer birthdays, anniversaries, weddings, and Mother's Day. In fact, any repetitive situation is worth examining to see if a standing plan will reduce planning time and increase efficiency of the operation.

Developing a standing plan often takes a great deal of time and thought. You will see in later chapters the amount of work that goes into developing a performance standard system or a training plan. But in the long run the time spent on planning is time saved for other managerial activities.

Standing plans have certain potential drawbacks. One is rigidity: If plans are followed without adapting and updating them, they may result in a stagnant operation lacking in vitality. People come to resist change: "We've always done it this way" is a common expression. A standing plan should be flexible enough to adapt to daily realities, and supervisors should have the imagination and daily initiative to do the adapting.

Another drawback of such plans is that changes often evolve in practice but written plans are not kept up to date. Then a new employee or supervisor following a written plan will not be using current procedures and will be out of step with everybody else. This often happens when a new cook comes in and follows an old standardized recipe from the file.

It is a good idea to review and update standing plans on a regular basis. This kind of review offers an opportunity for the supervisor to involve employees in generating alternatives or modifications to current plans. The people who do the work often see it more clearly than the supervisor does and are likely to have ideas about ways to do it better. Also, if employees are involved in making changes, they are more willing to adapt to change in general and more committed to making new plans work. It is good motivation all around.

✳ SINGLE-USE PLANS

single-use plan
A plan developed for a single occasion or purpose.

A **single-use plan** is a one-time plan developed for a single occasion or purpose. The nature and importance of the occasion or purpose will determine how much time you should spend on it. As in decision making, you have to keep your planning efforts in proportion to the consequences of carrying out the plan. If the occasion is an affair of no permanent significance and little risk, a plan can be made quickly, and if something goes wrong, there is little at stake.

But some single-use plans may have effects that last for years or forever, or could produce consequences that are immediate and disastrous. In such cases, every step of the planning process must be given the full treatment.

Often, the purpose of a single-use plan is a major change of some sort. For such changes the planning must be very thorough: The risks must be carefully assessed and the effects of each alternative weighed carefully. Such a plan might involve a change in the way the work is done, such as introducing tableside service in a restaurant, installing new kitchen equipment, putting in an automatic liquor-dispensing system at a bar, or using computerized recipes in a hospital kitchen.

You might plan a job-enrichment program or a performance standard system. You might have to plan how to carry out a companywide plan in your department—for example, computerizing all the transactions of the enterprise.

Sometimes a supervisor is required to make a departmental budget—another kind of single-use plan. A **budget** (see Figure 3.4) is an operational plan for the income and expenditure of money by the department for a given period: a year, six months, or a single accounting period.

budget

An operational plan for the income and expenditure of money for a given period.

Your Restaurant Budget 2016

	$	%
Sales		
Food sales	700,000	70
Beverage sales	250,000	25
Other sales	50,000	5
Total sales	$1,000,000	100
Cost of Sales		
Food	196,000	28
Beverage	55,000	22
Other sales	37,000	75
Total cost of sales	$ 288,500	28.85
Gross Profit	$ 711,500	71.15
Controllable Expenses		
Salaries and wages	240,000	24
Benefits	40,000	4
Direct operating expenses	60,000	6
Music and entertainment	10,000	1
Marketing	30,000	3
Energy and utility	40,000	4
Administrative and general	30,000	3
Repairs and maintenance	10,000	1
Total controllable expenses	$ 460,000	46
Rent and other occupation costs	50,000	5
Income before interest, deprecation and taxes	201,000	20.1
Interest	$ 15,000	1.5
Deprecation	20,000	2
Total	35,000	3.5
Net income before taxes	$ 166,000	16.6

FIGURE 3.4:   A sample budget.

Upper-level management usually sets the goal or limitation, such as the department's portion of the organizational budget, but sometimes the supervisor is included in the goal setting.

Preparing the budget requires forecasting costs of labor, food products, supplies, and so on. This, like other forecasting, is done on the basis of historical data plus current facts, such as current payroll and current prices of supplies and materials, plus estimated changes in conditions during the budget period—primarily, changes in needs and in costs. The completed budget is then used as a standard for measuring the financial performance of the department. Needless to say, this is one plan that you will make with great care.

❋ DAY-BY-DAY PLANNING

Planning the day's work has top priority for the first-line supervisor. As noted earlier, this planning is concerned primarily with the details of what is to be done and who is to do it and adjusting various standing plans to the needs of the day. Getting the day's work done also requires getting the necessary supplies, materials, and information to the people who need them in their work.

Some of this must be planned daily, some must be planned a day or more in advance, and some is planned by the week. Purchasing may be planned daily, weekly, or monthly, depending on the department, type of enterprise, or location of suppliers. Scheduling may be planned by the week and updated daily as necessary. But things change daily, hourly, minute by minute in this industry, and whatever planning has been done in advance must be reviewed each day to see that everything needed is on hand—enough people to do the work, enough food, linens, liquor, cleaning supplies—whatever your responsibility is.

Plan before the day begins. Make it a regular routine. Many supervisors come in early or stay a few minutes late the day before to have some peace and quiet for this task.

Established routines simplify planning but do not take its place entirely. Ask yourself what is different about today. Stay alert and aware: Nothing is ever the same in this business.

Wherever possible, reduce risks by increasing predictability (more facts) and flexibility (more options). The more ways a plan can go into action, the more emergency situations you can meet because you prepared for them by planning effectively.

❋ SCHEDULES

Gantt chart
A bar graph diagram showing the activity and the timing of each activity.

Hospitality supervisors regularly spend time scheduling the work to be done, the order it is to be done, and who will do each task. These functions are performed in conjunction with the normal schedule of shifts to be covered by associates in a week. There are two scheduling techniques used by the industry to help accomplish the goals. The **Gantt chart** is a bar graph diagram showing the activity and the timing of each activity. The chart shows both the actual progress and the goal over a period of time. By charting progress, supervisors can easily monitor progress toward the goal and see if it is on time or behind schedule.

Figure 3.5 shows a simple Gantt chart for planning a new menu. The time is expressed in weeks across the top of the chart and the activities are listed down the left side. The planning takes place in deciding what needs to be done and by when. The shaded box represents what has been accomplished, and a clear box represents a goal. In all, the project takes six weeks, and progress can be monitored each step or week.

For weeks one and two, the activity is to consult with the chef and service staff. The person responsible is the owner, general manager, or food and beverage director.

For weeks two and three, new menu items are proposed. The person responsible is the chef.

For week four, new menu items are tested. The persons responsible are the chef, manager, and service staff.

For week five, the menu is costed. The person responsible is the manager.

For week six, the menu is printed. The person responsible is the manager.

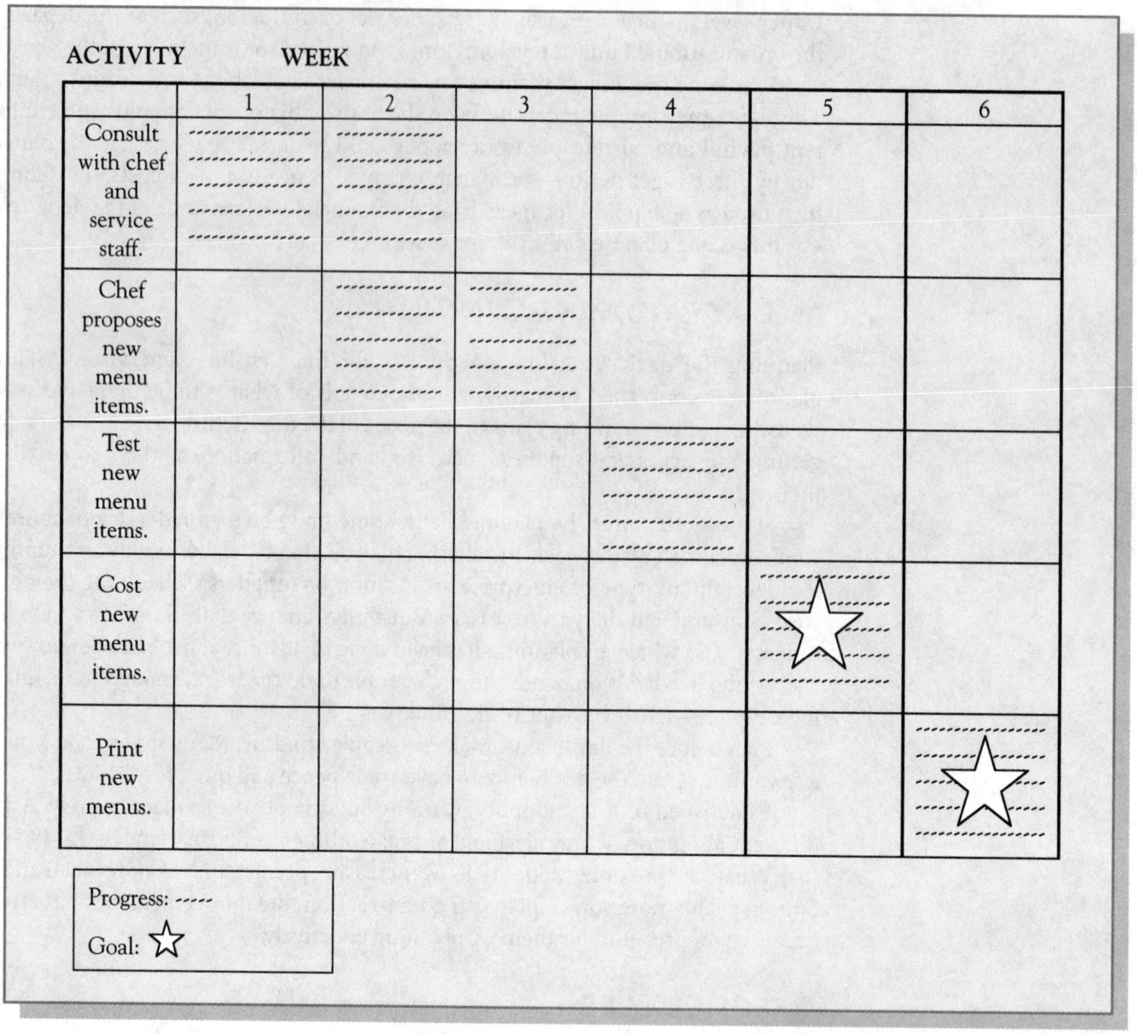

ACTIVITY	WEEK					
	1	2	3	4	5	6
Consult with chef and service staff.	----- ----- ----- ----- -----	----- ----- ----- ----- -----				
Chef proposes new menu items.		----- ----- ----- ----- -----	----- ----- ----- ----- -----			
Test new menu items.				----- ----- ----- ----- -----		
Cost new menu items.					----- ⭐ ----- -----	
Print new menus.						----- ----- ⭐ ----- -----

Progress: -----

Goal: ☆

FIGURE 3.5: A simple Gantt chart for planning a new menu.

program evaluation and review technique (PERT)
A diagram of a sequence of activities to complete a project.

critical path
The least amount of time for the activities to occur.

management by goals (MBG)
Employees jointly set goals for their departments and then plan strategies to meet or exceed the goals.

Another scheduling technique is called the Program Evaluation and Review Technique (PERT), which is a diagram of a sequence of activities to complete a project. It shows the time for each activity and plans for activities that can be done ahead of time or simultaneously. In using a PERT chart, a supervisor has to think of each activity required and how long each will take. A supervisor can also foresee any likely bottlenecks or other reason for slowdowns and plan to avoid them.

The PERT chart in Figure 3.6 indicates the time for planning a wedding function. It begins with setting up the room and allows two hours for that, and then it has another two hours to lay the tables, followed by one hour to decorate them. If we follow the chart all the way to the end and calculate the total time to complete the function, it comes to nine hours. This is the least amount of time required for all the activities to occur and is called the critical path.

Notice how some activities are done ahead of time to be ready when needed and the critical path begins with start and seeks to reduce the overall time to complete the project.

Supervisors and managers can use a planning technique called management by goals (MBG). With MBG, employees jointly set goals for their departments and then plan strategies as to how to meet or exceed the goals.

Management by goals = Planning goal setting

Progress toward the goals is monitored and rewards given for outstanding performance. MBG is a good motivational tool because associates are trying to meet or exceed the goals that they have been a part of setting.

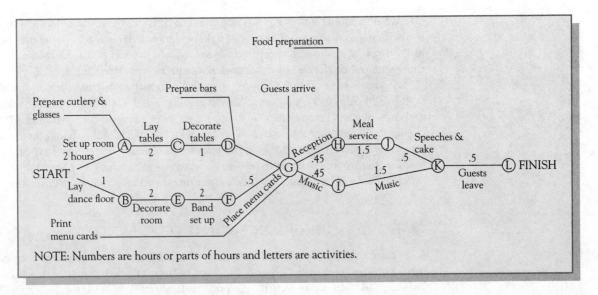

FIGURE 3.6: A PERT chart for a wedding function.

There are five key ingredients in an MBG program:

1. *Goal specificity*: Goals should be specific and measurable. Not just a desire to improve sales, but to increase sales from $4,000 per month to $5,000 per month.

2. *Participation*: With MBG, the goals are not set solely by top management; rather, supervisors and managers and their associates jointly set the goals so the participation makes for a synergy of effort that motivates all to achieve the goals.

3. *Time limits*: Each goal has a precise time limit by which it is to be accomplished and checkpoints so that the performance can be monitored.

4. *Who will do what*: Once the goals have been set, it is important to determine who will do what to ensure that the goals are met or exceeded.

5. *Performance feedback*: MBG needs feedback on a continuous basis so performance can be monitored. Self-evaluation is the best—just as athletes can look at the clock to check their time for a race, so too can hospitality associates monitor their feedback.

MBG works well in a variety of organizations and helps with other supervisory and managerial techniques such as TQM and motivation because it is participative in nature. Associates know what has to be done and can participate not only in the process but also in the rewards.

Planning for Change

LEARNING OBJECTIVE: Describe how to plan for change, identifying the various employee responses.

Every organization goes through big and little changes all the time, to adapt to new circumstances, to enhance competitive position, to be more cost-effective, and to improve products and services. If you stand still, you will soon be obsolete. Such changes are usually initiated at the top of the organization, but supervisors might be required to plan how to introduce these changes in their own departments. Sometimes, supervisors also initiate changes independent of what is going on in the rest of the company. Whether they can do this depends on the extent of their authority and responsibility.

When planning for change it is important to consider the stakeholders involved and affected by the changes. Also consider how the stakeholders will likely react—some may have the "what's in it for me" attitude, so be prepared for that. If you want to make changes you have to involve those who will be affected by the change in the process; it's that simple.

As suggested briefly, planning a change that affects the work must be done carefully and thoroughly. Although it is a single-use plan, it requires much time and thought because of the risks and consequences that a change involves. There are two sides to such planning. One is to plan the change itself; the other is to plan how to deal with the effects of the change on the workers. If you introduce a change without taking the workers into account, you may have real problems. The best way to plan for change is to involve those who will be affected by the change in the planning process. This will likely increase their commitment and involvement in ensuring that the change is successful.

✳ HOW WORKERS RESPOND TO CHANGE

Most employees resist change. Because any change upsets the environment, routines, habits, and relationships, it creates anxiety and insecurity in those affected. Even people who do not particularly like their jobs derive security from what is familiar. They know what to expect, life on the job is stable and predictable, and they have the comfort of belonging.

Change upsets all this, exchanging the known for the unknown, the familiar for the new, the predictable for the uncertain. People worry and feel threatened. They will have to adjust to new circumstances, learn new ways of doing things, work with different people, and readjust relationships. They don't know what to expect: Will they have to work harder? Will they be able to learn the new tasks? Can they get along with the new people? Will they lose their jobs?

People also resist change if it means a loss for them: less status, less desirable hours, separation from a friendly coworker, fewer tips, more work. This builds resentment, which is probably more difficult to deal with than insecurity and anxiety. If even a whisper of change is detected by a worker, rumors begin to fly, and fears are magnified and spread rapidly through the employee grapevine.

You cannot avoid dealing with them. It is best to do it sooner rather than later and to have a plan that will reduce resistance rather than to meet it head-on in battle or to steamroll the change through simply because you are the boss.

✳ HOW TO DEAL WITH RESISTANCE

resistance to change
A reaction by employees to changes in their work environment that may be accompanied by feelings of anxiety, insecurity, or loss.

The first essential for dealing with **resistance to change** is a climate of open communication and trust. You can reduce fears and stop rumors with facts: what the change will be, who will be affected and how, when it will take place, and why. Solid information will dispel much of the uncertainty, and it is always easier to adjust to something definite than to live with uncertainty and suspense, fearing the worst.

But the facts will not eliminate fear and uncertainty entirely. Employees must feel free to express their feelings, to ask questions knowing that they will get straight answers, to voice complaints. Venting one's feelings is the first step in adjusting to a new situation. Sometimes the change is advantageous to workers, and this should, of course, be emphasized. However, it is a mistake to oversell. If you try to persuade, or give advice, or become defensive about the change, or disapprove of people's reactions, or cut off complaints, or punish the complainers, you are shifting your ground from being helpful to being an adversary.

Your employees should feel that you want to make the change as easy for them as possible, that you understand their feelings and want to help them adjust. You can reassure them that their jobs are not at stake (unless they are), that they will have any additional training they need, that they are still needed and valued, and that you are all in the new situation together. You should avoid promising anything that might not happen—that would diminish their trust in you.

Besides reducing fears with facts, open discussion, understanding, and support, you have another very good way to approach resistance: You can involve your workers in planning and carrying out the change. Some of them will certainly have ideas and useful information, even though some of it might be negative. At the very least, they can give you another point of view that might help you in considering risks and consequences. Differing points of view are often more useful than agreement when you are trying to reach a sound decision.

Resistance often increases when employees feel they are not fully heard. Supervisors should avoid increasing resistance by avoiding threatening, criticizing, blaming, ridiculing, or being overly directive.

Actually, most people will respond positively to being included in planning changes that concern them. It gives them a sense of having some say in their own destiny. They can be very useful in providing information for forecasting, in generating alternative solutions, and in evaluating risks and consequences. If you can gain their commitment through participative planning, resistance is likely to melt away. Participation in planning also makes it far easier to carry out the plan, since people already know all about it and have a stake in making it succeed.

✳ EXAMPLE OF PLANNING FOR CHANGE

Let us run through an example of planning for change to see how it resembles ordinary problem solving and how it differs, and how it affects the workers and how you deal with that. Suppose that you are the owner of a 100-seat family-style restaurant. You have been in business six years and are well known for serving good food at affordable prices. But over the past year your sales have been dropping off slowly but steadily. You have missed some old familiar faces, and new people are not coming in to take their place. You have to do something to bring in new people—but what?

After some thought and study you decide to freshen up your decor, change two-thirds of your menu items, and use some special promotions to reach patrons moving into the new apartments and condominiums nearby. The first part of your planning concerns the new menu items. They are central to your overall plan and its success.

1. *Define your problem and set your goals.* Defining your problem is easy. Sales are dropping. Next, you try to formulate your goals. You write down: *Introduce eight new entrées on the dinner menu, retaining the four most popular items (prime rib, chicken Maryland, sole meunière, sirloin burger), with the goal of attracting new guests and retaining old ones.* You start a worksheet on which you list all the factors involved in planning new menu items (see Figure 3.7). After you have raised all the questions and answered several of them (columns 3 and 4), you write the following more-detailed objectives:

    a. Select eight new dinner entrées that meet the following criteria:

        • Must appeal to current and prospective clientele in type of food, price, quality, taste, appearance, and variety and be compatible with each other and with items retained from the current menu.

        • Must be preparable with present staff, equipment, and kitchen space, with minor adjustments. New equipment may be considered if it becomes an obvious need.

        • Must require only minor variations in current serving procedures, or none.

        • Cost of each entrée must be compatible with cost/price guidelines; price must fall in current price range.

    b. Test all entrées considered. For final choices, standardize recipe and design plate layout.

    c. Complete selection in six weeks.

    d. Train all kitchen personnel in preparation and all dining room personnel in service and in explaining new entrées to customers.

    e. Coordinate introduction of new menu with decorating and promotional efforts.

    f. Include workers in planning as follows:

        • Dining room supervisor and assistant manager in market research

        • Head cook in developing new menu items, other kitchen staff on voluntary basis

Factors to Consider	Questions	Must Do	Objectives and Limitations
Market, customer appeal	Type of food? Price? Quality? Taste? Appearance? Variety?	Define market. Survey guests, competition, neighborhood potential.	Final choices must meet guest tastes in all categories.
Preparation	Prep time? Skills required? Holding problems? Prep space? Ingredient availability? Ingredient storage? Compatible with other menu items (equipment, timing)? Extra personnel? Training time?	Test recipes for taste, yield, proportions. Standardize recipes. Check suppliers. Train cooks.	Items must be feasible with present staff and space.
Equipment	Need new? Cost? Space for? Delivery time? Utility lines? Train for use? Special dishes, utensils?	Avoid conflicting use.	No new, at least for now.
Service	More work? New ways to serve? New plating, garnishing? Training required? Training time?	Train for serving. Train for describing (taste, how cooked, ingredients, goes with . . .).	No major changes in service.
Cost, price	Cost/price ratio? Costing? Who will cost? Budget? Equipment cost? Testing cost? Training cost? Cost of extra help?	Costing. Pricing. Budgeting.	Cost must meet cost/ price guidelines. Menu price must fit present range.
Employee resistance	Bring workers into planning? When? Voluntary or required? Rewards? Anxiety? Competition? Who needs what information?	Must handle positively.	Voluntary participation. Rewards.

FIGURE 3.7: Supervisor's worksheet for planning new menu items.

- Servers to help in market research, search for new menu items, and evaluations on voluntary basis
- Rewards for participation to be given

2. Gather data from the past, present, and probable future in order to forecast what alternatives are most likely to succeed and to reduce the risks. Here is where your market research comes in. Only after you have a good picture of your clientele and their needs and desires and a good idea of what other restaurants are doing to attract that clientele can you decide

what kinds of entrées will please your customer group and how much they are willing to pay for them. Then you can begin your search for potential menu choices.

You will bring your employees into the picture as you begin your market research. You will begin with your supervisors: the assistant manager, the dining room manager, and the head cook. You will delegate responsibility to them for some of the market research, some of the search for menu items, and probably some of the training. You will also instruct them in handling worker resistance through information, listening, reassuring, and encouraging employees to contribute ideas.

Next, you will tell everyone else about the overall project. You will present it with enthusiasm as a way of increasing business that will benefit everyone. You will explain the procedures and invite questions and ideas. You will invite them to take part in market research by talking to customers, to contribute menu ideas, and to taste-test and help evaluate items being considered. You will make clear how valuable they will be to developing a good menu. You will assure them that they will have the training and tryout time they need and that their questions will be welcomed at all times. You will watch for continuing resistance and have the supervisors deal with it on a one-to-one basis.

3. *Generate alternatives, evaluate the pros and cons of each, and assess their risks and benefits.* This step will be put in motion by having the cook prepare various new items for testing and asking everyone to contribute comments. You will continue to involve the workers by having them take part in the evaluation.

Sampling the products will give them the experience of tasting and gain their cooperation in recommending items to customers. You might try out the most likely menu items one at a time as specials with your current menu and have your servers report on customer reactions. This would test the market and reduce the risk, and it would give both the kitchen and the servers a chance to experience the changes a little at a time.

You hope that by this time, everyone will be involved and that a team spirit will develop. But you know that some employees will do better than others, that insecurities and jealousies may develop, and that ongoing routines may tend to be slighted.

You hope to sustain enthusiasm for the new project while maintaining current product and service standards. You will do this through your own active leadership and through working closely with your supervisors. You will also introduce some sort of reward system.

A major part of planning for change is to define the goals or problem and set your objectives.

Andersen Ross/Getty Images

You may have to hire extra help for testing and costing recipes. You will make this decision later after you see how the kitchen staff responds to the project. This is a contingency plan.

4. Make the necessary decisions, after weighing each alternative in terms of the five critical decision-making questions: risk versus benefit, economy, feasibility, acceptability, and meeting the objectives. You will ask for input on making these final evaluations from your supervisors. You want to make sure you have considered not only the desirability of each item but the possible consequences of using it, such as difficulties in production, extra work, service tie-ups, and so on. In the end, you will make all the final decisions yourself. You will make this clear to your employees at the beginning of their involvement.

Having planned your decision making, you now make out a detailed but flexible schedule and turn to planning your new decor and your promotion. You will call on outside consultants for help.

5. *Implement the plan.* If the plan has been made with care and in detail, this should go smoothly. The plan should meet all the criteria for a good plan cited earlier in the chapter.

Carrying out your menu plan will include training both servers and kitchen personnel. You will finalize your training plans—the who, what, when, where, and how of carrying out the plan—after you have decided on the final menu items. By that time you will know how much training is needed, how your supervisors have responded to their extra responsibilities, and how closely the workers are involved.

You must also coordinate the changeover from one menu to another with the new purchasing requirements, the new menu card, and promotional plans. At this point you are satisfied that the

---

Supervisor at Taco Bell

Courtesy of Kiyoko Lawrence

Planning is essential to any business. Every day we plan goals for our business, we forecast, make schedules, make deployment charts, make truck orders, and so on. Each step is a piece of the guest service puzzle.

Forecasting is one of our biggest pieces. Without doing this properly, everything could be planned wrong, costing time and money. Let's say we forecast high but forget that one day is a holiday. We waste money with our scheduling by having too many people, we waste food by prepping too much, and we could infuriate our employees by sending them home early. Forecasting is the key to avoiding other planning disasters.

Scheduling and deployment charts are a way of planning the right people at the right time in the right places. During scheduling it's always good to plan one-on-ones with your employees to give them performance feedback. Once you plug this into your schedule, employees will be ready with any questions, and as the manager you won't "forget" to meet with them.

Truck orders might not seem like planning, but really, they rank right up there with forecasting. A busy spring break hotel must make sure to have enough toiletries and extra towels to accommodate all of the extra college kids. A restaurant also plans truck orders to ensure that it has enough food to make it to the next truck but not so much food that it expires. Restaurants have it tough with some holidays such as Lent, when some religious people eat fish instead of other meats.

Goals should be planned for each shift and daily with your team's input. This way, your goals are challenging but not so challenging that they can't be reached. Raise your goals as soon as you have exceeded the original bar so that in time, you can reach your overall goal.

Planning your business can be very detailed and involve a lot of computer/numbers work. Don't forget to plan for yourself as well. If enough time is not allotted in your day to get everything achieved, you might not be setting your business up for success. One thing I like to do is to plan from the task I hate to do to the task I love to do. Then there is a sense of accomplishment every time I cross something off my list. I don't like forecasting, so that is first on my list, then truck orders because they take so much time it's monotonous once you do it twice a week. By following my own plan, planning for my restaurant is easier. I set them up for success and not to fail.

plan for selecting and introducing the new menu items provides a flexible, workable program and that when the time comes, you can make everything fall into place.

You can see that in broad outline, planning for change is very similar to making other kinds of plans. The main differences are in the extent of the forecasting, the degree of risk, and providing for the impact of the change on the people it affects. Perhaps this example will give you some understanding of how necessary the groundwork is: the forecasting, the market study, the testing, the careful defining of criteria, the detailed planning, and the need to consider the consequences of decisions before taking action.

Change involves risks that cannot be avoided, but doing nothing is sometimes riskier than anything else is. Careful planning can minimize the risks of change.

Planning Your Own Time

LEARNING OBJECTIVE: Discuss how hospitality supervisors can best plan their time on the job.

The most relentless reality about working as a supervisor in the hospitality industry is the lack of time. There is never enough private time for planning and reflective thinking. There are never enough long blocks of time in which to plan your time. Furthermore, how can you plan your time when most of your managing consists of reacting to things as they happen? How can you plan for that?

Planning your time is not going to give you any more of it; there are just so many hours in your working day. What it will do is enable you to make more effective use of those hours you work.

Your job requires that you spend the time in your day in several different ways:

- Planning, organizing, and communicating the day's work
- Responding to the immediate needs, demands, and inquiries of others (customers, workers, bosses, salespersons, suppliers, health inspectors, and so on)
- Managing your people: hiring, training, directing, coaching, evaluating, disciplining
- Dealing with crises, solving problems
- Making reports, keeping records, enforcing rules (maintenance activities)
- Doing some of the work yourself if you are a working supervisor

There are certain parts of the day when the job controls your time, and when customer needs and demands are high. These include check-in and checkout times at the front desk, early mornings in housekeeping, food preparation for mealtimes in the kitchen, serving periods in the restaurant, and so on, depending on the function you supervise. At these times, you must be at the disposal of anyone and everyone who needs you to answer questions, settle disputes, deal with crises, make decisions, and observe your people in action.

It is important for you to be visible at such times, especially if your employees serve guests directly. It is important to greet guests and to let them see that you take a personal interest in them and the way they are being treated. And it is important to your people to have you out there with them: they feel your support and they see your example. It is management visibility.

These are segments of your time that you cannot plan for other than to set them aside and let nothing else intervene. Certain other responsibilities may also be pegged to fixed points in the day: the housekeeper's report, end-of-shift sales figures, cash deposits, and so on. The rest of the time is yours to fill with all the other duties you are required to do.

What *do* you do? If you analyze the ways in which you spend your time now, you can probably find ways to spend it better. The first step is to keep a running log for at least one typical day, several days if possible. Put down everything, including interruptions, with a beginning and an end time for each activity.

Figure 3.8 is a suggested format. To save making extensive notes, you can devise symbols such as those on the chart. But make your notes complete enough to make sense to you when you

Begin	End	Total (min.)	Activity	With Whom	Importance (Rate 1–5)
7:45	7:52	7	Plan for day	Self	1
7:52	7:52½	½	Q-Sched change	Sam	3
7:52½	7:53	½	Plan	S	1
7:53	7:53½	½	Tel	Alice	1
7:53½	7:56	3½	Tel-replace Alice	4 people	1
7:56	8:00	4	Plan	S	1
8:00	8:13	13	Howdy rounds	O	1–2

C-E Coaching-evaluating	Pl Planning	1 = Most important
CR Crisis	Pro Production	5 = Least important
Cus Customer	Q Answering questions	
Dec Decision	R Report	
Dir Giving directions	S Self	
Dis Discipline	Sol Solving problems	
Int Interviewing	Tel Telephone	
O Other(s)	Tr Training	

FIGURE 3.8: Format for keeping track of how you spend your time.

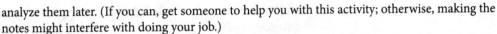

analyze them later. (If you can, get someone to help you with this activity; otherwise, making the notes might interfere with doing your job.)

The next step is to see what the record shows. Total the time you spent in each activity, and divide by the number of days to figure your daily average for each. Then ask yourself several questions:

- Is the amount of time spent per day appropriate to the activity? How can you reduce the time: Do it faster? Less often? Organize it better? Delegate it?
- How does the time spent on unimportant activities compare with time spent on highly important activities? Can you distinguish important from unimportant? How can you improve the ratio?
- Are you doing things that are not really necessary? Why are you doing unproductive work?
- Are you doing things that you could delegate to someone else?
- Can you group activities better as to time and place (e.g., make all your phone calls for purchases at one time during the day)?
- Was time wasted that could have been avoided by better planning? Standing plans? Better training? Better communications? Less supervision, more trust? Better records and better housekeeping? Saying no to unnecessary requests? Better decision making? An ounce of prevention?
- Does the log reveal that you do not spend any time at all on certain important but time-consuming activities you should be doing, such as making a list of sales prospects, developing holiday promotions, revising recipes, developing a procedures manual? Was it because you did not want to get started on a long project? When will you get started?

No doubt, your log will raise other questions, and no doubt you will be able to work out some good answers to all the questions.

First, get rid of activities that waste time or are not worth the time they take. Major wastes of time for supervisors include:

- Too much socializing
- Accepting drop-in visitors
- Allowing interruptions and distractions
- Not saying no often enough
- Poor organization of papers
- Procrastination and indecision
- Reading junk mail or checking e-mail
- Spending too much time online

Stephen Covey, author of *The Seven Habits of Highly Effective People*,[2] writes about how crucial it is to spend your time on important matters and to control the amount of time spent on urgent matters, such as when the cook announces that the kitchen is running out of a popular entrée. Just because a matter is urgent doesn't mean that it is automatically important, although you will certainly have to deal with the cook's problem in a timely but not time-consuming manner.

1. To *keep socializing to a minimum*, ask friends and family to call at work only if there is an emergency. Also make sure that you don't spend too much time socializing with your peers and employees. To cut down on outside visitors dropping in, make it a policy that visitors must have an appointment. When you find that certain employees are forever running to you with every problem and question, help them to find their own answers. If this doesn't help, ask them to save up several questions to go over at one time.

2. Today *more supervisors are using electronic communications*, as they are much quicker and easier to access for records purposes. Programs such as Microsoft Outlook and Go To

Meetings help facilitate scheduling, e-mailing, and actually conducting meetings at a distance.

3. Now that you have eliminated some of the time-wasters, *set priorities*. List all the things you want to get done today and divide them into musts and shoulds or number them 1 and 2. Do the must things (the 1s) first. It is said that 80 percent of your results can come from 20 percent of your efforts. Use a desk calendar (Figures 3.9 and 3.10) to plan and schedule the must items into the top 20 percent of your time.

A weekly planning sheet is a useful supplement to your daily calendar. At the end of the day, transfer what you didn't get done to tomorrow's list, add today's accumulation of

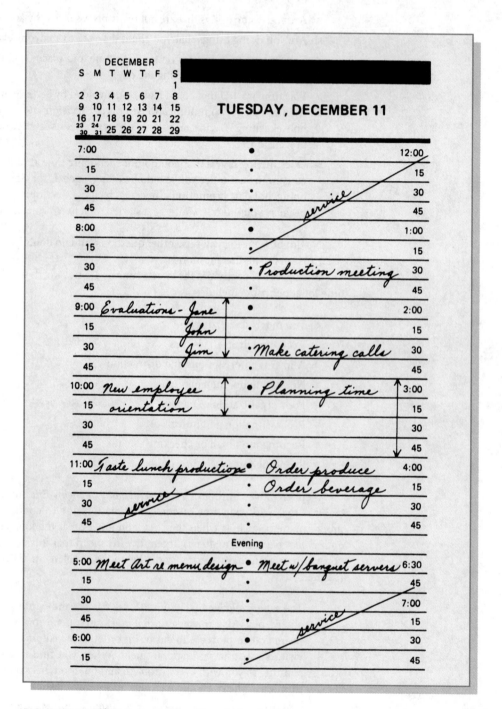

FIGURE 3.9: Planning your day with a desk calendar.

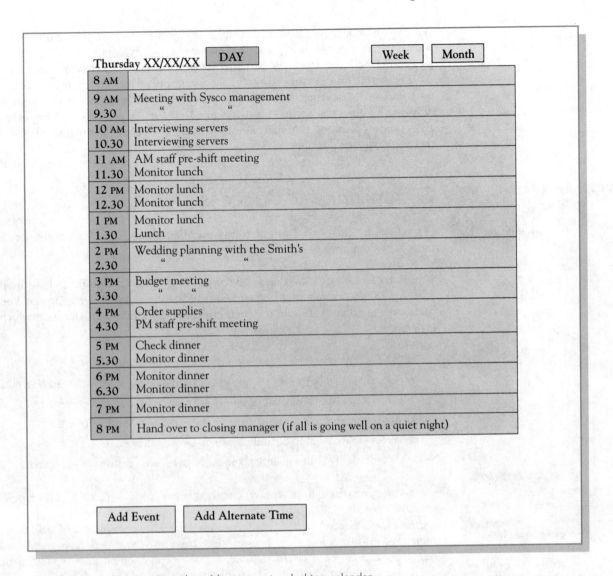

Thursday XX/XX/XX	DAY	Week	Month

8 AM	
9 AM	Meeting with Sysco management
9.30	" "
10 AM	Interviewing servers
10.30	Interviewing servers
11 AM	AM staff pre-shift meeting
11.30	Monitor lunch
12 PM	Monitor lunch
12.30	Monitor lunch
1 PM	Monitor lunch
1.30	Lunch
2 PM	Wedding planning with the Smith's
2.30	" "
3 PM	Budget meeting
3.30	" "
4 PM	Order supplies
4.30	PM staff pre-shift meeting
5 PM	Check dinner
5.30	Monitor dinner
6 PM	Monitor dinner
6.30	Monitor dinner
7 PM	Monitor dinner
8 PM	Hand over to closing manager (if all is going well on a quiet night)

Add Event   Add Alternate Time

FIGURE 3.10: Planning your day with a computer desktop calendar.

new musts and shoulds, reassess your priorities, and discard anything that no longer seems important. You will soon become adept at sorting out the time-wasters.

4. *Set aside regular periods of time without interruption* for interviews, problem solving, training, important decisions, and long-range projects. Every manager needs a quiet time each day for creativity and problem solving. Begin those time-consuming projects; divide them into manageable segments and get started on them. Getting them under way, and getting one or two of them accomplished, will renew your energy and confidence. Sometimes a small success is a big boost for self-motivation.

5. *Initiate long-range solutions to your time problems*—standing plans, better training, more delegation, reduced turnover, better communications—things that will eliminate the need for you to be in the thick of everything all the time (although still remaining visible). You will have fewer crises and fewer fires because everyone will know what to do and how to do it. You will have more employees taking more responsibility. You will have time to develop and motivate your employees, build trust, and exercise true leadership.

There will be many days when your plans are wiped out by unexpected events. When this happens, reschedule priority items for another day and go home and get a good night's sleep.

Hanging on to feelings of frustration will eat you up. But don't give up planning just because your plans don't work out some of the time. Adjusting your plans continuously is part of the planning process in your kind of job. And you can't adjust your plans if you haven't made any.

One way to improve time management is to record all thoughts, conversations, and activities for a week—or have someone do it for you—so you will see how productive you are. Another way of improving time management is to assign a time for completion of the important topics and allow time for planning, thinking, and strategizing. Even use a "do not disturb" sign if something must get done by a deadline.

Organizing for Success

LEARNING OBJECTIVE: Demonstrate how effective organization contributes to a department's maximum success.

The kind of long-range plans that will help you to solve your time problems have another, broader purpose: They will make your unit run more efficiently and effectively. One goal of this type of planning is effective organization. A well-organized and efficient unit is one in which the following are true:

- The organization's structure should be as simple as possible.
- Decisions that have a short-term effect should be made at the lowest levels.
- Lines of authority and responsibility are clearly drawn—and observed.
- Jobs, procedures, and standards are clearly defined—and followed.
- People know what to do and how to do it—and they do it.
- Standards of quality, quantity, and performance are clearly set—and met.

organizing
Putting together the resources—money, staff, equipment, materials—and methods for maximum efficiency to meet an enterprise's goals.

Setting everything up to run efficiently is **organizing**. Keeping it running efficiently and effectively is managing. Most hospitality supervisors move into a situation that has already been organized, whether well or badly, by someone else. If it has been organized well, their place in the company's organizational chart is clear, they know to whom they report, their authority and responsibility have been clearly defined, and they know the purpose (goal, function) of their department and their job. Each job classification is clearly defined, and there is a detailed procedures manual for each job. All necessary standards have been set, and people have been trained to meet them. Standing plans provide for repeating situations.

But many departments are not as well organized as this, and the lack of organization is a major contributor to crisis management. If you have inherited a badly organized unit, will you go on putting out fires, as your predecessor did, allowing yourself to be controlled by events instead of controlling them, developing the habit of nonplanning? It is an easy habit to develop and a seductive one: It gives you a sense of power, of being needed.

It is a short-run, shortsighted solution, and it will not contribute to your future in the industry. A better way to go is to set out to organize things better. How will you do it? One good way is to do critical tasks first and do them well if the goals are to be met or exceeded.

1. *Find out what you need to know about your own job.* Who is your boss—who directs you, and to whom are you accountable? What responsibilities and authority do you have, and what are you accountable for? How do your job and your department fit into the organization as a whole?

2. *Find out where poor organization is causing problems.* What you examine first depends on what is causing the most problems. Some areas to investigate follow.

   Chain of command (lines of authority and responsibility). Do all your employees know to whom they report—who has the right to tell them what to do and who holds them responsible for doing it? Does anyone take orders from more than one person? It is a primary principle of organization that each person should have only one boss (a principle known as **unity of command**). How many people do you supervise? Another principle of organization states that there is a limit to the number of employees one person can supervise effectively (known as **span of control**). One hundred people are far too many; 50 are too many; generally, 8–18 are considered the maximum number that one person can supervise or lead effectively. If you are given responsibility for supervising too many people, you may be able to delegate certain supervisory duties to key people in your department; in other words, create another level below you in the chain of command.

- *Job content and procedures.* Are jobs clearly defined in terms of duties, tasks, procedures, and acceptable performance standards? Is there a procedures manual for each job? Are the written materials kept up to date? Are there things people do—or should do but don't—that are not assigned to any job? Do any jobs overlap: Are the same responsibilities assigned to more than one job? Are people carrying out their assigned responsibilities and following procedures correctly? Do some people have too much to do and others too little?

- *Training.* Do employees know what to do and how to do it and what standard of performance is expected of them? Have they been well trained? Are they meeting the standards?

- *Evaluation and controls.* Have standards of quality, quantity, and performance been set, and are people meeting them? Are standards being used to evaluate and control products and services? Is there a system of performance review?

- *Standing plans.* Have standing plans been set up for repetitive tasks, routine reports, and emergency procedures? Are forms provided for reports, record keeping, information, and controls? Do people use them correctly?

Answers to such questions as these will give you a pretty good idea of the existing organization of your department and the degree to which it is functioning as it is supposed to.

Ensuring that you know who you report to, and what responsibilities and authority you have, is part of a good organization.

CandyBox Images/Shutterstock

Plan what you will do to improve the organization and efficiency of your operation. For this purpose many chapters of this book are at your disposal to help you along the way. Job descriptions will help you define jobs, set performance standards, develop procedures manuals, and provide the springboard for recruiting, training, and evaluation programs. The special chapters on these subjects will help you to develop and carry out such programs. Problem-solving and decision-making chapters will help you to define your problems, set your goals, and choose the best courses of action.

Delegation, job enrichment, and employee participation offer you ways of reorganizing at least some of the work for better results. You can relieve yourself of detail, broaden job responsibilities for employees, and provide more time for managing. Standing plans can reduce the management of repeating tasks to dealing with exceptional conditions and events.

CASE STUDY: Preparing Employees for Change

Michael is beverage manager at the principal hotel in a middle-sized manufacturing town. The hotel has a steady business trade and is now reaching out to attract local customers for its restaurant and bar. The restaurant is to be enlarged and the bar off the lobby is to be remodeled and expanded into a cocktail lounge with table service. Michael is in charge of planning all aspects of the change.

All of the bar equipment will be new, and Michael will work with the designer to determine what is needed. There will be a dozen tables served by waiters or waitresses whom Michael will hire and train. The bartenders will pour drinks for both the lounge and dining room servers in addition to customers at the bar, and Michael may have to add another couple of bartenders to the staff.

Michael knows the present bartenders will be worried about the change. They will be afraid they will work harder for fewer tips, since they will be competing with the table servers. They may also be concerned about the new layout and equipment. It is important for them to feel confident about working in the new setup and to be committed to the project and accepting of the new employees.

Questions
1. When should Michael break the news, and how?
2. How should he deal with their fears about new personnel?
3. What can he do to increase their confidence?
4. How can he gain their commitment to the new project?
5. Make a detailed plan incorporating your answers for Michael to follow in dealing with his present bartenders.

KEY POINTS

1. As a supervisor, you need to plan what is to be done, organize how it is to be done (this includes co-ordinating and staffing), direct the work being done, and control or evaluate what has been done.
2. There are various levels of planning, from strategic planning (or long-range planning) to short-term planning.
3. The planning process involves the following steps: define the purpose or problem/opportunity and set objectives; collect and evaluate data relevant to forecasting the future; develop possible courses of action; decide on the best course of action; and carry out the plan.
4. You can reduce risk when planning repetitive work by having a contingency plan, consulting people who have more experience than you do, or keeping records that add to your data for forecasting.
5. A good plan provides a workable solution and meets the stated objectives, is comprehensive, minimizes the degree of risk, and is specific and flexible.
6. A standing plan is an established routine, formula, or set of procedures designed to be used in a recurring or repetitive situation. Examples include policies and procedures.

7. A single-use plan is a one-time plan developed for a single occasion or purpose, such as a budget.

8. Most employees resist change. You can reduce their fear and uncertainty with facts, open discussion, understanding, and support, as well as involving them in planning and carrying out the change.

9. The first step in planning for change is to define your problem and set your goals. The second step is to gather data from the past, present, and probable future in order to forecast what alternatives are most likely to succeed and to reduce the risks. Next, generate alternatives, evaluate the pros and cons of each, and assess their risks and benefits. The fourth step is to make the necessary decisions, after weighing each alternative in terms of the five critical decision-making questions: risk versus benefit, economy, feasibility, acceptability, and meeting the objectives. The last step is implementing the plan.

10. To make good use of your time, eliminate time-wasters, set priorities, use a planning calendar, set aside regular periods of time without interruption for interviews, and initiate long-range solutions to your time problems.

11. Setting everything up to run efficiently is organizing. The first step in organizing is to find out what you need to know about your job. The second step is to find out where poor organization is causing problems, such as chain of command, job content and procedures, training, evaluation and controls, or standing plans. The third step is to plan what you will do to improve the organization and efficiency of your operation.

KEY TERMS

budget
contingency plan
critical path
forecasting
Gantt chart
management by exception
management by goals (MBG)
organizing
planning

Program Evaluation and Review
  Technique (PERT)
resistance to change
risk
single-use plan
span of control
standing plans
strategic planning
unity of command

REVIEW QUESTIONS

Answer each question in complete sentences. Read each question carefully and make sure that you answer all parts of the question. Organize your answer using more than one paragraph when appropriate.

1. Compare and contrast strategic planning to day-by-day planning.
2. Briefly describe the steps in planning. Why do they resemble the steps used in making a good decision?
3. How can you reduce your risk when planning repetitive work?
4. Discuss five qualities of a good plan, and explain why each is important.
5. Give five examples of a standing plan.
6. What is management by exception?
7. A budget is an example of what type of plan?
8. Why do employees resist change?
9. In the example of planning a new menu, what do you think of the manager's plan? What problems might there be? If you were the manager, would you have people participate even more, such as visiting competitors and reporting on their menus or contributing recipe suggestions? Why or why not? How would you reward the workers who take part?

10. List the five steps in planning for change.
11. What can you do to make more effective use of your time?
12. What areas of the operation might you investigate if poor organization is causing problems?
13. Give five examples of controls commonly found in hospitality operations.

 # ACTIVITIES AND APPLICATIONS

1. Discussion Questions
   - Because planning involves forecasting the future, so much of the supervisor's job is dealing with things that are unpredictable, and there is never enough time in the day anyway, do you think that planning is worth the time it takes? Defend your answer.
   - Juanita says that standing plans free the supervisor for other management tasks. Bruce says that standing plans are an excuse for laziness and a straitjacket on creative action. Which view do you agree with, and why? Can you somehow arrange to have the best of both views?
   - How does planning relate to all the other aspects of a supervisor's job? Give specific examples of how lack of planning can complicate getting the job done, starting new personnel on the job, or other examples from your own experience.
   - What responsibility does a supervisor have for departmental organization? If you were hired for a job in which organization was poor, how would you go about straightening things out? How would you deal with your own supervisor in reorganizing things?
   - Planning your personal time can be as useful as planning your work time. How good a planner are you? What are your time-wasters? Do you set priorities? Do you use a calendar for planning?

2. Group Activity: Brainstorming Plans and Controls
In groups of four, brainstorm examples of plans found in hospitality operations with which you are familiar. Have one person keep a list of plans and a separate list of controls. Share with the class.

 # ENDNOTES

1. www.foh.dhhs.gov/whatwedo/Training/planning.asp. Retrieved October 30, 2014.
2. Stephen Covey, *The Seven Habits of Highly Effective People* (New York: Free Press, 1989).

Chapter 4

Communicating Effectively

Human beings communicate all day, every day. We spend over 70 percent of our waking hours sending or receiving messages: speaking, listening, writing, reading, pushing keys on computers, watching the television screen. Since we communicate so much, we ought to be pretty good at it. But we're not. There are probably as many opportunities to be misunderstood as there are people with whom we communicate. Different people interpret what you say in different ways, and not necessarily in the way that you meant, and you do the same with what they say to you. Many of the problems we have on the job—and in our personal lives, too—involve some type of communication failure.

As a supervisor in a hospitality enterprise, you will be communicating constantly. You will be both a sender and a receiver of messages, and both roles will be very important. You must understand what comes down to you from the top so that you can carry out your supervisor's instructions and the policies of the company. You must communicate clearly with other supervisors to coordinate your work with theirs. You must communicate effectively with customers. Most important of all, you must communicate successfully with the people you supervise so you will have the power to get things done. You cannot supervise effectively if you cannot communicate effectively.

In this chapter we examine the communication process and its central role in managing people at work. After completion of this chapter, you should be able to:

- List the types of strong communication, assessing the value of each type.
- Explain why communication is important to the success of a hospitality supervisor.
- Describe the importance of listening as a supervisor.
- Examine the ways in which effective communication leads to effective management.
- Identify the best ways to improve your business writing skills.
- Discuss techniques to improve meeting productivity.

The Importance of Good Communication

LEARNING OBJECTIVE: List the types of communication, assessing the value of each type.

communications
The sending and receiving of messages.

Communications is the general term that sums up the sending and receiving of messages. The way employees communicate can make or break a company. Think of the difference between courteous and surly employees and the messages they convey to guests. We want to do business with people who can communicate the company philosophy to guests and give outstanding service. This all takes communication, which is the lifeblood of companies. It is critical that the front-line associates know the company mission and goals and how they are going to meet them. This information is a formal communication and is given via meetings, personal correspondence, e-mail, notice boards, and so on. Supervisors are vitally important, as they are the ones who explain the mission, goals, and company policy to their associates.

❋ TYPES OF COMMUNICATION

A communication may be a word-of-mouth message such as a verbal instruction given on the job or an announcement at a meeting. Or it might be a written communication: an e-mail, letter, memo, production sheet, housekeeper's report, or recipe.

A message might go from one person to another, as when the sous chef tells the soup cook what soups to prepare for lunch; when the housekeeping supervisor tells a housekeeper what

interpersonal communication
The sending and receiving of messages between people.

organizational communication
Communication down and up the organizational levels.

two-way communication (open communication)
Communication that moves freely back and forth from one person to another, or up and down the ladder.

interviewing
Conversation with the purpose of obtaining information, often used in screening job applicants.

small-group communication
Communication that takes place when two or more group members attempt to influence one another, as in a meeting.

mass communication
Messages sent out to many people through such media as Facebook, MySpace, Twitter, newspapers, magazines, radio, and television.

rooms she is to make up; or when one person says to another, "It's nice to have you back, we missed you." This is known as interpersonal communication.

A message might go down the corporate ladder from the president of the company to the general manager to the food and beverage director to the executive chef to the sous chef to the station cooks to the cooks' helpers. Such a message is likely to be a policy directive or some other matter affecting the organization as a whole. This is an example of organizational communication.

This type of message is likely to be reworded at each level, and there is little chance that much of the original meaning will survive the journey. One study of 100 companies showed that workers at the bottom of a five-rung ladder typically received only 20 percent of the information coming down from the top. The chances of messages going up the ladder from workers to top management are even less likely, unless it is bad news.

When messages move freely back and forth from one person to another, or up the ladder as well as down, we say that we have good two-way communication or open communication. Such communication contributes to a positive work climate and high productivity.

In addition to interpersonal communication and organizational communication, there are three other forms of communication that a supervisor may be involved in:

1. Interviewing is often defined as a conversation with a purpose. Supervisors use interviewing skills not only to screen job applicants but also to get needed information from employees and their own supervisors.

2. In small-group communication, three or more group members communicate in order to influence one another. Meetings are examples of small-group communication.

3. Mass communication refers to messages sent out to many people through newspapers, magazines, books, radio, television, and other media. Hospitality organizations often use mass communication to advertise for customers as well as job applicants. A restaurant newsletter is another example.

A communication is an interaction between a sender and a receiver.
Tyler Olson/Shutterstock

Although each context is different, they all have in common the process of creating a meaning between two or more people.

✳ THE COMMUNICATION PROCESS

A communication is an interaction between a sender and a receiver. In a successful communication, the sender directs a clear message to someone and the receiver gets the message accurately. It sounds simple enough. The problems lie in the words *clear* and *accurately*.

Let us take the process apart a bit, using Figure 4.1. The sender has something to tell someone—an idea in his mind that he needs to communicate. The sender knows what he means to say to the receiver, but he cannot transmit his meaning to her directly by mental telepathy. Therefore, he puts his meaning into words (symbols of his meaning) and sends the message by speaking the words to her or writing out the message. That is his part of this communication—conceiving the idea, expressing it, and sending it.

The receiver receives the message by hearing or reading the words, the symbols of the sender's meaning. She must translate or interpret the words in order to understand what the sender meant. Does she translate the words to mean what he intended them to mean? Does she then understand the message that was in the sender's mind before he put it into words? Receiving, translating, and understanding are her part of this communication.

These six processes happen almost simultaneously in spoken messages, but it is useful to break the process down because something can go wrong in any one of the six steps. From the beginning the message is influenced by the sender's personality, background, education, emotions, attitudes toward the receiver, and so on.

This, in turn, affects the sender's choice of symbols, how his meaning is expressed in these symbols, and whether he adds nonverbal symbols such as gestures and tone of voice. How he sends the message—whether he speaks or writes it and when and where—may affect its impact—how the receiver receives it or even whether she receives it. How the receiver translates the symbols, both verbal and nonverbal, will be affected, in turn, by her personality, emotions, and attitude toward the sender and so will her final understanding and acceptance of his meaning. We explore all this in detail shortly.

You can see that there are opportunities all along the way for things to be left out, misstated, missent, or misinterpreted. Sometimes, messages are not sent or received at all. The sender forgets or is afraid to send the message; the receiver does not hear or read or register it.

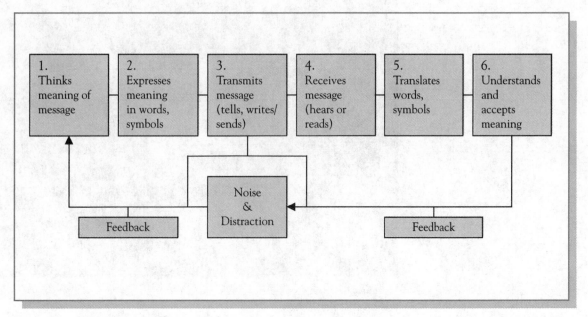

FIGURE 4.1: Six elements of a successful communication.

Often, people think they have communicated when in fact they have not. The sender thinks the receiver has understood. The receiver thinks so too, when in fact this hasn't happened.

When you see how complicated good communications really are and how easily they can go wrong, you might wonder how we ever get messages through. Yet nothing is more important to a supervisor both as a sender and as a receiver of messages.

Why Communication Is So Important

LEARNING OBJECTIVE: Explain why communication is important to the success of a hospitality supervisor.

Most supervisors probably think of themselves as senders rather than receivers, and most of the time they are. They spend the best part of their time at it. They direct the work of their subordinates, which is their major function as a manager. They give instructions and assign tasks: who will do this, who will do that, how they will do it, when it must be done. They provide information that their people need to do their jobs: how many people are guaranteed for this banquet, who is on duty, who is off, what room the banquet is in. They train people—communicate to them how to do the work their job requires. They give feedback on how well or how poorly people are doing. They recruit and interview and hire and fire. They discipline; they tell workers what they are doing wrong and how to do it right and what will happen if they don't shape up. Good supervisors also talk to their people informally to build working relationships, a positive climate, and a sense of belonging.

All these kinds of messages are vital to the success of your department. It is essential to send your messages clearly and explicitly to make sure that the meaning gets through. Only if your messages are understood and accepted and acted on can you get things done.

Messages that are garbled or misinterpreted or stalled along the way can make all kinds of trouble. They can waste time and labor and materials. They can cause crises, gaps in service, poor performance, and higher costs. Whenever they make something go wrong, they cause

Listening to guests is critical in any organization.
Minerva Studio/Shutterstock

frustration and usually hard feelings on both sides. If poor communication is habitual on the supervisor's part, it can build lingering resentment and antagonism and cause low morale and high employee turnover.

Not every supervisor realizes the importance of the other half of the communication process, the receiving of messages. Listening is probably the most neglected of communication skills. Moreover, in the hospitality industry, it is often hard to find time to listen. Yet it is very important to take that listening time with two kinds of people: the workers you supervise and the customers you serve. Your workers hold your success in their hands. Your customers hold the success of the enterprise in theirs.

The people you supervise want you to listen for many reasons. They want to give you their ideas and information. They want you to do something about a problem. They want to vent their anger and frustration. They want you to listen just for the sake of hearing them because they are human beings and they need to relate to people.

It is very easy to put aside these reasons for listening as being irrelevant to the job you are trying to do. You may think that you don't need people's ideas—they don't know anything you don't know. You certainly don't need their problems, their anger, their frustrations; and you don't have time to listen just for the sake of listening.

But the truth is that you really do need all these things. You need their ideas and their information because you might learn something new and valuable. You need to deal with the problems, the anger, and the frustrations even when you think you can't do anything about them. You need to listen for listening's sake because you need the good human relations that it builds. And you need to keep the door open all the time for upward communication, to build that positive work climate, you must have to lead effectively.

Listening to guests is important for similar reasons. Usually, they want to talk to you because they are complaining about something—the steak is tough, the pizza is cold. You don't want to hear about it, and you offer to get them another one. But wait! Maybe you have not received the message correctly. They really want to tell you how angry and disappointed they are and how they deserve better treatment, and they want you to listen. Certainly, they do not want another tough steak or another cold pizza. The amazing thing is that if you do hear them out with a sympathetic ear, 9 times out of 10, you will defuse their anger. Eventually, you can get them to suggest what they would like you to do about it—provide a tender steak, a hot pizza, or something entirely different—and you will have made a friend for your restaurant.

Obstacles to Good Communication

LEARNING OBJECTIVE: Discuss common obstacles to good communication and recommend tactics to avoid them.

We mentioned earlier the many problems in getting the sender's meaning through to the receiver. Figure 4.2 illustrates the most common barriers to good communication and shows how they can influence the message on its journey. In the next few pages, we discuss these problems in detail.

✳ HOW THE COMMUNICATORS AFFECT THE MESSAGE

Both sender and receiver can obscure or distort messages without being aware of it. In Figure 4.2, the two stages labeled 1 and 4 list various personal characteristics that can affect communication, especially if these characteristics are different for the sender and receiver. Differences in background, education, past experience, and intelligence can often cause communication difficulties. Big words, long sentences, and formal delivery may not get through to a school dropout or a person of limited background or intelligence. Today's slang may not reach someone older than you. Kitchen jargon may not be intelligible to a new employee. As a sender of messages, you need

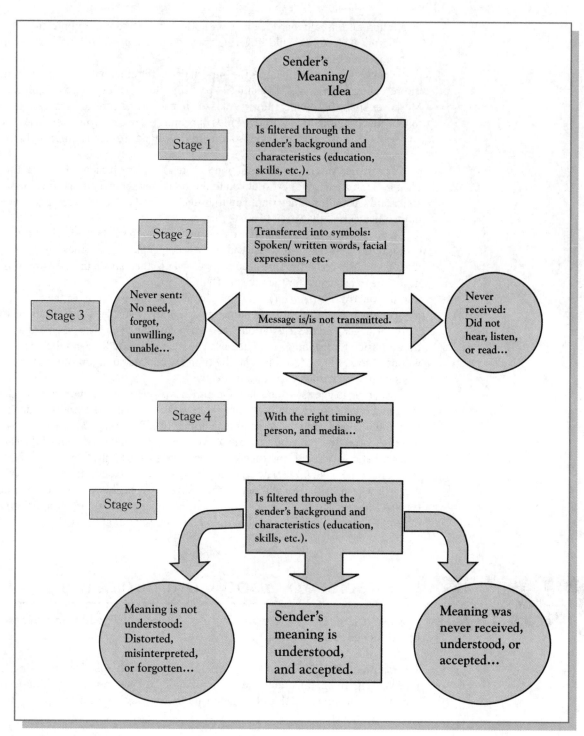

FIGURE 4.2: Barriers to communication.

to adjust to such differences. You must be aware of where you are and where the other person is so that you can make your messages understandable to that person.

Associates also differ in *attitudes*, *opinions*, and *values*, and these differences can inhibit communication or garble messages. Maybe you swear and tell off-color jokes as an informal and comfortable way of communicating. Some of your people are likely to find such speech obscure, offensive, or shocking, and might therefore miss your message entirely. Things that are important to you might mean nothing to your workers if it does not affect them. You care a lot about food costs and waste, but you have trouble getting that message through to your kitchen staff because they don't care—it's not their money going into the garbage can.

Prejudices can distort communication; this includes not only biases of the usual kind against women or ethnic groups but some intangible thing from your past that makes you sure that all men with beards or all tall women who wear thick glasses are not to be trusted. Prejudice can turn up in the words you choose in communicating or in your tone of voice, or it can make you leave something out of the message or forget to send it at all. If you let prejudice creep into something you say, it is likely to stir up anger and cause the message to be misinterpreted or rejected.

Sender and receiver may have different *perceptions* about the subject of a communication. Nobody can perceive reality directly; it is conveyed to us through our five senses, and its meaning is filtered through our minds and our emotions. Since we are all different, we do not always agree on what reality is. We do not see and hear things the same way. We do not agree on big and small, mild and hot, loud and soft, sooner and later. How many is several? How much is a lot? What does "season to taste" mean?

Everyone perceives things selectively and subjectively. If you are trying to calm down a couple of people who are having a fight, you will get two entirely different versions of the incident because each perceives and experiences it in his or her own way. They are not lying; they are telling you the truth as they see it.

As you give your people instructions and information, they all see and hear you subjectively, and they all tend to magnify what is pleasing to them in what you are saying and to play down what they do not want to hear or know about. Many times your message will be exaggerated out of all proportion to the point where you would not recognize what you said if they repeated it back to you.

Often, the sender and receiver of an instruction do not have the same perception of its importance. The supervisor may say, "Will you get this done for me?" and the employee may reply, "Yes, I will," and the supervisor may mean by noon today, whereas the employee is thinking in terms of the day after tomorrow or next week. They are focusing differently; each perceives things according to his or her own needs of the moment.

Assumptions and *expectations* distort communications. We assume that our listeners know what we are talking about, but sometimes they don't, and the entire message goes over their heads. We assume that our message has been received, and we are angry with people when they don't do what we tell them to do.

One supervisor told a cook, "Chiffonade purple kale and blanche prior to service so it can be sautéed for tonight's dinner party." What he meant was to cook the usual number of portions for a Thursday night dinner, and he assumed the cook would prepare about 5 pounds of kale. In the cook's mind, the message registered, "Cook all the kale that was delivered today." When the supervisor found 15 pounds of kale ready for service at 5:00, the predictable heated conversation ensued: "Why in the name . . .?" "But you didn't say that!" "Well, you should have known!"

When we make assumptions, we often jump to conclusions. If you see a bellperson sitting with his or her feet propped on a table and eyes closed, you assume that the bellperson is goofing off. If you see the director of marketing in the same pose, you assume that he or she is thinking. Your conclusions might be wrong; you do not know in either case what is really going on.

Often, we think we know what people are going to say, so that is what we hear, even when someone says something entirely different. Sometimes we even finish their sentences for them. As we already noted, often in listening, we extend the speaker's meaning far beyond what was said, and we answer inappropriately, as when the customer says, "My steak is tough," and you say, "I'll get you another one," when what the guest really wants is your attention to his or her feelings.

We make many assumptions about how people think and feel, what interests them, and what they value. If you assume that all people are lazy, are interested only in their paychecks, and have to be ordered around to get any work out of them (Theory X), you have probably closed the door to all meaningful communication and you will never find out what they are truly capable of.

One of the biggest troublemakers in communicating is the emotions of the people who are sending and receiving the messages. If you say something in anger, it is the anger that comes across, not the message, no matter what words you use. It always triggers emotions in the receiver: anger, hate, and fear. The message is buried under the emotions, and the emotions become the message. The receiver is likely to hear things that were not said, and the sender is likely to say things that were not intended. Communication becomes hopelessly snarled.

Sometimes underlying emotions color all communications between the supervisor and the people supervised. If there is contempt and suspicion on the part of the supervisor, there will be hatred, anger, and fear among the workers. The communications climate is thoroughly polluted, and messages are taken in the wrong way, or are rearranged according to suspected hidden motives, or are totally rejected. *The only healthy climate for communication is an atmosphere of trust.* If your people do not trust you, they will not be receptive to anything you have to say. Only if they trust you will they receive your messages willingly, understand them correctly, carry out your instructions, and feel free to send you messages of their own.

The *verbal skills* of the sender will have a lot to do with the clarity of the messages sent. Some people have the knack of saying things clearly and simply; others leave things out, or choose the wrong words, or tangle up the thought in long, strung-out sentences. Sometimes they mumble or speak too softly, or they write illegibly, and the receiver is not interested enough to ask the sender to translate or is afraid to ask.

On the receiving end, accurate reception depends in part on the *listening or reading skills* or sometimes even the hearing ability of the receiver. People who cannot read or hear and people who do not listen are not going to get the message.

✳ HOW SYMBOLS CAN OBSCURE THE MEANING

symbols
Words or pictures, spoken or written, to depict a message. International symbols are recognized by anyone, regardless of the language they speak.

Since we can't transmit messages directly by telepathy, we use various **symbols** to express our meaning (Figure 4.2, stage 2). Usually, the symbols are words, either spoken or written. Sometimes we use abbreviations (symbols of symbols). Sometimes we use pictures: diagrams showing how to operate equipment, posters dramatizing safety messages, movies or filmstrips demonstrating techniques or procedures. Graffiti in appropriate places are often used to send anonymous messages of anger or contempt. International symbols such as those in Figure 4.3 are used for instant recognition by anyone speaking any language.

The trouble with words is that they are often misinterpreted. Many words have several meanings. The 500 most often used words have an average of 28 different meanings apiece; the word *round* has 73 meanings. Many words and abbreviations are unfamiliar to inexperienced workers. Many words and phrases are vague (*they, that stuff, things, the other part,* and so on). Many words mean different things to people from different backgrounds. Slang expressions from yesterday may mean nothing to the teenager of today. You have to choose words that will carry your meaning to the people you want the message to reach.

If you have workers who do not speak English, you will have to speak in their language or use sign language, gestures, or pictures. Such situations require the sender of the message to watch carefully for signs of the receiver's comprehension. Actually, this is a good idea in all oral communication.

On the one hand, written words have the advantage that you can read them over to see if they express your meaning clearly. On the other hand, you have no feedback from the receiver unless you ask for a reply.

nonverbal communication
Communication without words, as in signs, gestures, facial expressions, or body language.

People also communicate without using words; this is generally referred to as **nonverbal communication**. They can deliberately use signs, gestures, and **body language** to convey specific meanings. Nodding one's head indicates, "I agree with you" or "I hear you." A smile says, "I want to be friendly," and invites a return smile. Amorous glances extend invitations. Shaking one's fist means, "I'm dead serious and don't you dare provoke me any further." A listener pays as much, if not more, attention to nonverbal as to verbal communication.

body language
Expression of attitudes and feelings through body movements, positions, and gestures.

People also convey feelings and attitudes unconsciously, through facial expression, tone of voice, intonation, gestures, and body language. Receivers often perceive them almost intuitively

FIGURE 4.3: Familiar symbols that convey meaning without words.

rather than consciously and respond with feelings and attitudes of their own. Figure 4.4 shows some typical bodily expressions of attitude and emotions.

Sometimes nonverbal messages run counter to the sender's words, and a mixed message is sent out. A speaker may tell new employees the company has their welfare at heart while frowning and shaking a fist at them about the rules they must follow (Figure 4.4e). Usually, actions speak louder than words in a mixed message, especially if the nonverbal message is a negative one. The receiver responds to the attitude or the emotion expressed nonverbally rather than to the spoken words.

✳ PROBLEMS IN SENDING THE MESSAGE

The simple mechanics of sending a message can often stop it in its tracks (Figure 4.2, stage 3). If you send it at the wrong time, to the wrong people, by the wrong means, it may never reach its destination.

Timing is important. For a message to get through, you have to consider the receiver's situation. The wrong time may be too soon or too late or a time when the receiver cannot receive it or cannot do anything in response to it.

If you send it too soon, it might not sink in or it might be forgotten. If you send it too late, there is no time for action. The sales manager, for example, must be able to tell the chef and beverage manager about the convention in time for them to order the food and liquor and hire the extra help and alert the station cooks.

Some people are more receptive at certain times of day. Their body clocks run fast in the morning and slow at night, or the other way around. Give an early bird messages in the morning and a night person messages at night.

Sometimes, the message is not received because you do not have the receiver's attention. There is no point in telling people anything when they are right in the middle of something else. A switchboard operator is not going to hear his boss tell him to postpone his lunch hour when he is handling 17 calls at once. A bartender is not going to remember what her boss said right after she dropped a jug of wine on her big toe and broke both the toe and the bottle.

To get a message through, you need to send it to the right person. Give it to the person directly; do not ask someone else to relay it. The "right person" always means everyone concerned. Leaving someone out can fail in two ways: He or she does not get the message and therefore does not carry out the instruction, and he or she then feels left out, put down, unimportant, and neglected. Don't let yourself be embarrassed by an employee saying, "No one told me."

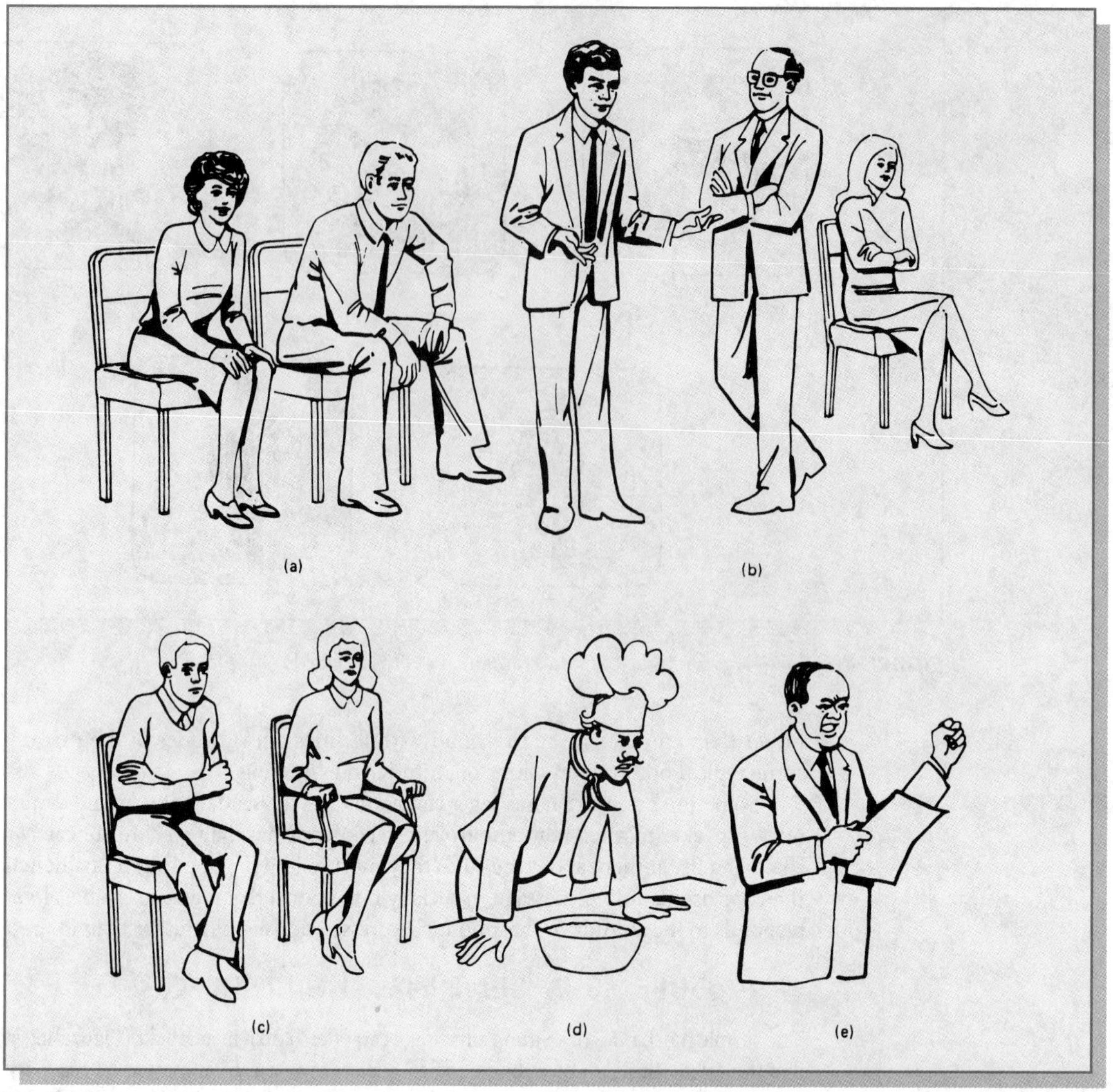

FIGURE 4.4: Body language: (a) accepting, ready to take action (leaning forward on edge of chair, arms open, legs and feet uncrossed; (b) (open hands, palms up, open coat) versus defensive, resting, rejecting (closed coat, crossed arms and legs, leaning backward); (c) nervous, anxious, upright, holding in emotions (crossed arms, fists clenched or hands gripping arms, locked ankles); (d) frustrated, angry, explosive (leaning far forward, head thrust forward, hands spread on table); (e) threatening (shaking fist).

To get a message through, you must choose the right means of sending it.

- If you announce it in a meeting, a few people will hear it correctly. Some people will hear it but will not understand it. Some people will hear you say something you did not say. Some people will not hear it at all.

- If you send the message in a memo, a few people will read it correctly, some will not understand it, some will misread it, some will read it two weeks later, some will not read it at all, and some will not even get the memo.

- If you post it on the bulletin board, no one will read it. Some people won't even know that you have a bulletin board.

- If you tell people individually, you may get your message through to most people. But some of them will be angry because they were the last to know.

You can't win. Your best shot is to tell each person individually, one to one, which is how most communication takes place at this level in this industry. Most hospitality and foodservice people are better at seeing and hearing than they are at reading, and individual contact reinforces the impact of the message.

Sometimes, messages are never sent at all. Sometimes, supervisors assume that communication is not needed. They assume that people know things: If they bused tables in another restaurant, they don't need to be told how to do it in their restaurant. They don't need to be told how to put paper napkins in napkin dispensers—even a five-year-old can do that—but on the other hand, it is their fault when guests can't get the napkins out.

There are other reasons why messages do not get sent. Sometimes the sender simply forgets to send the message. Sometimes he or she is unwilling, unable, or afraid to send the message because of the way the receiver might react. Supervisors who are uneasy in their relationships with their people sometimes avoid telling them things they know people will not like, even though the people need to know. This does no one any good. A supervisor might be unwilling to send a message when he doesn't want people to know as much as he does. In other words, he feels threatened. At other times, a supervisor might not send a message because he isn't really sure of the message himself. Sometimes, people do not communicate with the boss because they are unable, unwilling, or afraid or because they think the boss will not pay any attention. This also does no one any good.

✳ PROBLEMS IN RECEIVING THE MEANING

When the receiver hears or reads a message (Figure 4.2, stage 5), there might still be problems in understanding or accepting it. The meaning may be obscured by the way it is phrased or by something left out, assumptions, and so on. Sometimes the wrong message comes through; sometimes it is meaningless; sometimes nothing comes through at all.

Sometimes the receiver is preoccupied with something else or not interested enough to listen carefully. If you want people to listen actively and open themselves to receiving a message, there has to be something in it for them—information necessary to doing their job, or related to changes that affect them. It does not have to be pleasant; it just has to be important to them. If they think it does not affect them, they will not pay attention, or they will half listen and then forget.

Sometimes a message or the way it is delivered will trigger emotions that make it unacceptable, and people will either tune it out or react negatively. If a supervisor talks down to them, or talks over their heads, makes threatening or scornful remarks, speaks in a condescending tone of voice, or tells them to do something they do not consider part of their job, they are not going to accept the message as it was intended. They will misinterpret the instructions inadvertently or on purpose, or find other ways to withhold good performance—sometimes out of hostility, sometimes out of inertia. If people do not like the way they are being treated, it is quite literally hard for them to do a good job.

✳ REMOVING OBSTACLES TO COMMUNICATION

To summarize, let's list the many ways of removing obstacles to communication that have been mentioned in this section:

1. Build a climate of trust and respect in which communication is encouraged and messages are communicated with respect. Communicate to employees the way you would like them to communicate to you.

2. Send your messages clearly and explicitly, use language the receiver can understand, don't assume that the receiver knows anything, and take into account the receiver's ability to hear, read, and listen.

3. Send your message at the best time, and make sure that you have the receiver's attention.

4. Send your message to the right person(s): in other words, to everyone concerned.

5. Choose the best means of sending your message.

6. Check that your message has been understood, accepted, and acted upon.

7. Listen, listen, listen. This is discussed in detail next.

8. Be as objective as possible when communicating. Don't let any of your own stereotypes or prejudices shape what you say or how you send your message.

9. Avoid using slang names such as "Honey," "Babe," "Sweetheart," "Dear," "Guy," or "Dude." They are disrespectful and annoying. Also, don't tell jokes that poke fun at anybody.

10. Never communicate with someone when you are angry. Cool off first.

Listening

LEARNING OBJECTIVE: Describe the importance of listening as a supervisor.

If you want your people to listen to you, listen to them. If you want to be able to size them up, to figure out who has potential and who is a bad apple, listen to them. If you want loyal, willing, cooperative workers, listen to them. If you want to minimize conflict and complaints and to solve people problems, listen, and listen well. Listening means paying complete attention to what people have to say, hearing them out. It is the second half of the communication process, the most neglected half and sometimes the most important. It is a learnable management skill.

What can your people say that is worth listening to? They can keep you in touch with what is going on throughout your operation. They can tell you what guests think. They can suggest ways to make the work easier, improve the product, and reduce costs. They can clue you in on trouble that is brewing.

They want you to hear their problems and complaints. And what if you cannot solve them? Never mind; they still want you to listen. This may be the most significant listening you do.

❋ BAD LISTENING PRACTICES

Anyone as busy as a supervisor in a hospitality enterprise is going to have trouble listening. Your mind is on a million other things, and you go off on tangents instead of paying attention to the person trying to talk to you. And *going off on tangents* is the first bad listening practice. You must give the speaker your full attention.

It is hard to do this. You can think four times as fast as a person can talk. People talk at 100 to 125 words a minute and think at 500 words a minute, so you have three-fourths of your listening time for your mind to wander. You may be preoccupied with other things: the convention coming in next week, the new furniture that has not arrived, the tray carts that don't keep the food hot all the way to the last patient at the end of the corridor.

You may not really be interested in what the speaker is saying, or it may concern a touchy subject you would rather avoid, so you tune out. You may be distracted by your phone, your beeper, your unopened mail, or some unconscious habit of the speaker, such as pulling his beard or curling the ends of her hair around her fingers. Or maybe the speaker is following you around to talk to you while you do a dozen other things.

Another bad listening practice is to *react emotionally* to what is being said. If someone says something against your favorite person or cherished belief or political conviction, it is very easy to get excited and start planning what you will say to show that he or she is wrong. That is the end of the listening. The other person might, in fact, go on to modify the statement or to present evidence for what he or she is saying, but you will not hear it. You are too busy framing your reply. You might even interrupt and cut the communication short with an emotion-laden outburst and start arguing. The effect is to cheat yourself out of the remainder of the message and to antagonize the person who was talking to you, especially if you misinterpret the message. It is essential to hear the speaker out before you make a judgment and reply.

Sometimes there are words that hook your emotions and make you lose your composure, such as "baby" or "gal" or "dear" to a woman or "boy" or "buddy" to a man, or some of the more vivid four-letter words to some people. Once your emotions flood forth, you can no longer listen and the speaker can no longer speak to you. Chances are that the emotional cloud will also hang over future communications, inhibiting them and fogging the messages back and forth.

You have to stay calm and collected. On the one hand, maybe the speaker was deliberately trying to goad you. If so, the speaker has won and has found a vulnerable spot or word that can be used again. On the other hand, if the speaker used the fateful word without meaning anything by it, he or she will become embarrassed and defensive and will not try to communicate with you again.

---

listening
Paying complete attention to what people say, hearing them out, staying interested and neutral.

If certain words raise your temperature to the boiling point, try to find some way to word-proof yourself. It makes no sense to let a couple of words interfere with communication between you and your people.

Still another bad listening practice is to *cut off the flow of the message*. Certain kinds of responses on your part will simply shut the door before the speaker has finished what he or she needed to say. Suppose that one of your people is upset about a personal problem concerning a coworker. One way of shutting the door is to tell the person what to do. You may do this by giving orders, threatening consequences, preaching—"You ought to do this; you should have done that"—asserting your power and authority as the boss. Such negative responses not only end the conversation but arouse resentment and anger. They are bad for communication, bad for the work climate, and bad for the person's self-respect.

Other ways of telling people what to do might seem positive on the surface: giving advice, giving your opinion, and trying to argue them into accepting it. You may think you are being helpful, but you are really encouraging dependency along with feelings of inadequacy and inferiority. When you solve people's problems for them, you might be plagued with their problems from then on. If they reject your solution, they might resent you for having tried to argue them into something.

Probing, interrogating, or analyzing their motives only complicates your relationship to them. This is not what they came to you for; they do not want to expose their inner selves. They do not want you to see through them, and they resent your intrusion. Besides, your analysis could be wrong, and they will feel that you are accusing them unjustly. Even though they came to you with the problem, they might end up feeling that it is none of your business. They might find your probing scary and threatening, and from then on they worry about working for you when you know so much about them—or think that you do.

Still other ways of responding will also close the door and end the discussion—for example, blaming the person for having the problem in the first place or calling the person stupid or worse for getting into the predicament. Whatever the truth of the matter, this type of response solves nothing and only arouses negative feelings: anger and resentment toward you, plus feelings of self-doubt and self-reproach.

Responses at the opposite extreme, although not so destructive, still close the door on the communication and the problem. You might try to sympathize, console, and reassure the person in an effort to make the feelings go away, or you might belittle the problem by refusing to take either the problem or the feelings seriously. Neither response works. The problem and the feelings are just as big as ever, and you have, in effect, told that person you don't want to hear any more about it.

What about cheering this person up with something positive, such as saying what a good job he or she is doing? This sometimes improves the climate momentarily, but it will not solve anything unless it is related to the problem and *unless it is true*. If the employee knows it isn't true—and who knows this better than the worker?—your praise only raises doubts about your sincerity and increases the distance between you.

All these ways of responding to a worker who is communicating a problem are ways of refusing to listen anymore. They are different ways of saying, "Go away, that's all I want to hear." The employee stops talking to you, and the problem goes unsolved. Figure 4.5 lists these and other roadblocks to listening.

1. Withdrawing, distracting
2. Arguing, lecturing
3. Commanding, ordering
4. Warning, threatening
5. Diagnosing, analyzing
6. Judging, criticizing
7. Blaming, belittling
8. Interrogating, analyzing
9. Preaching, giving advice
10. Consoling, sympathizing

FIGURE 4.5:   Roadblocks to listening.

❋ HOW TO LISTEN

Good listening does not come easily to most busy people in charge of getting things done. You have to learn to listen, and you have to make a conscious and deliberate effort to discard all of the bad listening practices you may have been using. Here are five principles of good listening, along with a few techniques for putting them into practice:

1. *Give the other person your undivided attention.* Set aside everything you are doing and thinking and concentrate on what that person is saying. Don't answer your phone, don't open your e-mail or mail, don't look at your watch, and don't let other people interrupt. Don't make that person follow you around while you are tending to something else. Take whatever time is necessary, and take seriously the person's need to talk. Keep your mind on the message and don't go off on mental tangents. Look the person in the eye with an interested but noncommittal expression on your face.

2. *Hear the person out.* Don't stop the flow; don't tell the person what to do, or comment or argue or console or in any way take the conversation away from the person talking. Keep the door open: "I'd like to hear more. Tell me why you feel this way." In this way you acknowledge others' right to talk to you and let them know you really want them to.

   Encourage the person. At appropriate times, grunt ("unnh," "ummmm," "uh-huh"), say "OK" and "Yes," and nod your head. This lets the person know you really are tuned in and you really are listening. And you really do listen.

   This type of listening is known as **active listening**. It is most appropriate when a person is upset about something or has a complaint or a problem. It takes concentrated effort on your part. Suspend all your own reactions, make no judgments or evaluations, and do your best to understand how things look from the speaker's point of view and especially how he or she feels about them. The purpose of active listening is to find the ultimate solution to the problem. Active listening is listening for the complete meaning without interrupting the speaker or interpreting the meaning. Because, as we already noted, the average person speaks about 125–200 words a minute and the average listener can understand up to 500 words per minute, this leaves time for the mind to wander. The *active* part of active listening means empathizing attentively with the speaker and, to an extent, putting themselves in their position.

   You can raise the level of active but neutral listening by mirroring the speaker's words. When the speaker says, "I don't think you're being fair to me," you say, "You don't think I'm being fair to you." You can go further and paraphrase: "You feel I'm giving you more than your share of the work." You can take the process still further by mirroring the speaker's feelings as well as the words: "You feel I'm being unfair, and it's really making you angry, isn't it?" These techniques, as well as others described in Figure 4.6, used sensitively, will move the flow along until the speaker has said everything that he or she wants to say. Only then do you respond from your point of view as supervisor.

3. *Look for the real message.* It may not be "Solve my problem." Maybe it is, "Hey, I need to talk to someone who understands." Or maybe there is a message underneath the verbal message. Often, the first spurt of speech is not the real problem but a way of avoiding something that is difficult to talk about.

   Look for nonverbal cues: tone of voice, anxious facial expression, clenched fists, body tension. If the speaker is still tense and anxious when he or she stops talking, you probably have not heard the real problem yet. Wait for it; open the door for it; use active listening techniques. But remember that interrogating and analyzing are turnoffs, not invitations to go on.

4. *Keep your emotions out of the communication.* Stay cool and calm; don't let your own feelings interfere with the listening. Let remarks pass that you are tempted to respond to and keep your emotions in check and concentrate on the message. Remain inwardly calm all the time this person is talking, and do not get sidetracked planning a hard-hitting reply.

---

active listening
Encouraging a speaker to continue talking by giving interested but neutral responses that show that you understand the speaker's meaning and feelings.

Technique	Objective

1. Acknowledgment
Examples:
  "Uh-huh . . ."
  "Ummm . . ."
  "I see."
  "I understand."
  "Let's look at and discuss your last
    comment."
	To show interest, to encourage an employee to keep talking

2. Clarifying questions
Examples:
  "What exactly do you mean by . . .?"
  "Will you explain what you mean by . . .?"
  "What I understand you to say is . . ., is
    that right?"
	To clarify and/or confirm a message

3. Mirroring statements
Examples:
  "You feel it was unfair that Jimmy got
    Friday off instead of you."
  "You think someone else should help you
    in the dishroom."
  "You think you're being treated differently
    than the other people you work with."
	To keep the speaker talking

4. Summarizing check
Examples:
  "Let's hold on for a moment and review
    what we have discussed so far . . ."
  "These seem to be the key points you've
    expressed to me . . ."
  "To summarize, the key ideas as I
    hear them are . . ."
	To pull together important points in order to confirm understanding, review progress, and possibly lead to more discussion

FIGURE 4.6: Techniques listeners can use to increase understanding.

Concentrate on listening and staying neutral so that you can get the message and so this person can vent his or her emotions and clear the air, and maybe you can help keep your own emotions down by looking for the message behind the message that is causing this person to make such remarks to you. Don't let this person grab you with loaded words; don't react. Get the message straight and clear, and then, when the flow stops, you can respond appropriately.

5. *Maintain your role.* You can listen to personal problems, but do not try to solve them; do not get involved. Do not boomerang back over that line between manager and worker to help a buddy out. Do not let yourself be maneuvered into relaxing company policies or making promises you cannot keep. Stay in your role as manager; you accept the message with understanding and empathy, but you do not take on anything that is not your responsibility.

If the problem is job-related and within your sphere of authority and if you have listened successfully, you and the worker can talk about it calmly. Suppose, for example, that the real problem turns out to be something that makes this person—let us call him John—furious at you, and he has been seething about it for weeks. Listening successfully means that you are able to keep your own emotions out of it even though he ends up screaming at you and calling you choice names, and that you manage to refrain from making judgments and jumping to conclusions and telling him off.

Courtesy of
Holly Rielly

Throughout the past 15 years I have worked in the restaurant industry. My experience in the business ranges from the kitchen of a convent to casual and fine dining. Currently I am employed as a supervisor at Marina Jacks in Sarasota, Florida.

I think a good supervisor is someone who is fair to everyone and does not have "favorites." A good supervisor explains what needs to be done, requests input from the team members, and informs everyone of the decision. By gaining input from the team members it encourages their involvement in achieving the goals. Supervisors need to ensure that everyone has the necessary tools to do the job required; we all know how frustrating it can be when a restaurant runs out of supplies on a night when you are being "slammed"!

Good supervisors are also good leaders who gain respect not for what they are but for who they are. Leaders encourage superior performance just as a great coach does. They also praise and reward outstanding performance and create a fun atmosphere that makes employees look forward to going to work.

A good supervisor develops his or her team by arranging training sessions or offering wine appreciation sessions so that team members can learn more about wine and how to offer an appropriate wine to complement the meal to the guests. Supervisors should never lose sight of the fact that they need to delight the guest and staff while satisfying the owner. Now that's a tall order! At our restaurant we have preshift meetings to go over the menu and specials of the day, so we are all on the same page. We communicate with each other regularly throughout the meal service. On occasion we also have sales contests. The employees get rewards for the highest sales of a particular item.

In order for a supervisor to be a good leader and develop a successful team, he or she must know how to communicate well. Communication is essential in the hospitality industry due to the constant interaction with staff and patrons. When good supervisors communicate, they are precise and clear, and they check to be sure that the receiver of the information understands what is being conveyed to them. A team cannot be successful unless they communicate well and all work together toward the same goal. Therefore, in my opinion, an overall successful supervisor is fair, considers team input, communicates well, and has respectful relationships with staff.

Because you listen, John unloads his emotions and begins to simmer down, and when you don't get mad or shut him up he begins to appreciate that you are not using your position of power against him, and pretty soon he comes around to seeing that he has exaggerated a few things and is sorry about some remarks. Now you and John can begin to explore the causes of the problem and perhaps come up with a solution.

To demonstrate to the person that you have been paying close attention, try pausing before giving a response. When you pause, you convey that you have been giving careful consideration to his or her words. Pausing always helps you to hear more clearly what the other person has said. Asking questions is a way to prove that you have concern for the situation at hand. It is important that you ask questions to help you gain clarification. Think about asking open-ended questions such as, "What do you mean?" People will often rephrase their words so that you can better understand them. Questions such as *how, when, what, why, where, and who* will prove to be useful. Asking questions will help employees to expand their thoughts and also help you to gain a better understanding of what they want.

Of course, it doesn't always work. But if you stay in the boss's role and use a positive, person-to-person leadership approach, you have a chance of turning listening into two-way communication that works. Here, in a simplified list, are the five principles of good listening:

1. Give the other person your undivided attention.
2. Hear the person out: "A good listener listens 95 percent of the time."
3. Look for the real message.
4. Keep your emotions out of the communication so that you receive a clear message.
5. Maintain your role.

This kind of listening is not a skill you can develop overnight. Like everything else about leadership, it takes understanding and awareness and practice and maybe supervised training on the job. It is not common in our industry; there is too little time and too much to do, and we have other traditional ways of dealing with people. But it has tremendous potential for solving people problems if you can learn how to put it to work.

Companies that have trained their supervisory personnel in active listening techniques have found them extraordinarily effective. One manufacturing company that was losing money and suffering labor problems brought in a psychologist to train all its foremen in an intensive two-week course in listening. The investment paid off—grievances declined by 90 percent, and the company began making money again.

Many industries today are paying a great deal of attention to what workers have to say about their work and are finding it very profitable. In our time-pressured industry it is easy to think we do not have time for listening. But often it does not take long to receive a message. What it does take is an attitude of openness. A few minutes—a few seconds even—of total attention can pay off in countless ways.

Directing People at Work

LEARNING OBJECTIVE: Examine the ways in which effective communication leads to effective management.

directing
Assigning tasks, giving instructions, training, and guiding and controlling performance.

Effective communication leads to effective management: It gives you the power to get things done. One of your major functions as a supervisor is **directing**: assigning tasks, giving instructions, telling people how to do things, and guiding and controlling performance. If you can give your people direction in a way that will get them to do things as you want them done, you will be an effective communicator and an effective manager.

Most chefs have a directing or briefing staff session in the kitchen before they start their shift. This is very common in most professional kitchens. The executive chef has morning briefings with his or her staff and speaks with each department head as to what needs to be done and to confirm that all are on the same page before day begins.

Your effectiveness as a communicator depends on several positive things that you can do to avoid the many obstacles to good communication. The most important are to send clear messages, to get people to accept your messages, and to make a positive impact. Let us look at each of these in detail.

✳ SENDING CLEAR MESSAGES

A clear message is one that is specific, explicit, and complete. It tells everything the other person needs to know: the who, what, when, where, how, and why of the information to be given or the task to be done. Most of the communicating you do in directing your people will be giving instructions and information on the job. It will be very informal, and it will likely be one of those 48-second contacts that make up your day. It will be in the middle of the kitchen, at the front desk, out in front of the hotel or the country club, or in the storeroom or the bar or the laundry or wherever the person you need to talk to is working. It will probably be a fragment of conversation, and it may take place under severe time pressure, so it will be very easy to run into all the common obstacles to good communication. But brief as it is, and however difficult the circumstances are, you need to give each person the entire message.

"Cook the chicken" is not an entire message. "Cook 30 dinner portions of fried chicken to be ready at 5:00 P.M." is specific and complete. It makes no assumptions, takes nothing for granted. It tells all. Say it clearly and distinctly, or write it down clearly and distinctly. Or do both. It takes only 25 of your 48 seconds.

A clear message is also one that is understandable and meaningful to the person to whom it is sent. It must be phrased in terms that the person can understand. It must be delivered on that person's level. It must be meaningful within that person's experience.

Making messages understandable and meaningful requires awareness on your part. It takes awareness of the other person's background and experience and ability to comprehend, awareness of your own assumptions about this person and how he or she regards you, awareness of your tone of voice and choice of words and how you come across to other people. It means deliberately adapting your message to your audience. It means knowing your people and knowing yourself.

✳ GETTING YOUR MESSAGES ACCEPTED

Of the six steps of communicating (see Figure 4.1 to refresh your memory), you as the sender control only three. The other three are up to the receiver. How can you influence that person to come through on the other three steps, to receive and understand and accept your message?

The first essential here is trust. If your employees trust you, they will have a built-in attitude of acceptance, of willingness to do as you say and do a good job for you. If they do not trust you, the message probably will not come through to them clearly. Their opinion of you or their feelings about you are likely to distort facts and meaning and are also likely to lessen their desire to carry out your instructions. Professional chefs today communicate with their staff as a coach-type leader, with a clear voice and passion that conveys upbeat energy.

Building trust takes time. Meantime, in dealing with someone you know does not trust you, you should do your best to maintain a pleasant atmosphere and a calm and confident manner. Be extra careful to send clear, simple, and very explicit messages and to explain why the task or information is important. Then follow up to make sure that the person in question is carrying out the task correctly.

A second essential for acceptance is the interest of the receiver in the message: People have to see *what's in it for them*. Make sure that they understand what your messages have to do with their work, as well as how and why the information or instructions affect them. Perhaps something of value for them is involved: better tips, more satisfaction in the work itself. Or perhaps it is something less pleasant, yet something they must adjust to. Whatever it is, people will pay careful attention to anything affecting them. Look at your messages from their point of view and emphasize whatever is important to them.

A third essential for acceptance is that your instructions must be reasonable. The task to be done must be possible to do within the time allowed. It must be within the ability of the person and within the scope of the job as the employee sees it. It must be legally and morally correct and

Emotional Intelligence is the ability to manage ourselves and our relationships effectively.
wavebreakmedia/Shutterstock

compatible with the needs of both the person and the organization. If you violate any of these criteria, the worker is likely to balk openly or to do the task grudgingly or leave it unfinished to prove that it is impossible.

When instructions that seem unreasonable to a worker seem perfectly reasonable to you, you need to discuss them—find out why they seem impossible and explain how and why it really is possible to do what you want done. It may be that your communication was poor the first time.

❋ MAKING A POSITIVE IMPACT

If you want people to get your messages clearly and do willingly what you want them to do, your messages must have a positive impact. People must feel like complying. They must not be put off by the way you have delivered your messages.

Many supervisors make the mistake of talking to their people from a position of power, authority, or status. In effect, they are commanding, "You'd better hear what I say because I am your boss, and what I say is important because I say it is, and you'd better pay attention and do what I tell you to do."

In every instance where you are directing people, you must remember to think, "How are the people I am talking to going to hear me, and how will they feel about it?" Your purpose as a manager, as a supervisor, is not to impress them but to get across the message that you want to give them. So put yourself on their level and talk to them person to person.

The situation affects your style of communication. Sometimes, in pressure situations, it is all too easy to let your emotions take over and to yell and scream orders when the food is not going out fast enough or customers are waiting and the tables are not being cleared. Unfortunately, yelling and screaming often stops the action entirely. Remember that emotion takes over the message: People react to the emotion rather than the instruction, and anger and fear drive out good listening and good sense. In a real emergency a sharp, controlled command may be appropriate, but never one expressing anger, fear, or loss of control.

❋ EMOTIONAL INTELLIGENCE

emotional intelligence (EI)
The ability to manage ourselves and our relationships effectively.

Emotional intelligence (EI), often measured as the EI quotient (EQ), popularized by the work of Daniel Goleman, is defined as "the ability to manage ourselves and our relationships effectively."[1] According to his research, emotional intelligence is an important influence on leadership effectiveness, especially in more senior management positions. In Goleman's words: "The higher the rank of the person considered to be a star performer, the more emotional intelligence capabilities showed up as a reason for his or her effectiveness."[2]

The critical components of EI are the following:[3]

- *Self-awareness*: Ability to understand our own moods and emotions, and to understand their impact on our work and on others
- *Self-regulation*: Ability to think before we act and to control otherwise disruptive impulses
- *Motivation*: Ability to work hard with persistence and for reasons other than money and status
- *Empathy*: Ability to understand the emotions of others and to use this understanding to better relate to them
- *Social skill*: Ability to establish rapport with others and to build good relationships and networks

How many emotions are there? Although there are several emotions we could all name, research has identified six universal emotions: anger, fear, sadness, happiness, disgust, and surprise.[4] Are any of these emotions used in the workplace? Absolutely! I get *angry* after a bad experience with a guest. I *fear* that I could receive a poor evaluation. I am *sad* about a coworker who was victimized in an accident. I'm *happy* about being promoted. I'm *disgusted* with the way my manager treats women on our team. I'm *surprised* to find out that management plans to reorganize our department.[5]

We all know that a hospitality business is made up of people with varying emotional dispositions, so it is especially important for supervisors to recognize individual emotions and adapt to situations to maintain a harmonious workplace. Personally, we should all assess our own EI and then see what others think about our EI because if we want to advance in our careers, we should have good EI. It is good to express emotion in critical situations as long as the person directing is in control and is respective. Positive emotion with seriousness in their tone and facial expression resonates with employees, much like a coach giving direction to his players.

Supervisors and managers work with people. Yet, as simple and obvious as this sounds, it is not always easy to do well.[6] The emotional awareness of *"reading people"* is a good start; being able to zero in on people and identify emotions is vital to success in interpersonal relationships in the fast-paced hospitality industry. The ability to identify emotions consists of a number of different skills, such as accurately identifying how you feel and how others feel, sensing emotion in art and music, expressing emotions, and reading between the lines. Perhaps most critical is the ability to detect real versus fake emotions.[7] Hospitality companies are using EI in their hiring decisions.

✳ GIVING INSTRUCTIONS

Here is a detailed set of directions for giving instructions or orders. Not every step will apply to every kind of instruction or every circumstance, but it is a good standard you can adapt to your own needs. There are five steps:

1. Plan what it is you are going to say, whom you will say it to, when you will give the instruction, where you will give it, and how.
2. Establish a climate of acceptance.
3. Deliver the instructions.
4. Verify that the instructions were understood.
5. Follow up.

Let's look at these in more detail.

Plan

The first step is to plan. You plan what it is that you are going to say (*the who, what, when, where, how, and why of the task*), whom you will say it to, when you will give the instruction, where you will give it, and how (orally or in writing or both, with maybe additional materials such as diagrams, recipes, or a manual of procedures).

Generally, an oral order is best suited to simple tasks, to things that people have done a hundred times before, to filling in details, to explaining or amplifying written orders, to helping someone or showing somebody how to do a task (show and tell). It is also appropriate to something requiring immediate action, such as an emergency, or to instances where a written order is not likely to be read or understood. As mentioned earlier, most people in the hospitality business are not great readers; there is nearly always a time pressure; and communication tends to be oral, for better or worse, and with all its attendant risks.

Written instructions are best when precise figures or a lot of detail is involved, such as specifications, lists of rooms to be made up, recipes, production sheets, and specific needs. You might need 150 salad plates and 250 dessert plates in the banquet kitchen by 4:00 P.M., and you would write this order out for the dishwashers so that there will be no chance of mistake.

It is best to use written instructions when the details of the task are very important, when mistakes will be costly and there is no margin for error, when strict accountability is required, when you are dealing with a slow or forgetful or hostile person with a poor track record, or when you are repeating orders from above or are enforcing company policy.

Written instructions should be short, complete, clearly written or typed, and clearly stated in simple words. You should write them with the reader in mind, and you should read your message over to be sure that it includes everything and says it clearly.

Written instructions are not appropriate when time is short, when immediate action is called for, or when it is likely that people will not read them or will not grasp the meaning. They can cause problems when they are not written clearly, when they are incomplete, when they are too long or too complicated, or when they are poorly organized (such as a recipe whose final instruction is to soak the beans overnight).

In some cases, it is best to give instructions both orally and in writing. Then one method reinforces the other. People receive the impact of the oral directive and have the written instructions for confirmation and reference.

Establish a Climate of Acceptance

The *second step* in giving instructions is to establish a climate of acceptance. Often this is something as simple as making sure that people are not preoccupied with something else and are ready to listen. This is the point at which you explain the why of the task (if people don't know) and what's in it for them, to involve their interest, and, if possible, secure their commitment.

Quiet surroundings free of distractions help to establish a climate of acceptance. But your surroundings and conditions will probably be less than ideal—the typical fragmentary 48-second time-pressured conversation—so you have to make the most of it.

Among the different types of people who work for you, there will be those who are cooperative and enthusiastic, with whom you have a relationship of trust and goodwill. They will always be receptive and accepting. There will be other people who just plod along doing whatever they are told to do—no less, no more. You must be sure that you have their full attention and that everything is included in the instructions because that is exactly what they will do—just enough to get by.

There will be a third type, the hostile employees, the ones who do not trust you and whom you do not trust. There are always a few of these, and they are looking for ways to beat the system, to do as little as possible, to challenge you, to show you up if they can. If you make a mistake in your instructions to them, they will follow that mistake to the letter and take delight in the trouble it makes for you.

An incident at a riverboat restaurant in St. Louis illustrates the point. The manager, appalled at the total disorder of the dishroom, told the two dishwashers, "Get this place cleaned up now! I'll be back in 30 minutes to check it out!" When he came back 30 minutes later, the room was totally clean, not a dish in sight—cleaned out. The dishwashers had simply pushed everything into the Mississippi River.

With this type of person, spell everything out—why they are given the task, exactly what they must do, how, why, and what results you will hold them responsible for. If it is something important, put it in writing. You cannot expect a spirit of acceptance from these people; you must force acceptance of the instructions to whatever degree you can, using your powers of enforcement as necessary.

Deliver the Instructions

The *third step* is to deliver the instructions. Your manner of delivery is critical. Gestures, inflection, tone of voice, facial expressions, all the nonverbal ways of communicating come into play here, as well as what you say. Give your instructions calmly and confidently. An air of confidence is critical to giving the order. You can appear confident even when you are not by acting calm, competent, and collected, speaking lower and slower than you normally do, and talking without hesitation or groping for words. Your image as a leader is involved here, and this is one of the things that make people listen and take your directions seriously.

Where you stand or sit in relation to people you are directing can sometimes have an effect on the communication. Research has established that there are unexpressed zones of comfortable communication between people (**communication zones**). Two to 3 feet is **personal space**—don't come any closer unless invited. Four to 7 feet is **social distance**—it's okay to give instructions from this distance. Seven to 25 feet is **public distance**, and that's too far away—it is not a public meeting you are addressing. These are American zones. People from some other cultures, such as Latin Americans, have much smaller zones, and you can have communication problems when one of these people tries to get close enough to speak comfortably and the other person keeps moving away to maintain personal space.

communication zones
Unexpressed areas in which people are comfortable communicating with each other under various situations.

personal space
The distance within which a person feels uncomfortable allowing others to come closer unless they are invited to approach (for Americans, about 2–3 feet).

social distance
The distance that is comfortable for having a conversation (for Americans, about 4–7 feet).

public distance
The distance at which conversation is impersonal; it is suited for public speakers, not for one-on-one conversations (for Americans, about 7–25 feet).

There are several ways of issuing instructions. You can request people to do things. This is an easy method to use, and it works well with most people—cooperative types, plodders, long-time employees, older people, and sensitive individuals. You can suggest actions to certain kinds of people when you want something done and there is no set way you want it done. This is a subtle and gentler form of direction that you cannot use with everybody because you are leaving up to the worker not only how something is done but whether it is done. It is a method to use with smart, ambitious, experienced people; they will jump on your suggestion and run with it because they want to please or impress you. It is not a good technique to use with inexperienced people or with plodders or those who are hostile.

Verify That the Instructions Were Understood

The *fourth step* in giving instructions is to verify that the instructions have been understood. There are various ways of doing this. You can watch for spontaneous signs: the look of comprehension in the eyes, nodding of the head, a verbal okay. This means that the person thinks that he or she understands or at least wants you to think so. By contrast, a glazed look in the eyes or a lost expression on the face can tell you that you have not gotten through.

If you ask people whether they understand, they are likely to say yes whether or not they do understand. A better way to check understanding is to ask whether they have any questions. The trouble with this is that people don't like to admit that they do not understand, especially in group situations. Sometimes they do not even understand enough to formulate a question that makes sense. They must feel at ease with you before they can handle asking you questions. Sometimes if you ask them, "Can I clarify anything for you?" they will admit there is something they have not understood.

One way to test understanding is to ask people to repeat your instructions back to you. Some people are insulted or embarrassed by this. Sometimes you can take the edge off this impact by presenting the repeating as a way of checking up on yourself: "Have I told you everything necessary?" People who know they have trouble getting things straight generally do not mind repeating things back to you. It is a technique best used selectively according to the person you are dealing with. The best proof of understanding is seeing people carry out your orders correctly. But it is risky to wait and see, and it is a bit late in the game for corrections.

Follow Up

The *fifth and final step*—following up—deals with just this problem. You should not consider that you have carried out your direction-giving responsibilities fully until you find out how your instructions are being carried out. Observe your employees at work. Measure results where they can be measured. Give assistance and further direction where they are needed. And check back on your own performance: Did your instructions do the job? Can you do even better the next time you give directions to these unique and diverse people who carry out the work of your department?

❋ COMPUTER- AND TELEPHONE-AIDED COMMUNICATIONS

Technology and, in particular, information technology has totally changed the way supervisors communicate in the hospitality industry. Today, vital information can be communicated to and by supervisors about sales and recordkeeping far more quickly than previously. For example, a chain of restaurants' daily sales and other relevant data may be automatically sent to the corporate office rather than being typed up, faxed, and then entered into the computer system; schedules can be updated with a few clicks of a mouse; and training programs are now available via CD-ROM; and policy and procedures manuals can be checked online instead of printing and distributing copies.

Networked computer systems link corporate and independent hospitality businesses to one another, the supply chain, and various information sources via the World Wide Web. For instance, if you wanted to find some new recipes, it's easy to surf the Web to find several to choose from for any dish.

E-mail is a quick and convenient way for supervisors and employees to share information and to communicate to one or several people. Some companies use *instant messaging (IM)*, which

A server records an order electronically. The order will reach the kitchen before the server does.

Tyler Olson/Shutterstock

is interactive, real-time communication that takes place among computer users who are logged onto the computer network at the same time. With IM, there is no waiting for someone to read the e-mail, but users must be logged onto the network at the same time.

Voice mail allows information to be stored and retrieved later. Voice mail is also very useful in the hospitality industry because much of supervisors' time is spent away from their desks, so messages can be checked periodically. However, for fast-paced operations it is necessary to have immediate voice communications, which improve guest service and satisfaction. For example, when a guest checks out of a New York hotel and needs a cab, they don't have to wait around outside in the winter cold—the front desk agent or the bellhop can speak into a microphone to request the doorperson to hail a cab.

Texting enables text messages to be instantly sent from one hand-held device to another.

Fax machines and e-mailing PDFs permit the transmission of text or graphics over telephone lines. The sending fax machine scans and digitizes the document and the receiving machine reads the scanned information and reproduces it on paper. All-in-one devices can print/scan/copy and fax; however, sending PDFs via e-mail has become the most efficient and effective way of sending documents from one computer to another.

❄ SOCIAL MEDIA

Twitter is a communication platform that helps businesses stay connected to their guests. As a business, you can use it to quickly share information with people interested in your company, gather real-time market intelligence and feedback, and build relationships with guests, partners, and other people who care about your company. As an individual user, you can use Twitter to tell a company (or anyone else) that you've had a great—or disappointing—experience with their business, offer product ideas, and learn about great offers.[8]

Twitter lets you write and read messages of up to 140 characters, or the very length of this sentence, including all punctuation and spaces. The messages are public, and you decide what sort of messages you want to receive—Twitter being a recipient-driven information network. In addition, you can send and receive Twitter messages, or tweets, equally well from your desktop or your mobile phone.[9]

Facebook is the world's largest social utility that connects more than 400 million people with friends and others who work, study, and live around them. Users can join networks organized by workplace, school, or college.[10]

Instagram is a fast, fun, and free way to share photos and video with family and friends by posting them.

An *intranet* is a private, organization-wide networks that acts like a website but to which only people in the organization have access. Intranet is being increasingly used in the hospitality industry, especially with WiFi. This voiceover allows associates to make and receive calls on the same wireless broadband network that the company uses for Internet access. WiFi is good for making employees available no matter where they are on the property. Resorts, amusement parks, conference centers, sports complexes, hospitals, and cruise ships all cover huge areas, and yet with WiFi everyone is easily and almost instantaneously accessible.[11]

An *extranet* is a company communication network that uses Internet technology, which allows approved users in the company to communicate with key people outside the company such as guests or suppliers.

Business Writing

LEARNING OBJECTIVE: Identify the best ways to improve your business writing skills.

As a supervisor you are involved in all types of writing: job descriptions, policies and procedures, memos to employees and managers, performance appraisals, disciplinary action notices. It is crucial for you to be able to express yourself effectively in these different types of written documents as well as write in a clear, organized manner. Let's first take a look at some common problems, or pitfalls, in business writing.

- Too long, too wordy
- Too vague
- Too much jargon
- Too many hard-to-understand words
- Poorly organized
- Purpose not clear
- Sloppy, misspelled words; incorrect grammar and punctuation
- Too negative
- Indirect, beats around the bush

You have surely read letters, memos, and policies with some, or even most, of these problems.

Following are 10 tips for better business writing:

1. Pay attention to who the reader will be and write from his or her perspective.
2. Organize your thoughts so that your writing is then better organized and better able to communicate.
3. Use simple words to communicate your message. Stay away from jargon, slang, and big words, as they only clutter up your message and the reader might not understand them.

4. Get to the point quickly. Use only as many words as you need to get your point across. Trim all unnecessary words. Be specific about what you want to communicate. Avoid vague terms and expressions (e.g., *somewhat, sort of, rather*).

5. Be positive and upbeat. Even when you have to give bad news, provide some good news.

6. Write as though you were talking. Be natural.

7. Write clearly. Proofread your writing for clarity.

8. Show how the reader will benefit from reading your written communication.

9. Keep the document as short as possible.

10. Always check your document for correct grammar, spelling, punctuation, and neatness.

Meetings

LEARNING OBJECTIVE: Discuss techniques to improve meeting productivity.

Mention the word *meeting* to a group of supervisors and you might hear a few groans and comments such as, "What a waste of time!" Meetings have a bad reputation. Surveys help tell us why. Surveys show that meetings often have problems such as people getting off the subject, no set goals or agenda, they last too long, people are not well prepared, and important issues are not resolved. Some things you can do to make meetings effective are:

1. Be prepared for all meetings. If you are calling the meeting, plan out exactly what you want to accomplish, and formalize your list of topics to be covered in an agenda, a written statement of topics to be discussed.

2. Start on time and first review the agenda.

3. Summarize at appropriate points and move on.

4. When the conversation goes off in a different direction, bring it back to the matter at hand.

5. Make sure that meeting minutes are kept and followed up.

6. Have some rules of order so that everyone has a fair chance of speaking.

7. Handle differences of opinions with respect.

When run well, meetings really can accomplish a lot, without too many groans!

 # CASE STUDY: The Meeting

Once a month, Sue, a district manager for a foodservice contract company, holds a meeting at a single location that is accessible to all the foodservice directors at her six hospital accounts. The meeting is supposed to start at 2:00 P.M. but Sue waits until the last manager arrives at 2:25.

The meeting starts as Sue thanks everyone for coming and asks how everyone is doing. Three managers start in at the same time with problems they want Sue's advice on, until one manages to outshout the others and gets Sue's attention. He complains about problems using the new computers in his hospitals. Once he's done, each manager takes a turn asking questions about concerns in his or her own operation. Their questions are all directed to Sue. By the time Sue has given each question her attention, it is almost 3:30. Since the meetings are only supposed to go until 4:00, Sue tries to share her news with everyone in 30 minutes.

After the meeting is over, Sue realizes that she forgot to mention something very important. She becomes angry with herself because she once again ran out of time and didn't write an agenda. The managers grumble and groan about wasted time as they run to their cars to return to their accounts.

Questions
1. What went well with this meeting?
2. What went poorly?
3. How could Sue make her management meetings more productive?
4. What could the managers do to make the meetings more productive?
5. Did Sue really not have enough time to write an agenda, or did she not make it a priority?

 ## KEY POINTS

1. Communication is the transference of understanding and meaning between two or more people.
2. Supervisors are involved in interpersonal communication, organizational communication, small-group communication, and sometimes mass communication.
3. Figure 4.1 shows the six elements of a successful communication.
4. Communication is important because supervisors spend most of their time communicating, as when directing people at work, giving instructions, training, interviewing, hiring, firing, and so on.
5. Figure 4.2 describes many of the obstacles to communication.
6. The following can be done to remove many obstacles to good communication: build a climate of trust and respect; send your messages clearly; use language the receiver can understand; don't assume anything; take into account the receiver's ability to receive; send your message at the best time; send your message to the right person; choose the best means of sending your message; check for understanding; listen; be objective; avoid slang and disrespectful terms; and don't communicate when you are upset.
7. Listening is the second half of the communication process—the most neglected half and sometimes the most important.
8. Bad listening practices include going off on tangents, reacting emotionally, and cutting off the flow of the message.
9. Five principles of good listening include giving the other person your undivided attention, hearing the person out, looking for the real message, keeping your emotions out of it, and maintaining your role.
10. To direct work effectively, a supervisor must send clear messages, get people to accept the messages, and make a positive impact.
11. The steps in giving instruction include planning, establishing a climate of acceptance, delivering the instructions, verifying that the instructions have been understood, and following up.
12. Tips for effective business writing include paying attention to your reader, organizing your writing, using simple words, getting to the point quickly, being upbeat, writing as though you were talking, writing clearly, showing how the reader will benefit from reading your communication, keeping it short, and checking your document for grammar, spelling, punctuation, and neatness.
13. For meetings to work for everyone, you need to be prepared, have an agenda, follow the agenda, and respect differences of opinion.

 ## KEY TERMS

active listening

body language

communication zones

communications

directing

emotional intelligence

interpersonal communication

interviewing

listening

mass communication

nonverbal communication

open communication

organizational communication

personal space

public distance

small-group communication

social distance

symbols

two-way communication

REVIEW QUESTIONS

Answer each question in complete sentences. Read each question carefully and make sure that you answer all parts of the question. Organize your answer using more than one paragraph when appropriate.

1. Define *communication* and the five types of communication discussed in this chapter.
2. List five obstacles to communication and how to overcome each.
3. Describe briefly the five principles of active listening.
4. You need to give new instructions to the late-shift supervisor about how to close the operation for the night. Describe how you might do so.
5. Discuss briefly three problems you have seen in business writing and how you would overcome them.
6. What can be done to prevent lengthy meetings that don't accomplish anything?

ACTIVITIES AND APPLICATIONS

1. Discussion Questions
■ Explain why two people describing the same object or the same situation might give different accounts of it. How does this interfere with communication?
■ Comment on the statement, "When emotions are involved, the emotions become the message." Do you agree? Give examples to back up your answer. How does emotion block communication on the job?
■ Give several examples of words that mean different things to different people, such as *fun, soon, a little,* or *pretty.* Pick two or three words and ask several people to describe what each means to them. Compare the meanings and discuss how the differences could complicate communication.
■ The supervisor is responsible not only for giving an instruction but also for making sure that it is received and carried out, yet it is the worker who controls the reception. How can a supervisor deal with this problem?
■ Do you know people who are really good listeners? What do they do that makes them good listeners? Do you think a first-line supervisor would be able to become an active listener? What is the best way to do this?

2. Group Activity: Communication Methods
Complete a chart listing the advantages and disadvantages of face-to-face communication, telephone communication, memos, meetings, and e-mail. Compare your chart with those prepared by other groups.

3. Memo Critique
Critique the memo in Figure 4.7 using the criteria in the text. What could you do to make this memo more effective?

4. Are You Listening?
With the students working in pairs, one student draws two or three shapes (such as squares, rectangles, or triangles) on a blank piece of paper. The other student should not see it. Next, the first student describes what he or she drew to the second student, who tries to draw the same picture on a blank sheet of paper. The second student is not allowed to ask questions or say anything. Once completed, reverse roles. What did you learn?

MEMO

TO: Jill

FROM: Pete

DATE:

I can't stop wondering why we always have problems when you are not at work. Today, while you were off, I had a visit from the local health department. They came to do an inspection because of a complaint from a customer about one of YOUR cafeteria employees not wearing a hairnet which is against the rules and regs! During this inspection, the inspector found several of your employees (I forget who) without hairnets and also found someone with their bare hands in the tuna fish salad! Then, he found a reach-in refrig. with pies in it not working. When I looked at the daily temperature log for the refrigs, I found that no one had filled it in for a while. If someone had done so, we might have fixed the refrig. in time. Please speak to your employees about these problems.

FIGURE 4.7: A sample memo.

 # ENDNOTES

1. Daniel Goleman, "Leadership That Gets Results," *Harvard Business Review*, (March–April 2000), pp. 78–90. See also his books *Emotional Intelligence* (New York: Bateman Books, 1995) and *Working with Emotional Intelligence* (New York: Bateman Books, 1998). As cited in John R. Schermerhorn Jr., *Management* 11th ed. (Hoboken, NJ: Wiley, 2012), p. 339.
2. Daniel Goleman, "What Makes a Leader?" *Harvard Business Review* (November–December 1998), pp. 93–102, As cited in John R. Schermerhorn Jr., *Management* 11th ed. (Hoboken, NJ: Wiley, 2012), p. 339.
3. Ibid.
4. H. M. Weiss and R. Cropanzano, "Affective Events Theory," in B. M. Staw and L. L. Cummings, *Research in Organizational Behavior*, vol. 18 (Greenwich, CT: JAI Press, 1996), pp. 20–22. As cited in Stephen P. Robbins and Mary Coulter, *Management* 12th ed. (Upper Saddle River, NJ: Pearson Prentice Hall, 2013), p. 355.
5. Adapted from Stephen P. Robbins and Mary Coulter, *Management* 12th ed. (Upper Saddle River, NJ: Pearson Prentice Hall, 2013), p. 355.
6. J. R. Schrock, former dean of the School of Hospitality Management, University of South Florida, personal communication, February 16, 2014.
7. Ibid.
8. https://business.twitter.com/basics. Retrieved December 18, 2014.
9. Ibid.
10. Retrieved May 13, 2010.
11. www.statisticbrain.com/social-networking-statistics/. Retrieved December 18, 2014.

Chapter 5

Equal Opportunity Laws and Diversity

Promoting equal opportunity, diversity, and inclusion in the workplace sounds simple enough, but we all know it simply isn't. For years, women and minorities were not—and in some cases, still are not—treated equally. Lack of equal opportunity was such a pressing concern that in June 1963, President John F. Kennedy sent comprehensive civil rights legislation to Congress. Later that summer, in front of the Lincoln Memorial, Dr. Martin Luther King, Jr. gave his famous "I Have a Dream" speech that came to symbolize the insistence for meaningful legislation to address the demand for racial equality and justice.[1]

The Equal Employment Opportunity Commission (EEOC) was established in 1978 as a central authority, responsible for leading and coordinating the efforts of federal departments and agencies to enforce all laws relating to equal employment opportunity without regard to race, color, religion, sex, national origin, age, or disability.

Today, a visit to the Equal Employment Opportunity Commission's website at www.eeoc.gov will likely have an example of a hospitality company being sued by the EEOC for violation of equal opportunity laws. As a hospitality leader, you will be responsible for equal opportunity in the workplace, for employing and supervising people from cultures different from your own. Applying a "standard" approach to employees—one that does not consider each employee's cultural background—will often create communication barriers. Culturally appropriate communication strategies are needed. But how does that conform to equal opportunity in the workplace? What role does culture play? What is diversity? Why should we want equal opportunity, diversity, and inclusiveness?

Our culture is defined as our values, which are manifested in the way we behave, speak, think, and dress; our religious beliefs; the music we like and the food we eat; and the way we interact with others. Culture strongly influences behavior. On the one hand, failure to understand and respect the diverse cultural backgrounds of your employees, and the differences among them, can result in misunderstandings, tension, poor performance, poor employee morale, and higher rates of absenteeism and turnover. On the other hand, when differences are respected, the working environment is richer, much more fun, and even more interesting. Employee satisfaction and performance improve because of this. Equal opportunity, diversity, and inclusiveness in the workplace are critically important in the hospitality industry.

In this chapter you will learn to:

- Define equal opportunity in the workplace.
- Discuss legal issues related to managerial decisions about employment.
- Choose the most EEO-compliant practices when evaluating job applicants.
- Identify discriminatory employment practices as a hospitality supervisor.
- Describe the importance of understanding and promoting diversity in the hospitality workplace.
- Employ cross-cultural interaction skills.
- List ways to increase personal cultural awareness.
- Describe the process of management through effective communication skills.
- Explain how hospitality supervisors can promote cultural diversity in the workplace.
- List the steps to establishing a diversity and inclusion program.
- Compare and contrast ways of positively managing diversity issues in the workplace.

Equal Opportunity in the Workplace

LEARNING OBJECTIVE: Define equal opportunity in the workplace.

Today, whenever a job is advertised and candidates are recruited, interviewed, tested, and selected, it is necessary to take equal opportunity into account. Progressive corporations create offices and programs responsible for planning, developing, implementing, and evaluating a comprehensive equal opportunity and diversity program with multifaceted initiatives to support the company's commitment to equal opportunity, diversity, and inclusiveness. Many large hospitality companies have an office of equal opportunity and diversity (EO&D). They may also be called by similar names, such as diversity and equal opportunity (DEO).

❋ EEO AND DIVERSITY

diversity
Physical and cultural dimensions that separate and distinguish individuals and groups: age, gender, physical abilities and qualities, ethnicity, race, sexual preference.

The equal opportunity and diversity office provides effective leadership to ensure that diversity and equal opportunity are a thriving part of the fabric of your company. This department provides an array of services:

- Education and training the public about equal opportunity and diversity
- Advocacy for diversity
- Support for companies' initiatives toward equal opportunity and diversity
- Consultation on best strategies for equal opportunity and diversity recruitment
- Conflict mediation and resolution
- Monitoring employers' equity and affirmative action goals
- Reviewing compliance with state and federal regulations
- Processing and resolving complaints

Applebee's is one of the restaurant industry's most progressive companies. Former CEO Lloyd Hill took a stand on racial and sexual orientation issues by saying, "There have been too many 'no comments' on these matters." Operations, finance, and marketing have been the "big three" of the industry for years and something crucial has been left out of the equation: human resources. Applebee's, for example, has a chief people officer, Lou Kaucic, who says that it is critical for human resources to have a seat around the executive table.[2]

Sodexo, which is rated one of the top 50 employers for diversity, says that it is committed to respecting, leveraging, and celebrating the diversity of its workforce, its clientele, and the community in which they live, work, and serve.[3]

Marriott International, one of *Fortune* magazine's 100 best companies to work for, says that its commitment to diversity is absolute. It is the only way to attract, develop, and retain the best talent available.[4]

❋ INCLUSION

inclusion
To include, to make a person feel welcome.

Inclusion in the workplace means exactly what it says: to include everyone regardless of gender, marital status, race, national origin, religion, age, disability, sexual orientation, weight, or physical appearance.[5] People who are overweight or less beautiful should not be discriminated against—the only employment criterion should be the ability to do the job. Likewise, a person's sexual orientation should be immaterial: the person's job performance is the only factor that matters.

The restaurant industry's 10-year effort to improve diversity and inclusion in all aspects of the business has been average at best and failing in some areas, according to Garry Fernandez, president of the MultiCultural Foodservice Alliance. Fernandez urges the pursuit of diversity in four areas: workforce, customer, community, and suppliers.[6]

Equal Employment Opportunity Laws

LEARNING OBJECTIVE: Discuss legal issues related to managerial decisions about employment.

A number of laws have been passed to ensure that no individual or group is denied the respect deserved. Understanding the legal requirements of equal opportunity in the workplace is important for three reasons. It will help leaders to (1) do the right thing, (2) realize the limitations of your company's HR and legal departments, and (3) minimize your company's potential liability. Equal employment opportunity is a concept that means that people should be treated equally in all employment matters. Figure 5.1 lists important federal laws commonly referred to as equal employment opportunity (EEO) laws. In general, EEO laws make it unlawful for you to discriminate

Federal Laws	Type of Employment Discrimination Prohibited	Employers Covered
Equal Pay Act of 1963	Gender differences in pay, benefits, and pension for substantially equal work	Almost all companies, private and government
Title VII, 1964 Civil Rights Act (amended in 1991)	Discrimination in all human resource activities based on race, color, gender, religion, or national origin; established Equal Employment Opportunity Commission to administer the law	Companies with 15 or more employees
Age Discrimination in Employment Act of 1967 (amended in 1986)	Age discrimination against those 40 years of age or older	Companies with 20 or more employees
Pregnancy Discrimination Act of 1978	Prohibits discrimination in hiring, promoting, or terminating because of pregnancy; pregnancy to be treated as medical disability	Same as Title VII
Immigration Reform and Control Act (1986 and 1990)	Prohibits discrimination on the basis of citizenship status and nationality	Companies with 4 or more employees
Americans with Disabilities Act of 1990	Bars discrimination against disabled persons in hiring and employment	Businesses with 15 or more employees
Family and Medical Leave Act of 1993	Mandates 12 workweeks of leave for husband or wife upon birth or adoption of a child or sickness in the family	Companies with 50 or more employees
Fair employment practice acts of states and local governments	Bars discrimination; varies	Varies

FIGURE 5.1: Equal employment opportunity laws.

Diversity in the workplace is on the increase.
Andrey_Popov/Shutterstock

Equal Pay Act of 1963
A law that requires equal pay and benefits for men and women working in jobs requiring substantially equal skills, effort, and responsibilities under similar working conditions.

Civil Rights Act of 1964, Title VII
An act that makes it unlawful to discriminate against applicants or employees with respect to recruiting, hiring, firing, promotions, or other employment-related activities on the basis of race, color, religion, gender, or national origin.

Age Discrimination in Employment Act of 1967 (ADEA)
An act that makes it unlawful to discriminate in compensation, terms, or conditions of employment based on a person's age. The ADEA applies to everyone 40 years of age or older.

The Pregnancy Discrimination Act of 1978
An act that makes it unlawful to discriminate against a woman on the basis of pregnancy, childbirth, or related medical conditions.

against applicants or employees with respect to recruiting, hiring, firing, promotions, compensation, or other employment-related activities, on the basis of race, color, religion, gender, nationality, age, or disability. Discrimination in the workplace can be thought of as making employment decisions based on factors that have nothing to do with a person's ability to do the job.

The starting point for EEO laws was probably passage of the Equal Pay Act of 1963. This law requires equal pay and benefits for men and women working in jobs requiring substantially equal skills, effort, and responsibilities under similar working conditions. Congress passed the Civil Rights Act of 1964, Title VII (amended in 1991) to bring about equality in employment decisions. The act makes it unlawful for you to discriminate against applicants or employees with respect to recruiting, hiring, firing, promotions, or other employment-related activities on the basis of race, color, religion, gender, or national origin. Other employment-related activities include, but are not limited to, wages, overtime pay, job assignments, training opportunities, leaves of absence, and retirement plans. Title VII does not require you to hire, promote, or retain employees who are not qualified. The law does provide for you to hire a person of a particular gender if it is based on what is called a bona fide occupational qualification (BFOQ). For instance, it is permissible to hire a man to clean lounges and restrooms reserved for men. The Equal Employment Opportunity Commission (EEOC) was created by the Civil Rights Act of 1964, and it is responsible for enforcing the employment-related provisions of that act as well as other EEO laws. Employees with the EEOC, which also develops and issues guidelines to enforce EEO laws, can file complaints of discrimination.

Age discrimination was addressed in the Age Discrimination in Employment Act of 1967 (ADEA), amended in 1978 and 1986, which makes it unlawful for you to discriminate in compensation, terms, or conditions of employment because of a person's age. The ADEA applies to all people 40 years of age and older. This act also bans forced retirement. The Pregnancy Discrimination Act of 1978 makes it unlawful to discriminate against a woman on the basis of pregnancy, childbirth, or related medical conditions. You cannot refuse to hire (or promote) a woman just because she is pregnant. According to this law, pregnancy is a temporary disability and women must be permitted to work as long as they are physically able to perform their jobs. Employers cannot determine the beginning and ending dates of a pregnant employee's maternity leave.

Immigration Reform and Control Act
A federal law that requires employers to verify the identity and employment eligibility of all applicants and prohibits discrimination in hiring or firing due to a person's national origin.

The **Immigration Reform and Control Act** (IRCA; written in 1986 and amended in 1990) was prompted by problems associated with the increasing numbers of immigrants living in the United States. This act makes it illegal to discriminate in recruiting, hiring, or terminating based on a person's national origin or citizenship status. In these kinds of cases, fines can be charged and judges can order employers to provide back pay, pay court charges, and reinstate an employee. Although Title VII of the Civil Rights Act of 1964 has long prohibited this type of discrimination, IRCA covers employers with 4 or more employees, and Title VII covers employers with 15 or more employees.

The only people you can discriminate against are those you are not legally allowed to hire (or continue to employ): illegal aliens. IRCA imposes penalties for hiring unauthorized aliens. To help ensure that you don't hire an illegal alien, IRCA requires employers to verify that the people they hire are eligible to work in the United States. This is done by completing an I-9 Employment Eligibility Verification form within three days after hire. Using this form, the employer may ask for certain documents that establish the person's identity (such as a driver's license) and employment eligibility (such as a U.S. birth certificate or valid Immigration and Naturalization Services' Employment Authorization Card). To be fair and nondiscriminatory, you cannot request certain work status documentation from some applicants but not others.

Americans with Disabilities Act (ADA)
An act that makes it unlawful to discriminate in employment matters against the estimated 43 million Americans who have a disability.

The 1990 **Americans with Disabilities Act (ADA)** makes it unlawful to discriminate in employment matters against the estimated 43 million Americans who have a disability. Under the ADA, a person has a disability if he or she has a physical or mental impairment that substantially limits one or more major life activities, such as hearing, seeing, speaking, or walking. It also covers recovering alcohol and drug abusers (as long as they are in a supervised treatment program) and people infected with the HIV virus.

It is unlawful to ask an applicant whether he or she is disabled or about the disability itself. You can ask an applicant questions about his or her ability to perform job-related functions as long as the questions are not phrased in terms of a disability. You can also ask the applicant to describe or demonstrate how (with or without reasonable accommodation) he or she will perform job duties. The ADA does not interfere with your right to hire the best-qualified applicant, and a disabled applicant must satisfy your job requirements and be able to perform essential job functions.

Reasonable accommodation, which is legally required, refers to any change or adjustment to a job or the work environment that will enable someone with a disability to perform essential job functions. For example, a worktable might be lowered to enable someone to work while seated, a work schedule might be modified, or a job might be restructured. Employers are not required to lower quality or quantity standards to provide an accommodation, nor are they required to make an accommodation if it would impose an undue hardship on the operation of the business. Undue hardship is defined as an "action requiring significant difficulty or expense" and is determined on a case-by-case basis.

In addition to the federal EEO laws, state and local governments have fair employment practices (FEPs) that often include further conditions. For example, some states forbid employment discrimination on the basis of marital status. It is important to learn about EEO laws because you need to be able to select applicants in a fair and nondiscriminatory manner.

Family and Medical Leave Act of 1993
An act that allows employees to take an unpaid leave of absence from work for up to 12 weeks per year for the birth or adoption of a child or a serious health condition of the employee or his or her spouse, child, or parent.

The **Family and Medical Leave Act of 1993** allows employees to take an unpaid leave of absence from work for up to 12 weeks per year for any of the following reasons:

- Birth or adoption of a child
- Serious health condition of a child
- Serious health condition of a spouse or parent
- Employee's own serious health condition
- Military leave

When the employee returns from leave, he or she is entitled to his or her former position or an equivalent position. To be eligible for a leave of absence, the employee must have worked for the employer for at least 12 months. If it was provided before the leave was taken, the employer is obligated to maintain group health insurance during the leave.

EEO Laws and the Hiring Process

LEARNING OBJECTIVE: Choose the most EEO-compliant practices when evaluating job applicants.

Figure 5.2 lists recommended ways to ask questions of job applicants, whether on job applications or during interviews, to avoid charges of discrimination. The kinds of questions that are not allowed

Subject	Inappropriate Questions (May Not Ask or Require)	Appropriate Questions (May Ask or Require)
Gender or marital status	• Gender (on application form) • Mr., Miss, Mrs., Ms.? • Married, divorced, single, separated? • Number and ages of children • Pregnancy, actual or intended • Maiden name, former name	• In checking your work record, do we need another name for identification?
Race	• Race? • Color of skin, eyes, hair, etc. • Request for photograph	
National origin	• Questions about place of birth, ancestry, mother tongue, national origin of parents or spouse • What is your native language? • How did you learn to speak [language] fluently?	• If job-related, what foreign languages do you speak?
Citizenship, immigration status	• Of what country are you a citizen? • Are you a native-born U.S. citizen? • Questions about naturalization of applicant, spouse, or parents	• If selected, are you able to start work with us on a specific date? If not, when would you be able to start? • If hired, can you show proof that you are eligible to work in the United States?
Religion	• Religious affiliation or preference • Religious holidays observed • Membership in religious organizations	• Can you observe regularly required days and hours of work? • Are there any days or hours of the week that you are not able to work? • Are there any holidays that you are not able to work?

FIGURE 5.2: Equal employment opportunity: Appropriate and inappropriate questions sometimes used in hiring a new employee.

Age	• How old are you? • Date of birth	• Are you 21 or older? (for positions serving alcohol)
Disability	• Do you have any disabilities? • Have you ever been treated for (certain) diseases? • Are you healthy?	
Questions that may discriminate against minorities	• Have you ever been arrested? • List all clubs, societies, and lodges to which you belong. • Do you own a car? (unless required for the job) • Type of military discharge • Questions regarding credit ratings, financial status, wage garnishment, home ownership	• Have you ever been convicted of a crime? If yes, give details. (If crime is job-related, as embezzlement is to handling money, you may refuse to hire.) • List membership in professional organizations relevant to job performance. • Military service: dates, branch of service, education, and experience (if job-related)

FIGURE 5.2: *(continued)*

relate to race, gender, age (except to make sure that the applicant's age meets labor laws), family and marital status, religion, national origin, appearance, and disabilities unrelated to the job.

Job requirements or qualifications, such as those regarding education and work experience, must be relevant to the job, nondiscriminatory, and predictive of future job performance. Although requiring a high school diploma for an entry-level foodservice job, such as server, seems to be acceptable, there are certainly many servers who do their jobs well without a diploma. The requirement of a high school diploma when it is not related to successful performance of the job can be viewed as discriminatory.

Any type of preemployment test must be valid, reliable, and relevant to the job. To be valid, tests must be related to successful performance on the job. To be reliable, tests must yield consistent results. Tests should be given to all applicants, with a single standard for rating scores, and must be given under the same conditions. Even when a test is given to all concerned, it might be considered discriminatory if the test eliminates members of protected groups (the groups protected or covered by EEO laws) more frequently than members of nonprotected groups.

A good way to check yourself to ensure that you are not discriminating when evaluating job applicants is to be sure you can answer yes to the following five questions:

1. Are the qualifications based on the actual duties and needs of the job, not on personal preferences or a wish list?

2. Will the information requested from the applicant help me to judge his or her ability to do the job?

3. Will each part of the selection process—including job descriptions, applications, advertising, and interviews—prevent screening out those groups covered by EEO laws?

4. Can I judge an applicant's ability to do the job successfully without regard to how he or she is different from me in terms of age, gender, race, color, nationality, religion, or disability?

5. Is the selection process the same for all applicants?

Equal Opportunity in the Workplace: What Leaders Need to Know

LEARNING OBJECTIVE: Identify discriminatory employment practices as a hospitality supervisor.

The following is excerpted from the U.S. Equal Employment Opportunity Commission "Training and Technical Assistance Program."[7]

✻ Q & A: RACE, ETHNICITY, COLOR—WHAT PRACTICES ARE DISCRIMINATORY?

Title VII of the Civil Rights Act of 1964 prohibits employment discrimination based on race, color, religion, sex, or national origin. It is illegal to discriminate in any aspect of employment including:

- Hiring and firing
- Compensation, assignment, or classification of employees
- Transfer, promotion, layoff, or recall
- Job advertisements
- Recruitment
- Use of company facilities
- Training and apprenticeship programs
- Pay, retirement plans, and disability leave
- Terms and conditions of employment

✻ INTERVIEWING

Questions you can and cannot ask at an interview are discussed in more detail in Chapter 6 on recruiting and selecting applicants. But we should mention here that there are several inappropriate questions that should be avoided (also refer back to Figure 5.2): How many children do you have? What country do your parents come from? What is your height? What is your weight? What church do you go to? Are you a U.S. citizen? Do you have any disabilities? Are you dating anyone right now? When did you graduate from high school? A simple rule to follow is if it's not job related—don't ask. When facing charges of discrimination, the employer bears the burden of proving that answers to all questions on application forms or in oral interviews are not used in making hiring and placement decisions in a discriminatory manner prohibited by law. The guiding principle behind any question to a job applicant is: "Can the employer demonstrate a legitimate job-related or business necessity for asking the question?" Both the intent behind the question and how the information is to be used by the employer are important for determining whether a question is an appropriate preemployment inquiry.[8]

Diversity

LEARNING OBJECTIVE: Describe the importance of understanding and promoting diversity in the hospitality workplace.

One of the biggest business drivers is the changing demographics. The ever-changing face of the U.S. population continues to reflect an increase in women, racial and ethnic groups, immigrants, older workers, and individuals with disabilities, as well as changing family structures and religious diversity. At Darden restaurants approximately 46 percent of employees are minorities and about 51 percent are women. Both percentages rank above average in the industry.[9]

Understanding and embracing diversity is of critical importance in today's increasingly multicultural and diverse society. The term *diversity* is often used when discussing people of different cultures. Diversity refers to the following cultural as well as physical dimensions, which separate and distinguish us both as individuals and as groups. This list is not meant to be all-inclusive of all groups. Diversity is so much more than just what is listed:[10]

- Culture
- Ethnic group
- Race
- Religion
- Language
- Age
- Gender
- Physical abilities and qualities
- Sexual orientation

culture
The socially transmitted behavior patterns, art, beliefs, institutions, or all other products of human work or thought characteristic of a community or population.

Culture, ethnic group, and *race* are related terms. **Culture** is a learned behavior consisting of a unique set of beliefs, values, attitudes, habits, customs, traditions, and other forms of behavior. Culture influences the way that people behave. Cultural behavior varies from culture to culture. *Culture* refers to the behaviors, beliefs, and characteristics of a particular group, such as an ethnic group. *Ethnic groups* share a common and distinctive culture, including elements such as religion and language. *Race* refers to a group of people related by common descent.

The population of the United States is becoming more multicultural, and diverse, every day. Almost one in four Americans has African, Asian, Hispanic, or Native American ancestry. It is estimated that by 2020, the number will rise to almost one in three, and by 2050, the number will be almost one in two.

As the United States becomes more diverse, so does the workplace. The hospitality workplace employs a particularly diverse group of employees. A restaurant's staff often resembles a miniature United Nations, with employees from all around the globe. According to the U.S. Department of Labor, 12 percent of foodservice employees are foreign-born, compared to 8 percent in other occupations. The National Restaurant Association's website states the following:[11] The restaurant and foodservice industry is one of the most diverse in the United States. It employs more minority managers than any other industry. Women represent 55 percent of the restaurant workforce, and more than a fourth of all foodservice managers are foreign-born.

Up until the late 1980s, white males made up the majority of the U.S. workforce. Many of the workers now entering the labor force are minorities, such as Hispanics, Latinos, and Asians, and many are immigrants. The reasons behind these trends include a young, growing minority population and a continuing high rate of immigration.

The market-savvy businesses of today are responding to the changing demographics by targeting diverse consumers, employees, and supply partners in ways that build meaningful and reciprocal relationships. If companies' marketers and service providers do not reach out to minority communities in a holistic way, they're setting themselves up for failure in the long run.

Promote inclusion in the supply chain by partnering with minority-owned firms that support businesses as patrons. This demonstrates a commitment to inclusion and creates jobs in the very communities that support businesses as patrons. If companies' marketers and service providers do not reach out to minority communities in a holistic way, they're setting themselves up for failure in the long run. The market-savvy businesses of today are responding to the changing demographics by targeting diverse consumers, employees, and supply partners in ways that build meaningful and reciprocal relationships.[12]

PROFILE   Gerry Fernandez

Courtesy of Gerry Fernandez.

Gerry Fernandez began working as a cook at Royal's Healthside Restaurant in Rutland, Vermont. Ernest Royal, owner of the restaurant and a noted New England restaurateur, was the first African American board member of the National Restaurant Association. Royal had experienced considerable racism, and in his honor, Gerry Fernandez conceived the Multicultural Foodservice and Hospitality Alliance (MFHA; *www.mfha.net*). The MFHA is dedicated to promoting diversity within the foodservice industry.

Since his first position as cook at Royal's, Gerry has had a successful career, beginning with earning a bachelor of science in food service management from Johnson and Wales University, where he also earned a degree in culinary arts. His stint as a cook was followed by terms as sous chef, manager, and general manager of various New England restaurants. Gerry spent more than 10 years as a senior manager opening and operating fine-dining restaurants.

In 1995, Gerry moved to General Mills as a technical service specialist in foodservice research and development. He provided support and training to sales, marketing, and product development teams. Additionally, he evaluated current new competitor products, and he conducted recipe development, concept testing, tolerance testing, photo shots, and product presentations. Gerry has received numerous awards, including the General Mills "Champion's Award" and *Nation's Restaurant News* "50 Power Players."

In a recent interview, Fernandez spoke about the vision of the MFHA organization and its intended impact on foodservice operations worldwide.

How does diversity affect foodservice in general? "Comprehensive diversity issues exist in many large companies, and they are asking foodservice operators, 'Where do you stand?' It is coming, whether or not you like it. Diversity is not simply a social agenda issue; it is a bottom-line issue. When people talk about diversity, they think about inclusion of more women and more people of color. They tend to think

only about the soft issues, the green issues. Diversity really is a green discussion more than it is any other color. MFHA is striving to make this an economic discussion rather than a social discussion."

What is the object of MFHA? "To be the solutions bridge for multiculturalism in the foodservice industry so that operators can leverage diversity as a positive influence on the bottom line. We think of multicultural diversity as a way to improve the foodservice business in all aspects: human resources, marketing, training, community relations, and so on. We are the multicultural Yellow Pages for the industry. We are solution focused: a connector of people to issues and people to information."

What does MFHA offer operators? "This is a place to start the diversity process. If operators are looking for opportunities, recipes, programs, or qualified diversity experts, they can call on us. Whether you are an on-site operator—self-op or contract managed—there is a concern regarding bids for city, state, and federal contracts. These potential clients are inquiring directly as to what percentage of your business purchases are from women and minority-owned businesses. Companies are realizing that they need a way to address this issue. We offer a context in which such issues can be explored constructively."

What kinds of services does MFHA have available? "We help identify qualified women and minority-owned business operators who can do business with the big boys. Additionally, MFHA can provide in-house solutions in the form of awareness and skills training, recruiting and retention, marketing, purchasing, and referral services."

How is MFHA evolving? "The last three years have been internally focused as we have developed the infrastructure. Now we are focused externally on our members. We will be able to provide more research and, through focus groups and benchmarking, help operators by training in ways to be more strategic in their diversity effort and to recruit better talent. It's not about one company or one ethnic segment, it's about our industry as a whole reaching out and recruiting from and to every segment of the population."[13]

In the hospitality workforce it is vital that multicultural management recognizes and respects cultural differences among employees and allows and encourages them. Cross-cultural awareness in today's hospitality industry promotes harmonizing with other cultures and prevents misunderstandings. By allowing and encouraging variation, blends of people from all different kinds of backgrounds are able to learn from one another and grow in aspects of the workplace.[14]

ARAMARK's definition of diversity is, "The mosaic of people who bring a variety of backgrounds, styles, perspectives, values, and beliefs as assets to ARAMARK and our partners."

Kaleidoscope Vision states that ARAMARK is composed of unique individuals who, together, make the company what it is and can be in the future. Only when all individuals contribute fully can the strength and vision of ARAMARK be realized. Here are ARAMARK's guiding principles for diversity:

> *Because we are committed to being a company where the best people want to work, we champion a comprehensive diversity initiative. Because we thrive on growth, we recruit, retain, and develop a diverse workforce. Because we succeed through performance, we create an environment that allows all employees to contribute to their fullest potential.*[15]

Why Does Cultural Diversity Matter?

LEARNING OBJECTIVE: Employ cross-cultural interaction skills.

Cultural diversity matters to every single one of us, both professionally and personally. When a group or segment is excluded or oppressed, all of us are denied. For businesses and communities not only to survive but to thrive, each of us must be aware and sensitive to all members of the community. Our communities are rich in resources. When all segments are respected and utilized, it benefits everyone involved. America is the most diverse nation in the world. Our ethnicity, religion, life experience, and so on make each one of us unique.

Diversity in itself is not a challenge but, in fact, an opportunity. It is an opportunity for us to build diverse teams, diverse knowledge perspectives, and experiences that can solve business problems and create value for our shareholders and guests. The main thrust of the initiative is moving beyond awareness training and toward diversity skills training, which helps to enhance the skills of managers and supervisors in communication across lines of difference. This enhances the ability to recognize and respond to the needs of our diverse customers. Every successful business needs to practice sensitivity to diversity.

By developing cross-cultural interaction skills, you will be better equipped to do your job and to motivate diverse employees to accomplish company goals. But don't think you will be able to develop these skills overnight or, for that matter, even over a few months. By considering the major steps listed in Figure 5.3, you will better appreciate that this process is complex and will take time to master. The effective supervisor is aware that employees come from different cultural backgrounds, and so the supervisor learns about how their cultures differ and works with employees without passing judgment about their cultures.

Shifting demographics make practicing diversity more than just a politically correct idea in the hospitality industry. Diversity is anything that makes people different from each other, such as gender, race, ethnicity, income, religion, and disabilities. Foodservice has welcomed minorities for a long time, and minorities make up the largest percentage of workers in the foodservice industry. For supervisors in the hospitality industry it is important to encourage minority talent. Promoting people based solely on their abilities, skills, and job performance into supervisory positions helps promote minority advancement in the foodservice industry.

1. Increase personal awareness.
2. Learn about other cultures.
3. Recognize and practice cross-cultural interaction skills.
4. Maintain awareness, knowledge, and skills.

FIGURE 5.3: Developing cross-cultural interaction skills.

How to Increase Personal Awareness

LEARNING OBJECTIVE: List ways to increase personal cultural awareness.

Without realizing it, it is possible to become culture bound, meaning that you believe that your culture and value system are the best, the one and only. You think your way of speaking, perceiving, thinking, valuing, and behaving are normal and right. For example, when you hear someone talking with an accent, you are likely to think how strange it sounds, or even how wrong or abnormal it is. How many of us realize that each of us has an accent, which probably sounds strange to those of different backgrounds? The first step in developing your cross-cultural skills is to examine how your own culture has influenced who you are. Consider, for example, how your culture has influenced your attitudes toward the following:

- Education
- Work
- Family
- Self-sufficiency
- Money
- Authority
- Expression of emotions

An activity at the end of this chapter helps you look more deeply at your own cultural attitudes and compare them to others.

After becoming more aware of your own culture, the next step is to learn various facts of other cultures. As a supervisor, it is crucial to see other cultures as objectively as possible and not pass judgment. By learning about another culture, it is hoped that you will be better able to understand people from that culture, as well as to be understood better in turn. Some aspects of another culture that are interesting to learn include verbal and nonverbal language differences, values, customs, work habits, and attitudes toward work.

A danger in learning about any culture is that the information may be overgeneralized, thereby promoting stereotypes. It is important to keep in mind that regardless of cultural background, a person is still an individual and needs to be treated and respected as someone with a unique personality, wants, and needs.

You can learn about other cultures in various ways: reading about them in books and magazines, attending cultural fairs and festivals, and interacting with individuals from other cultures. By learning about other cultures and interacting with people of varying backgrounds, you can work on valuing your differences as well as uncovering and overcoming any of your own fears, stereotypes, and prejudices.

How to Recognize and Practice Cross-Cultural Interaction

LEARNING OBJECTIVE: Describe the process of management through effective communication skills.

A person's nationality, culture, race, and gender affect how he or she communicates. However, communication between people of different cultures can often be difficult when neither person is familiar with the other's style of communicating. There are three specific problem areas that supervisors must take steps to overcome:

1. The tendency not to listen carefully or pay attention to what others are saying.
2. Speaking or addressing others in ways that alienate them or make them feel uncomfortable.

3. Using or falling back on inappropriate stereotypes to communicate with people from other cultures.

To be an effective supervisor in a culturally diverse workforce, you must be able to recognize the different ways that people communicate, be sensitive to your own employees' cultural values, and adapt your own supervisory style accordingly.

For example, in some cultures, people rely primarily on verbal communication. In other cultures, the spoken word is only part of communication; people express themselves "in context"—language, body language, the physical setting, and past relationships are all parts of communication.

Eye contact and facial expressions are two other nonverbal communication techniques that vary among cultures. Whereas in North America it is common to maintain good eye contact and employ facial cues such as nodding the head when listening to someone speak, not all cultures share those practices. In many other cultures, people will make strong eye contact when speaking, but when listening, make infrequent eye contact. They also might not use facial expressions when listening to others. These nonverbal communication differences may lead to misunderstanding. A supervisor may wrongly misinterpret that an employee who does not make eye contact or nod in response is simply not listening or doesn't care, when in fact the employee is listening in a respectful manner.

Cultural differences also affect other areas of communication, such as the rate at which people speak, the volume, speech inflections, and the use of pauses and silence when speaking. In some cultures, silence is regarded as an awkward thing that must be avoided; and so it may be common to speak loudly and often to break any silences. This is not always the case in other cultures. Silence is not regarded as an interruption or indication that the conversation has ended but is often considered as much a part of conversation as speaking is. Silence is also often used as a sign of politeness and respect for elders rather than a lack of desire to keep talking. Whereas loud speech in many countries is often interpreted as being aggressive or even angry, a soft-spoken voice elsewhere might be seen as a sign of weakness or shyness.

Another communication difference is the tendency in some cultures to be direct in conversation and get to the point. In many other cultures, this practice is considered impolite and rude. To some, being direct might be interpreted as being insensitive to the feelings of others. Cultures that place emphasis on respect and harmony will use indirect speaking methods to achieve those ends.

Creating opportunities to learn about other cultures is helpful in creating cultural harmony.
auremar/Shutterstock

As a leader, you should be sensitive to your employees' cultural values and understand their different communication styles. Always be open for feedback when communicating. Feedback can tell you how you are perceived by others as well as how well you are getting your point across. Also, keep in mind that it is only natural that people from other cultures speak with a different tone of voice, rhythm, and pace. Finally, as a supervisor, you can also focus on core values that transcend cultural boundaries by creating a workplace where all employees feel valued, safe, and respected.

Leading Cultural Diversity in the Workplace

LEARNING OBJECTIVE: Explain how hospitality supervisors can promote cultural diversity in the workplace.

In the twenty-first century supervisors and managers, in order to be effective, have to handle greater cultural diversity. Supervisors and managers who are not able to handle diversity in the workforce are a liability. Poor supervision can cost companies dearly in the following ways:

- Discrimination lawsuits
- Litigation time and money
- Legal fees/settlements
- High employee turnover rates
- Negative community image

Understanding what cultural diversity is, why it matters, and how to effectively manage your diverse team of associates can minimize risks.

Leading diversity in the workplace means to recognize, respect, and capitalize on the different backgrounds in our society in terms of race, ethnicity, gender, and sexual orientation. Different cultural groups have different values, styles, and personalities, each of which may have a substantial effect on the way they perform in the workplace. Rather than punishing or stifling these different management styles because they do not conform to the traditional white (male) management methods, employers should recognize these differences as benefits. Not only can diverse leadership

More women are members of the executive committee or guidance team.

michaeljung/Shutterstock

styles achieve the same results as traditional methods, but a diverse workforce can also help improve the company's competitive position in the marketplace.

Diversity, or sensitivity, training is now commonplace in the corporate world. However, small businesses need to be aware of these issues just as well. As a small business owner, your awareness and respect of diversity truly matters to your employees and client base.

You must create a balance of respect and understanding in the workplace to have happy and optimally productive workers. An organization's success and competitiveness depends on its ability to embrace diversity and realize the benefits. When organizations actively assess their handling of workplace diversity issues and develop and implement diversity plans, multiple benefits are reported.[16]

Rohini Anand, senior vice president and chief diversity officer, Sodexo USA, says that diversity has an extremely broad definition at Sodexo. It includes all those differences that make us unique, including race and gender, ethnicity, sexual orientation, religion, class, physical and mental abilities, language abilities, and so on. Sodexo also has a very clear mission statement about diversity being the right thing to do and making business sense for the company. It's about creating a work environment for their employees as well as giving back to the community and providing socially responsible services to their clients.

There must be a real commitment for top management to "walk their talk." Plus, senior management must establish goals and monitor their accomplishment; otherwise, they won't be met. In some companies people only do things because the president says so, but what if the president leaves? The next president may not have the same agenda. It is better to have diversity, inclusion, and multiculturalism in the mission statement and in the departmental goals with objectives as to how the goals will be met.

Sodexo strives to be the best in class in the hospitality industry and is rapidly becoming a benchmark for corporate America. This is being achieved by six strategic imperatives, diversity being one of them. Alongside financial results the company reports on how it is doing on diversity and inclusion. The company also has an incentive program where bonuses of 10 to 15 percent are linked to the diversity scorecard. Twenty-five percent of the executive team's bonus is linked, and the CEO has guaranteed that this bonus will be paid out regardless of the financial performance of the company. Now that's putting your money where your mouth is!

Sodexo recommends having someone whose sole job is to attend to diversity and inclusion because it has such a broad scope in terms of the kinds of things that are involved in a diversity effort. Plus, it really shows the commitment. The symbolic aspect is as important as the actual work. According to Anand:

> There are so many pieces included in a diversity effort. It ranges from recruiting and sourcing to retention . . . my recommendation is that somebody be responsible for diversity and inclusion and report to senior management—preferably the president. You want to have influence at the top. A diversity effort can only be successful if you get top level buy-in along with grass roots efforts.[17]

An example of managing cultural diversity in the workplace is that many hospitality organizations are hiring more Hispanics because they have proven themselves capable and enthusiastic. However, this is often accompanied by new communications problems. Whatever your approach, patience is a virtue—and a necessity. Establish workplace policies and resources, then recognize and encourage a family mentality. Provide easy and affordable access to make long-distance calls home; supply phone cards as incentives and gifts; ask about the well-being of their relatives; arrange for easy and affordable transmission of money home.[18]

Hospitality human resource professionals and leaders should be careful to avoid problems associated with requiring employees to speak English at all times in the workplace. This is in conflict with Title VII, which prohibits workplace discrimination based on national origin. Clearly, it is important for hotel owners, operators, and managers to balance how to best deliver a unique brand experience along with the rights of their workforce. The increasingly diverse U.S. workforce makes English-only communication difficult. This potential conflict is best reflected in a recent settlement between the EEOC and The Melrose Hotel Co. of New York. In that case, 13 employees received an $800,000 settlement for complaints of a hostile work environment—which included allegations that Hispanic employees were subjected to an English-only rule.[19]

Establishing a Diversity and Inclusion Program

LEARNING OBJECTIVE: List the steps to establishing a diversity and inclusion program.

The following five steps are the how-tos for establishing a diversity and inclusion program:

1. Develop a mission statement that includes diversity and inclusion.
2. Develop goals for diversity and inclusion for each key operating area.
3. Develop objectives/strategies to show how the goals will be met.
4. Develop measurements to monitor progress toward the goals.
5. Monitor progress toward goal accomplishment.

Gerry Fernandez, president of the Multicultural Foodservice and Hospitality Alliance (MFHA), says that with more than $1 trillion in buying power, the minorities of today will be the majority of tomorrow. That's not a philosophical argument; it's based purely on facts and data found in the U.S. Census. Focusing on multiculturalism isn't about doing the right thing; it is about doing the right thing for your business.[20]

Some might be concerned about the costs of promoting diversity and education about other cultures. According to Salvador Mendoza, director of diversity at Hyatt Hotels, it is possible to leverage diversity to the bottom line. Multicultural initiatives are bottom-line issues. You have to be sensitive to the needs of employees, and you have to make money. There's a way to do both while promoting a multicultural environment.[21]

Many of the larger, well-known hotel and restaurant corporations have a diversity initiative to encourage minority ownership of franchised hotels and restaurants as well as becoming suppliers to those companies. Cendant's Keys to Success program aims to give minorities a leg up in the hotel business by offering an allowance of $1,000 per room for properties up to 74 rooms and $1,500 per room for larger ones. At Starwood Hotels and Resorts, which *Fortune* magazine recently named among the top 50 companies for minority employees, 80 percent of the associates are women and minorities. Starwood seeks out differences and strives for variety in every aspect of their business, so much so that diversity and inclusion are key components of their overall business strategy.[22]

The W Hotel on Union Square in New York celebrates the holidays of many religions and countries, theme days in the employee recreation room and coffee breaks, and provides opportunities for more formal communication by an employee survey index, which measures how successful it was in creating a multicultural workplace of excellence. Supervisors and managers take this seriously because one-third of their bonus is based on the survey results.[23]

Managing Diversity Issues Positively

LEARNING OBJECTIVE: Compare and contrast some ways of positively managing diversity issues in the workplace.

The following list of tips and suggestions will help you to remember that your staff is made up of individuals, which is important to keep in mind, no matter how diverse your staff.

✳ GENERAL GUIDELINES

- Get to know your employees, what they like about their jobs, what they do not like, where they are from, what holidays they celebrate. Listen to their opinions. Help to meet their needs.
- Treat your employees equitably but not uniformly. Do not treat everyone the same when, after all, they are all different. Of course, there must be some consistency to what you do,

but as long as you apply the same set of goals and values to each situation, you can treat each employee individually and consistently.

- Watch for any signs of harassment, such as employees telling jokes that make fun of a person's cultural background, race, sexual orientation, religion, and so on. Know your company's policies on harassment.
- Foster a work climate of mutual respect.
- Encourage the contributions of diverse employees at meetings, in conversations, and in training. Recognize their valuable contributions. Also, allow differences to be discussed rather than suppressed.

❋ GENDER ISSUES

- Make sure that you are not showing favoritism to males or females, by, for instance, granting time off more readily or allowing certain employees to come in late or leave early.
- Show the same amount of respect and listen actively to both genders.
- Know your company's policy on sexual harassment and take seriously any charge of misconduct.

❋ CULTURAL ISSUES

- Learn some of the foreign language phrases that are used by your employees. It shows respect for the employees who speak that language and improves communication.
- Find out how your employees want to be addressed in their own language and how to pronounce their names correctly. Avoid using slang names such as "Honey," "Sweetheart," "Dear," "Fella," and so on. They are disrespectful and annoying.
- Give rewards that are meaningful and appropriate to all employees.
- If employees are having trouble with English, be careful when speaking to them. Speaking a little more slowly than usual might be helpful, but speaking too slowly might make your employees feel that you think they're stupid. Speaking very loudly will not make things any easier to understand. Make sure that slang terms and idioms are understood. It's always important that employees (even those who speak English well) know and are comfortable enough to tell you when they don't understand what you're saying.
- Be cautious about the use and interpretation of gestures. Gestures such as a thumbs-up are by no means universal. For example, in the United States, a customer may gesture "one" to a server in a restaurant by putting up an index finger. In some European countries, this gesture means "two." If you are not sure what someone's gesture means, ask for the meaning.

❋ RELIGIOUS ISSUES

- Be consistent in allowing time off for religious reasons.

❋ AGE ISSUES

- Both the young and the old sometimes feel that they do not get the respect they deserve. They need to know what is going on in the department and how well they are doing their jobs, just like anyone else. Make them feel like part of the team.
- Young workers want to do work they consider worthwhile and to have fun doing it. They want their supervisors to listen to them, to let them participate in decision making. Not surprisingly, they do not want supervisors to bark orders military-style. They like to have time and money invested into their training and development.
- Do not have higher expectations of older adults than of their younger coworkers, and don't patronize them.

❋ PHYSICALLY AND/OR MENTALLY CHALLENGED ISSUES

About 43 million Americans have a physical disability. At work, people with disabilities might feel that supervisors do not see beyond their disabilities and do not think they are capable.

- Look at the differently abled employee the same way you look at other employees: as a whole person with likes, dislikes, hobbies, and so on, and encourage the employee's co-workers to do so.
- Speak directly to the differently abled employee.
- The hospitality industry has a responsibility to provide job opportunities for all.
- Employees with disabilities are just as productive as other employees. You might have to make some adjustments for disabled employees, but none of these adjustments affect the quality of their work.

Even with the greatest cross-cultural interaction skills, you might do something that offends an employee. When this happens, do the common-sense thing: listen closely, apologize sincerely, and avoid making the same mistake in the future.

CASE STUDY: Culture Clash

As the new head of housekeeping, Nancy, a white, middle-aged woman, oversees a staff made up of a diverse group, mostly minorities. She is not sure that she is getting her messages through to her staff because they do not respond the way she expects them to when she gives directions. She is starting to wonder if they are really listening. To solve this problem, Nancy decides to spend a half-hour each day supervising the employees directly. She accompanies them on their cleaning assignments and employs a hands-on approach to analyzing their tasks and assessing their proficiencies.

One week later, she comes to the conclusion that just about everyone is doing the job the way they should, but she hears from another department head that housekeepers are grumbling about being watched over. Later that day, one of the housekeepers complains in a very emotional way, and Nancy isn't sure what to do.

Case Study Questions

1. What did Nancy do right?
2. What did Nancy do wrong?
3. In what ways was Nancy insensitive to cultural differences?
4. What should Nancy do now? How can she learn more about the cultures represented in her staff to help her to do her job better?

KEY POINTS

1. Equal employment opportunity was denied to so many for so long that eventually in 1963 Congress passed the Equal Pay Act and, in 1964, the Civil Rights Act, Title VII, which established the Equal Employment Commission.
2. Progressive companies embrace equal opportunity, diversity, and inclusiveness. Many have EEO/diversity officers who plan, develop, implement, and monitor EEO, diversity, and inclusion programs.
3. The equal employment opportunity laws are reviewed.
4. When hiring, it is important to know the questions you can or cannot ask and an outline of what supervisors need to know.

5. On the one hand, failure to understand and respect the differences, or the diversity, of your employees can result in misunderstandings, tension, poor performance, poor employee morale, and higher rates of employee absenteeism and turnover. On the other hand, when differences are respected, the working environment is richer, more fun, and more interesting, and employee satisfaction and performance improve.

6. Steps to develop cross-cultural interaction skills are:
   - Increasing personal awareness
   - Learning about other cultures
   - Recognizing and practicing cross-cultural interaction skills
   - Maintaining awareness, knowledge, and skills

7. The chapter lists tips that can be used to manage diversity issues positively.

KEY TERMS

Age Discrimination in Employment Act of 1967 (ADEA)

Americans with Disabilities Act (ADA)

Civil Rights Act of 1964, Title VII

culture

diversity

Equal Employment Opportunity Commission

Equal Pay Act of 1963

Family and Medical Leave Act of 1993

Immigration Reform and Control Act (IRCA)

inclusion

The Pregnancy Discrimination Act of 1978

REVIEW QUESTIONS

Answer each question in complete sentences. Read each question carefully and make sure that you answer all parts of the question. Organize your answer using more than one paragraph when appropriate.

1. Outline equal opportunity in the workplace.
2. List the laws that affect equal opportunity in the workplace.
3. Identify the important things that every supervisor needs to know about diversity.
4. Describe the process of developing cross-cultural interaction skills.
5. Identify two diversity-related problems that could come up in a day-to-day job scenario, and give tips for managing these situations.

ACTIVITIES AND APPLICATIONS

1. Discussion Questions
   - Have you ever been discriminated against in the workplace, or observed when someone else was discriminated against? Describe what happened and how it could have been handled better.
   - Do you know a supervisor/manager who handles diverse employees skillfully? If so, what skills does he or she have?
   - What are some traditions that are important in your family? After those of several families have been discussed, look for similarities and differences.
   - What are some foods and dishes native to your culture?

2. Group Activity
In groups of two (or four) students, each student writes down two adjectives (positive or negative) that come to mind for each of these groups: New Yorkers, Californians, Latinos, lesbians, whites, Asian

Americans, women, teenagers, older adults, men, Catholics, African Americans, and disabled. Discuss your reactions with the group using the following discussion questions.

- Were there any groups you know so little about that you felt uncomfortable writing about them?
- Were any of your adjectives used because you had limited personal experience?
- Were any of your adjectives reflective of stereotypes?
- Do you think the way you described certain groups would affect how you would communicate with them?
- How quick are we to prejudge others when we know little about them?

3. Group Activity: The Cultural You

In groups of two, each student should first write down his or her attitudes toward each of the following: money, expression of emotions, time, religion, education, authority, family, independence, work, children, competition, and use of alcohol. Next, discuss your attitudes with your partner. Identify to what extent your culture has influenced your attitudes.

ENDNOTES

1. www.eeoc.gov/abouteeoc/35th/pre1965/index.html. Retrieved December 18, 2014.
2. Charles Bernstein, "The Missing Piece," *Chain Leader*, vol. 9, no. 7 (June 2004), p. 10.
3. Sodexo, "Workplace Diversity," http://sodexousa.com/usen/corporate-responsibility/responsible-employer/diversity-inclusion/thought-leadership.aspx. Retrieved December 18, 2014.
4. http://www.marriott.com/diversity/diversity-and-inclusion.mi. Retrieved December 18, 2014.
5. Dr. Chad Gruhl, personal communication, February 16, 2014.
6. Dina Berta, "MFHA: Industry Gets a Barely Passing Grade on Diversity Issues," *Nation's Restaurant News*, vol. 40, no. 34 (August 21, 2006), p. 14.
7. www.eeoc.gov/facts/qanda.html. Retrieved on December 18, 2014.
8. www.isis.fastmail.usf.edu/eoa/interviewfaq.asp. Retrieved on December 18, 2014.
9. www.darden.com/diversity/workforce.asp. Retrieved on December 18, 2014.
10. Dr. Chad Gruhl, personal communication, July 16, 2014.
11. www.restaurant.org/Industry-Impact/Employing-America/Faces-of-Diversity. Retrieved on December 18, 2014.
12. Gerry A. Fernandez, "Multicultural Diversity: It's the Right Thing to Do for Your Business," *Nation's Restaurant News*, vol. 37, no. 20 (May 19, 2003), p. 42.
13. Texas Center for Women's Business Enterprise, Austin, TX. SBA Online, August 10, 2004.
14. John R. Walker, *Introduction to Hospitality Management*, 4th ed. (Upper Saddle River, NJ: Prentice Hall, 2012), p. 532.
15. www.aramark.com/search.aspx?searchtext=diversity. Retrieved December 18, 2014.
16. www.multiculturaladvantage.com/recruit/diversity/diversity-in-the-workplace-benefits-challenges-solutions.asp. Retrieved December 18, 2014.
17. Rohini Anand, "Make Diversity Part of the Business Plan," *Restaurants and Institutions*. vol. 114, no. 10 (May 2004), p. 22.
18. Davis E. Morrison & Michael L. Sullivan, "English-Only Can Be Discriminatory," *Lodging Hospitality*, vol. 62, no. 13 (September 1, 2006), p. 28.
19. Gerry A. Fernandez, "Multicultural Diversity: It's the Right Thing to Do for Your Business," *Nation's Restaurant News*, vol. 37, no. 20 (May 19, 2003), p. 43.
20. John P. Walsh, "Putting It Together," *Hotel and Motel Management*, vol. 218, no. 15 (September 1, 2003), p. 4.
21. Carlo Wolf, "Hotel Companies Diversify their Diversity Targets Even as Growth Lags," *Lodging Hospitality*, vol. 60, no. 10 (July 15, 2004), pp. 46–48; www.starwoodhotels.com/corporate/about/values/diversity.html. Retrieved December 18, 2014.
22. Arash Azarbarzin, personal communication, June 23, 2014.
23. Ibid.

Chapter 6

Recruitment, Selection, and Orientation

You've run a social media ad for a weekend housekeeper, but the only person to put in an application is a high school student looking for her first job. You interview her, and you look at the housekeeper's work schedule and realize that if you don't hire her today, you'll have to spend the weekend doing housekeeper duties yourself. So you hire her and the next day, when she starts, you ask an experienced (but not very friendly) housekeeper to get the new hire started. By next Saturday, she has quit, so you put another ad on the hotel Web page and you think, "There's got to be a better way."

How do you find the people you need? How can you choose people who will stay beyond the first week, do a good job, and be worth the money you pay them? Does it always have to be the way it is today? No, it doesn't.

There is no foolproof system: Human beings are unpredictable, and so is the day-to-day situation in the typical hospitality operation. But the knowledge and experience of people who have faced and studied these problems can be helpful to you, even though you must adapt it to your own situation.

In this chapter, we examine the processes and problems of recruiting and selecting hourly employees for hospitality operations. After completion of this chapter, you should be able to:

- Describe the characteristics of entry-level jobs in the hospitality industry.
- Recall the factors that must be considered when hiring new employees.
- Identify the various recruitment methods.
- List the best ways to find qualified employees, assessing the pros and cons of each method.
- Describe safeguards against negligent hiring.
- Explain the two primary purposes of employee orientation.

The Labor Market

LEARNING OBJECTIVE: Describe the characteristics of entry-level jobs in the hospitality industry.

labor market
In a given area, the workers who are looking for jobs (the labor supply) and the jobs that are available (the labor demand).

The term **labor market** refers to (1) the supply of people looking for jobs as well as (2) the jobs available in a given area. When you need people to fill certain jobs, you are looking for people with certain characteristics—knowledge, abilities, skills, personal qualities—and you have a certain price you are willing or able to pay for the work you expect them to do. The people who are in the market for a job are looking for jobs with certain characteristics—work they are qualified to do or are able to learn, a place they can get to easily, certain days and hours off, a pleasant work environment, people they are comfortable working with and for, and a certain rate of pay (usually, the most they can get). The trick is to get a good match between people and jobs. A challenge in the best of times, recruiting and retention has again emerged as one of the most critical issues facing the hospitality industry. The average U.S. unemployment rate fluctuates; it was close to 10 percent in 2010, but in the spring of 2015 is 5.9 percent.[1] The National Restaurant Association estimates that the number of jobs in the industry will grow by about 15 percent over the next decade to 14.8 million by 2024.[2] Operators across the country are offering higher hourly wages, and may have to offer benefits such as healthcare.

When jobs are plentiful and few people are unemployed, employers have a harder time finding the people they want, and workers are more particular about the jobs they will accept. When many people are looking for jobs and jobs are scarce, employers have a better choice and workers will settle for less. The number of employers looking for the same kinds of people also affects the labor market. You are always in competition with hospitality operations like your own, as well as retail stores, which also offer many part-time, entry-level jobs.

Hospitality companies identify where they are in the marketplace for employees, meaning the Ritz-Carlton will likely attract a different person than a Motel 6. Companies assess the need for additional employees for a brief period of a "full house" versus some overtime being worked by existing staff.

The following comment from Jim Sullivan, a seasoned hospitality consultant, gives us something to discuss: "Human resource professionals and supervisors spend too much time on dealing with difficult employees. If you do not terminate people who are not working out, you increase the possibility of having to let go of the people who are."[3]

❋ JOBS TO BE FILLED

Many of the jobs in food and lodging operations demand hard physical labor. People are often on their feet all day doing work that is physically exhausting. About the only people who sit down are telephone operators, cashiers, reservationists, and many clerical employees. Kitchens are hot and filled with safety hazards. At busy times, pressure is intense and tension is high. Many jobs are uninteresting and monotonous—eight hours of pushing a vacuum cleaner, making up guest rooms, polishing silver, setting up function rooms, washing vegetables, spreading mayonnaise on bread, placing food on plates, washing dishes.

In many of these jobs the pay is entry level, but there is the possibility of promotion. It is not surprising that the duller and more demanding a job is, the harder it is to fill it with a good employee and the more often you have to fill it. The main attraction of such jobs is that they are available, and you are willing to take people with no experience and no skills. For example, operators may offer starting positions to employees whose English communication skills need improving.

These individuals can, once they are more proficient in English, advance to other positions within the operation. Examples of this in a hotel would be in housekeeping and stewarding. For certain jobs you must look for specific skills and abilities. Front desk clerks, servers, and bartenders must have several kinds of skills: verbal, manual, and social. Cooks must have technical skills, varying in complexity with the station and the menu. All these jobs require people who can function well under pressure. The rate of pay goes up for skilled employees, except for servers, who are usually paid minimum wage or less and make most of their money in tips.

❋ DAYS AND HOURS OF WORK

In the hospitality industry, there is a pattern of daily peaks and valleys, with the peaks forming around mealtimes and the valleys falling between. This makes for some difficulty in offering the regular eight-hour day that many people are looking for. You also have some very early hours if you serve breakfast, evening hours if you serve dinner, and late-night hours if you operate a bar or feature entertainment or serve an after-theater clientele.

This irregular kind of need encourages split shifts, part-time jobs, and unusual hours, which can work both for you and against you in finding employees. Sometimes you cannot guarantee a certain number of hours of work per week: employees are put on a call-in schedule and must simply take their chances of getting as many hours as they want. But if they cannot count on you, you may not be able to count on them.

You also have varying needs according to days of the week. These form a fairly predictable pattern, predictable enough for you to plan your hiring and scheduling. In restaurants, staff needs are lighter during the week and heavy on weekends, which closes your doors to people looking for a Monday-to-Friday week. In business hotels the pattern is the reverse, heavy during the week and light on weekends; however, resorts are busier on weekends.

Restaurant employees typically work when other people are playing—evenings, weekends, and holidays—which complicates finding people to fill jobs. Restaurants may also have urgent temporary needs for parties and promotions and emergencies when regular employees are out sick or leave without warning. This requires a banquet server call-in system or overtime for regular employees.

In some facets of the foodservice industry, the timing of people needs is regular and predictable. In hospitals and nursing homes, the population is generally steady seven days a week, and the only variation in need comes with the daily peaks and valleys of mealtimes. Schools have steady Monday-to-Friday patterns, with short days built around lunch, and they follow the school calendar, closing down for vacations, when they lose many people. Business and industry feeding follows the workweek of the business or plant.

In hotels the pattern of need is likely to be irregular but fairly predictable. Reservations are typically made ahead except in the restaurants, and need is generally geared to coming events in the community or in the hotel itself, or to predictable vacation and travel trends. Often, a hotel will require large numbers of temporary workers for single events such as conventions and conferences. Temporary extra help is often supplemented by having regulars work overtime. Where needs vary widely and frequently, leaders can spend a great deal of time on staffing and scheduling alone. Hospitality operators normally have a number of "on-call employees" who are called on to work banquets and catering functions as required.

The types of jobs, unusual working hours and days, minimum wages, and the up-and-down character of the need for workers limits the appeal of hotel and foodservice jobs to people who can fit this pattern or can slip in and out of it easily. Accordingly, it attracts people who are looking for short-term jobs, part-time work, or jobs requiring no skills or previous experience. Some people deliberately seek the unusual hours to fit their own personal schedule: people going to school, moonlighters, parents who must be at home to take care of the kids. Many people are looking for temporary work and have no interest in long-term employment or a career in the industry. "I am only working here until I can find a *real* job" is a common attitude.

❄ SOURCES OF EMPLOYEES

The source of workers continues to change as the composition of the U.S. labor force changes. The majority of new workers entering the hospitality workforce are women, minorities, and immigrants. Why is this? It is due to the combination of a shrinking, older, white U.S. population; a younger, growing minority population; recent easing of immigration restrictions; and increasing numbers of women entering or returning to work.

If the job you need to fill is anything above the lowest level in terms of pay, interesting work, and decent hours, *the first place to look for someone to fill it is inside your own operation.* Upgrading someone whose attitudes and performance you already know is far less risky than hiring someone new and will probably assure you of a good, loyal worker. You will spend less time in training, and the adjustment will be smoother all around.

Social media have become free and increasingly popular ways for hospitality companies to advertise and recruit employees. Operators are using sites such as Facebook, Twitter, LinkedIn, and YouTube. The popular seven recruiting methods are:[4]

1. In-house job referral
2. Company website
3. Social media
4. Online résumés
5. Job fairs
6. Schools
7. Professional/industry associations

Consider also how people would feel if you brought someone in from outside to fill a job or a shift they would like to have. It is important for morale to give your employees first chance, even when you might find it easier to fill the vacant job from outside than to fill the job your current employee will vacate. It is part of being a good leader to consider your own people first and to move them along and develop their capabilities for better jobs.

As an industry, we are always looking for people, and we are among the few employers who will hire people without experience. Usually, first-timers want the jobs for the money, the experience of working, and the advantage it gives them in getting their next job. A few, but not many, apply because they think the work will be interesting. Often, they choose a particular place because a friend is working there or because it is close to home.

Many are looking for part-time work because they are students. Many are working *until*— until school starts or until they get enough money to buy a car. Some hospitality companies are now helping new employees with English classes so they can become more valuable employees. One resort even offered a quick course overview of the hotel to recently graduated but unemployed former high school students and ended up hiring several of them.

Another group of potential hospitality employees is *women* who want to go to work to supplement the family income or simply to get out of the house. A woman with children may be very happy with part-time work, three or four hours spanning the lunch period while the kids are at school, or an evening shift when her husband can take care of the children.

Another group of part-time employees is interested in evening work: *moonlighters*, people looking for a second job. This is not ideal for either you or them, since they are often tired from working their first job. However, students and homemakers also carry a double load, so perhaps moonlighting is no more difficult.

Another source of employees is the *unemployed*. If they have worked in an operation like yours, they may have skills and experience useful to you. If they were in another line of work, you may be competing with unemployment compensation, which is often more than the wages you pay. Workers from the automobile industry, for example, may have been making $52 an hour in wages and benefits, and although their unemployment compensation is not as high as that, it is still above hospitality wages.

If compensation runs out and they go to work in a hotel or restaurant, workers from higher-paying industries rarely find satisfaction in their jobs. They are likely to see both the pay and the work as a step down from the jobs they lost. They are truly *until*-type employees. Yet, some welfare-to-work programs are having success with companies such as Marriott.

Some people seek work in hotels or restaurants just to get away from what they have been doing. Sometimes, recent college graduates find that they are not happy with the jobs they have taken or the field they prepared for, and they just want to get out. Sometimes these people just want a breather, some time to think things over and make new plans. Sometimes they are thinking of switching to the hotel or restaurant field and want to experience it from the inside before they make up their minds. A number of people today are interested in learning professional cooking because the pay at the top is high and a certain glamour accompanies the profession.

Some first-time job hunters apply for lodging and foodservice jobs because the jobs are available.

Hiring *retired people* is becoming more commonplace, although the number of retired people who do return to work is still quite small. The over-65 group is growing and will increase to 20 percent of the population in 2030. Retirees often want to work to fill some empty time or perhaps to supplement their income. Although some of the jobs are not suitable because of physical demands and odd hours, this is not a labor source you should dismiss routinely. Not only is it against the law to discriminate based on age, but also older workers often have stability and an inner motivation that younger people may have not yet developed.

One national fast-food chain has made a special effort to develop jobs and hours that fit the availability and skills and talents of the retired. It has found this group to be an excellent source of employees: they are dependable, work-oriented people who are happy to have jobs. In general, retirees have proven to be loyal, willing, and service-oriented workers. They come to work on time, have much prior work experience on which to draw, and do their jobs well.

Another group of people who might be interested in working in the hospitality industry is the *disabled*. A disabled person has a physical, mental, or developmental impairment that limits one or more of life's major activities. For example, a disabled person may have a visual or hearing impairment.

There are disabled employees doing many different hospitality jobs. For instance, a cashier or payroll clerk can work from a wheelchair, a person who has a hearing impairment can do some food preparation tasks, or a person with mental development disabilities can wash dishes and pots. It is illegal not to hire a disabled worker unless the disability would directly interfere with the person's ability to perform the work.

It is your job to build a supportive environment in which the disabled employee, and your other employees, will work well together. Discussing with your employees ahead of time what the new employee will be doing and what to expect can do this. Encourage your employees to talk honestly about how they feel and about any of their concerns in a respectful and succinct manner. Be positive about the placement of disabled people in the workplace and what they can accomplish.

Often, we set up qualifications for jobs we want to fill that are totally unrealistic (and quite possibly illegal), and if we get what we say we are looking for we will have overqualified and unhappy people. Setting such requirements, in fact, can be interpreted as discriminatory. For some jobs, people do not even need to be able to read and write. All they need is the ability to perform the required tasks. The requirements we set up for a job must be based on the requirements of the work.

It can be difficult to fill certain types of jobs, so recruiters working with limited resources must decide whom to target, what message to convey, and how to staff the recruitment efforts.[5]

First, a concept must consider carefully its brand and its work culture in crafting a message that will stand out from a crowd of hiring notices. Second, target potential talent where they spend their spare time—in person and online. Motivated job seekers, especially those in Generations X and Y, respond well to employers who can showcase flexibility, a sense of humor, and a commitment to social values such as sustainability and community involvement.[6]

Some operators do not accept applications; they direct job seekers to the Internet, where their applications are automatically assessed for attitude and availability. Online recruitment is now the norm, which not only opens up the talent pool exponentially, but also provides a wealth of resources that can help with recruitment efforts.[7]

Being green is important to an increasing number of applicants. Dan McGowan, president of the Chicago-based pan-Asian chain Big Bowl, had a group of 10 applicants, eight of whom applied because they had read about some of the chain's green efforts and virtual interviews can be done via Skype—for free![8]

✻ CHARACTERISTICS OF YOUR LABOR AREA

You will find it helpful to know something about the labor market in your own area: such things as prevailing wages for various kinds of jobs, unemployment rates for various types of workers, the makeup of the labor force, and the kinds of enterprises that are competing with you for workers, both in and out of your own industry.

demographics
Characteristics of a given area in terms of the data about the people who live there.

You should know something of the demographics of your area: ethnic groups, income levels, education levels, and where in your area different groups live. Where do low-income workers, young married, immigrants, and the employable retired typically live? Employers sometimes note the zip code of the area in which the majority of their employees live to know where to do community advertising.

There are other useful things to know about your community. Where are the high schools and colleges that can provide you with student workers? What agencies will work with you to find suitable disabled workers? What are the transportation patterns in your area? Are there buses from where your potential workers live that run at hours to fit your needs? Can workers drive from their homes in a reasonable length of time?

Operations such as airports or in-plant cafeterias in outlying areas often find transportation the greatest single problem in finding employees. In a large organization your human resource department may have such information. In fact, they often take care of much of the routine of recruiting. But the more you participate and the better you know the labor resources of the area, the more likely you are to know how to attract and hold the kind of people you want.

Determining Labor Needs

LEARNING OBJECTIVE: Describe the factors that must be considered when hiring new employees.

If you are a busy hospitality leader and you see a heading like this, your first reaction may be to laugh. You need people all the time. You've got no time to make out lists; you need whoever walks in the door, and you are just afraid nobody will walk in.

But what if you could turn things around and avoid panic and crisis by hiring employees who are right for the job and will not walk off and leave you in the lurch? And do you realize the hidden costs when you hire unqualified people or people who are wrong for the jobs you ask them to do?

Hiring such employees is worse than useless. Either you will keep those employees and suffer their shortcomings, or you will have to fire them and start all over—and perhaps make the same mistakes. If you train those workers and the ones you replace them with, your training costs will skyrocket and the work will suffer until you get them trained. If you do not train them, they will not do their jobs correctly, and they will waste things and break things and turn out inferior products and give inferior service.

If they are unhappy or incompetent, they will be absent or late a lot, and their morale will be poor and will negatively impact everyone else's morale, too. They will not get the work done on time, and you will have to pay overtime. They will give poor service and drive customers away, and your sales will dwindle. When you finally do fire them, your unemployment compensation costs will go up and you will have to hire people to take their places—and the next people you hire may be even worse. It is a very, very costly way to choose people, and in time it could cost you your reputation as a good employer, your job, or your business. There are better ways to go about hiring people based on the thinking and experience of experts, and the place to start is to figure out exactly what to look for.

❋ DEFINING JOB QUALIFICATIONS

job specification
A list of the qualifications needed to perform a given job.

To define a job's qualifications, you need to list the knowledge, skills and abilities, work experience, and education and training required. This is known as a job specification. Figure 6.1 shows a sample job specification. Note that there is the heading "Preferred Qualifications"—the reason for this is to avoid any problems with affirmative action. If some applicants do not have the preferred qualifications then they are not as qualified for the position as those who do have the preferred qualifications. Training and certifications may also be added to the specification.

Job Specification: _____Server_____

Department: Dining Room

Grade   6

Job Qualifications:

KNOWLEDGE ___Basic knowledge of food and cooking.___

SKILLS AND ABILITIES ___Present a good appearance—neat and well-groomed,___

___interact with guests in a courteous and helpful manner, work well with other___

___personnel, write neatly, perform basic mathematical functions (addition,___

___subtraction, multiplication, and division), set tables, serve and clear.___

WORK EXPERIENCE ___Six months satisfactory experience as a server required.___

___One year preferred.___

EDUCATION AND TRAINING ___High school graduate and/or service___

___training preferred.___

PREFERRED QUALIFICATIONS ___a) 1 year in a fine dining restaurant___

___environment. b) must be able to work on weekends.___

FIGURE 6.1: Job specification.

Knowledge consists of the information needed to perform job duties. For example, a cook must know that one cup holds eight ounces, and other measurements, just as the dietary manager in a hospital kitchen must know which foods are not allowed on modified diets. You can use verbs such as _knows, defines, lists,_ or _explains_ to begin a knowledge statement.

Skills and abilities refer to competence in performing a task or behaving in a certain manner. Must a person be able to lift 100-pound bags and boxes? Add, subtract, and multiply? Convert recipes? Mix x number of drinks per hour? Cook eggs to order at a certain rate? Have a responsive, outgoing approach to people? Be as specific as possible.

Performance standards, if you have them, will tell you the specific skills you are looking for. You must decide whether to buy these skills or do your own skills training. If you plan to train, you need to define the qualities that will make people trainable for a given job. A bartender, for example, needs manual dexterity. Desk clerks and serving personnel need verbal skills.

The qualifications that you list in your job specification must not discriminate in any way on the basis of race, national origin, gender, age, marital or family status, religion, or disability. The place to begin in avoiding discrimination is with your job specifications. It is important that you phrase them in concrete terms of what each job requires and that you think in these terms as well. A job specification cuts to the quick with your requirements whereas the job description defines the duties and requirements of an employee's job in detail. The job specification provides detailed characteristics, knowledge, education, skills, and experience needed to perform the job, with an overview of the specific job requirements.[9]

❋ FORECASTING STAFFING NEEDS

Anticipating your needs for staff will give you time to look for the right people. If you need extra people for holiday and vacation periods, hire them ahead of time or else your competitors will beat you to the best people. Records of past sales or occupancy or special events may indicate trends in people needs. Look ahead to changes in your business: Is your employer planning to expand? If so, how will it affect your department's need for people?

scheduling
Determining how many people are needed when, and assigning days and hours of work accordingly.

Scheduling is a key factor. Your work schedules form a day-to-day forecast of the people you need at each hour of the day. Plan them in advance. Make sure that your workers are aware of any changes you make, and make sure that they tell you well in advance of any changes that they have in mind.

Employees need an environment that motivates them and offers benefits. Let employees know that you value their opinion. If at all possible, allow schedules to be flexible. This gives employees a feeling of control and the comfort of knowing that if something comes up, they will not be criticized. Today, more people are demanding that their personal lives be taken as seriously as their work lives.[10]

As an employer it is important that you try to meet the needs of both your employees and the company. Examine your scheduling as a whole. First, does it provide efficiently for your needs? Second, are there ways of organizing the shifts that would be more attractive to the type of person you would like to hire? Do you ask people to work short shifts at unattractive hours, such as early in the morning or late at night? A country club advertised a split shift of 11 to 3 and 5 to 11, three days a week—who is that likely to appeal to? That's a 10-hour day with hardly enough time between shifts to go home, yet it is not a full 40-hour week.

Consider revamping your schedules with people's needs and desires in mind. Look at the hours from their point of view. How far do they have to travel? How much useful personal time does your schedule leave them? How much money do they make for the time involved in working for you, including travel times? Ask your workers how they feel about their days and hours, and try to devise schedules that will not only fill your needs but will be attractive to new people as well.

Your employees will appreciate it if you give them a chance to move to a shift they like better before you hire someone new to fill a vacancy. Often, before making decisions it is important to implement new plans or policies with the staff. Include staff in the decision-making process and find out how they feel on certain policies; you may be surprised at what they have to say.

Scheduling is an important task that, when done well, helps ensure a smooth-running operation.

Another key factor in forecasting employee needs is *downtime*, the length of time that a position is vacant until a new employee who can fully perform the job fills it. Let's consider how long downtime might normally be: An employee resigns and gives you only two days' notice. It's not unusual, particularly if you don't make a point of requiring proper notice (usually two weeks) and withhold something of value to the employee, such as accrued vacation time, if proper notice is not given.

Once the employee resigns, depending on your employer's procedures, you might have to fill out an **employment requisition form** (see Figure 6.2). A requisition is something like a purchase order that must be signed by the appropriate person before you can begin the recruiting process.

Let's say that this takes one week. If you want to advertise the job, you will probably have to wait another week before the ad appears and you get responses. Now you can probably plan on one to two weeks to screen applicants, interview and test applicants, check references, and make a final selection. Often, the person you hire must give his or her current employer two weeks' notice, so you wait a little more.

employment requisition form
A standard form used by departments to obtain approval to fill positions and to notify the recruiter that a position needs, or will need, to be filled.

Employment Requisition

Department: _____

Position: _____

Reason for vacancy: _____ Incumbent leaving the company

                               Name: _____

                               Separation Date: _____

          _____ New Position

Is position budgeted? _____ Yes _____ No

Is position temporary _____ or permanent _____?

Is position full-time _____ or part-time _____?

Hours of position/days off: _____

When needed? _____

Job qualifications: _____

_____

_____

_____

_____

Approvals   Department Head _____

               General Manager_____

               Director, Human Resources _____

FIGURE 6.2: An employment requisition is completed by the department head and given to HR either electronically or by hard copy.

STAFFING GUIDE

Department: _____ Date: _____

Positions	Number Full Staff	Staff on Hand	Current Openings	Anticipated Openings	Total to Be Hired	Time Required to Recruit and Train

FIGURE 6.3: Staffing guides help supervisors determine when to hire new employees.

Now if you believe in magic, when the new employee shows up for the first day of work, you will think your problems are over and put the new employee right to work. Wrong! Now it will take at least one week, probably more, before your new employee gets up to speed in the new position. It has now been about six to seven weeks since your former employee resigned. One way to help reduce downtime is to forecast your personnel needs periodically. Figure 6.3 shows a staffing guide form that can be used every two months to help determine when to hire new employees so that downtime is minimized. Staffing guides are based on the budget and expected volume of business.

❋ TRAINING VERSUS BUYING SKILLS

In determining your staffing needs, you must decide whether to buy skills or to train new people yourself. Most managers will tell you they simply don't have time to train people—they are too busy with the work itself. They look for people who have experience in the jobs they are hiring for, even when they have to pay a higher wage.

There is no security in hiring experience, however. You might pay more to break someone of five years of forming bad habits than it would cost you to train an inexperienced person from scratch. For exactly this reason, a number of corporations hire only people with no experience for certain jobs. If you do hire experience, it is important to verify it by checking references and to evaluate it by testing performance.

Training takes the time of both trainer and trainee, and that is expensive. But putting people in jobs without enough training is likely to be more costly in the end. The worker does not perform well and is unhappy, the customer suffers and is unhappy, and you will suffer and be unhappy, too. You really don't have time *not* to train people. There is more on this subject in Chapter 10.

Recruiting

LEARNING OBJECTIVE: Identify different types of recruitment methods.

Recruiting is finding the most suitable employee for an available position. The process begins with announcing the opportunity; sometimes this is done first within the organization, then outside. Applications are received from a variety of sources: internal promotion, employee referrals, applications filed from people who have sent in résumés, online applicants, company website applicants, transfers within the company, advertising, colleges and universities, and government-sponsored employment services.

✳ GENERAL RECRUITING PRINCIPLES

Since the legal aspects of recruiting and selection were covered in Chapter 5, we now move on to recruiting. **Recruiting**—looking actively for people to fill jobs—is a form of marketing. You are in the labor market to sell jobs to people who might want them. Because your need is constant and urgent, because you have many competitors, and because many of your jobs are not the most exciting ways of making a living, you really need to work at making your recruiting effective.

The first word to keep in mind is *appropriate*. You must put out your message in appropriate places and aim it toward people you would like to hire. Use techniques appropriate to your image and to the kinds of people you want to attract. A help-wanted sign in a dirty and fly-specked window is going to reach only people who pass by and attract only people who reflect that image themselves—if it attracts anyone at all.

recruiting
Actively looking for people to fill jobs. Direct recruiting: going where the job seekers are, such as colleges, to recruit. Internal recruiting: looking for people within a company to fill jobs. External recruiting: looking for people outside a company to fill jobs.

Recruiting excellent candidates is requisite to ensuring excellent morale and guest satisfaction.
wavebreakmedia/Shutterstock

"Now Hiring" hanging in a clean window is only one step up. Take a look at some of the classified ads found in Figure 6.4. Which one might you respond to? Can you decipher what all the abbreviations mean? Which advertisement tells you the most about the restaurant and the nature of the jobs available? If you project an image of being a desirable employer through your advertisements, you are probably going to attract desirable applicants.

Your message must be appropriate: Tell them what they want to know. They want to know (1) what the job is, (2) where you are, (3) what the hours are, (4) what qualifications are needed, and (5) how to apply. "Bartender Wanted" and a phone number will not pull them in until after they have tried everyone else. They are also interested in (6) attractive features of the job, such as good wages and benefits.

It is also essential to use channels of communication appropriate to the people you want to reach, the same channels that they are using to look for jobs. You must get the message to the areas where they live and use the media of communication they see and hear.

The second word to keep in mind is *competitive*. You are competing with every other hotel and foodservice operation in your area for the same types of people. For unskilled labor you are

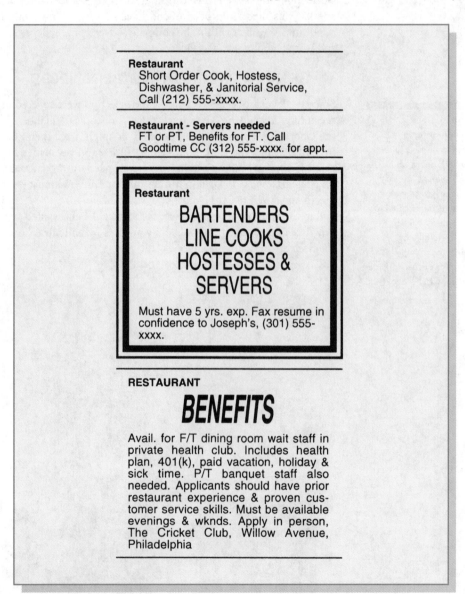

Restaurant
Short Order Cook, Hostess, Dishwasher, & Janitorial Service, Call (212) 555-xxxx.

Restaurant - Servers needed
FT or PT, Benefits for FT. Call Goodtime CC (312) 555-xxxx. for appt.

Restaurant

BARTENDERS LINE COOKS HOSTESSES & SERVERS

Must have 5 yrs. exp. Fax resume in confidence to Joseph's, (301) 555-xxxx.

RESTAURANT

BENEFITS

Avail. for F/T dining room wait staff in private health club. Includes health plan, 401(k), paid vacation, holiday & sick time. P/T banquet staff also needed. Applicants should have prior restaurant experience & proven customer service skills. Must be available evenings & wknds. Apply in person, The Cricket Club, Willow Avenue, Philadelphia

FIGURE 6.4: Classified advertisements.

competing with other types of operations as well: retail stores, light industry, and so on. You must sell your jobs and your company at least as well as your competitors sell theirs, if not better.

The third word to remember is *constant*. It is a good practice to be on the lookout for potential employees all the time, even when you have no vacancies. Even the best and luckiest of employers in your field will probably replace at least 6 out of every 10 employees in a year's time, and many operations run far higher than that. Keep a file of the records of promising people who apply each time you fill a job, and look through them the next time you need to hire.

You will also have drop-in applicants from time to time. Pay attention to them; they have taken the initiative to seek you out. Ask for a résumé and let them know that you will add it to the talent bank. Give them a tentative date to call back, and be cordial. They should leave with a feeling of wanting to work for you; remember that you are marketing yourself as a good employer, and you may need them tomorrow.

❋ ONLINE APPLICANTS AND SELECTION TESTS

Today, many hospitality companies have a space for employment opportunities on their websites. This free advertising is attracting an increasing number of applicants. Applications can be completed online, saving both time and money. Applicants may be asked to complete selection tests online. Here are some examples:

- *Cognitive ability tests* measure intelligence.
- *Aptitude tests* measure the ability of an individual to learn or acquire new skills.
- *Personality tests* measure an individual's basic characteristics, such as her or his attitudes, emotional adjustment, interests, interpersonal relationships, and motivation.
- *Honesty/integrity tests* are designed to measure an applicant's propensity toward undesirable behaviors such as lying, stealing, and abusing drugs or alcohol.
- *Substance abuse tests* are intended to ensure a drug-free workplace. Concern about workplace safety issues, alcohol, and illegal and unsafe drug use has prompted many employers to require employees and applicants to submit to substance abuse testing.

There are some who do not agree with applicant/employee testing. However, if employers keep selection tests in context and perspective, and if they are used in their proper job-related manner and are nondiscriminatory to protected-class members, then they are permitted under the law in your state, county, and city.[11] Employment testing is discussed later in this chapter.

The final words of wisdom are *use multiple approaches*. Do not depend on a single resource or channel; try a variety of methods to attract people. There are many channels: social media, Craigslist, schools and colleges giving hotel and foodservice and bartending courses, well-chosen word-of-mouth channels such as current employees, notices on the right bulletin boards (the student union, the school financial aid office), newspaper and radio ads, online job resources, trade unions, employment agencies, community organizations, summer job fairs, and organizations working to place certain groups of people such as immigrants or minorities or disabled persons. You can also go out into the field and recruit workers directly wherever they are.

Let us look at some of these resources and channels in more detail.

internal recruiting
Searching for job applicants from within an operation.

❋ INTERNAL RECRUITING

Internal recruiting is the process of letting your own employees know about job openings so that they may apply for them. Often, the most successful placements occur through people who already work for you. Internal recruiting often results in **promoting from within**, a practice in which current employees are given preference for promotions over outside applicants with similar backgrounds. Promoting from within has several advantages: It rewards employees for doing a

promoting from within
A policy in which it is preferable to promote existing employees rather than filling the position with an outsider.

Courtesy of
Sharon Morris.

As a restaurant supervisor, when recruiting I always prepare myself by reviewing the applications of candidates selected for interview. Our restaurant requires all applicants to go online and complete a talent assessment questionnaire to assess their aptitude for a service position, back or front of the house.

I use a structured format with some flexibility. For example, if I ask about a success the candidate had at a previous job, the applicant's response might lead to a follow-up question. Always the same questions are asked of each candidate. A skills test is also given in the relevant job area. If it's in the kitchen, then a kitchen-related skills test is given; the applicant is asked to prepare a certain food item on the menu.

Before offering applicants a position, we must do a background check. We use an outside company and request them to do as extensive a background check as the law permits and to do a drug test—the reason being that we do not want to be held liable for negligent hiring—meaning, if an employee were hired and does something serious to a guest or another employee, then we could be held liable. Finally, we need to ensure that applicants can do the job description that they were hired to do. We get their signature on the job description saying that they can do the job.

job posting
A policy of making employees aware of available positions within a company.

employee referral programs
A program under which employees suggest to others that they apply for a job in their company. If a person referred gets a job, the employee often receives recompense.

good job, it motivates employees and gives them something to work toward, and it maintains consistency within the enterprise.

Now how can you be sure of letting all employees know about open positions? Using a practice called **job posting**, a representative (usually from the human resources department) posts lists of open positions (see Figure 6.5) in specific locations where employees are most likely to see them, including on the company website.

Usually, employees are given a certain period of time, such as five days, in which to apply before applicants from the outside will be evaluated. In most cases, employees must meet certain conditions before responding to a job posting. For instance, the employee may be required to have a satisfactory rating on his or her last evaluation and have been in his or her current position for at least six months. These conditions prevent employees from jumping around too often to different jobs, a practice that benefits neither the employee nor the employer.

When you can't find a current employee to fill an open position, your employees may refer their friends and acquaintances to you. Some employers give a cash or merchandise reward to employees who bring in somebody who works for at least a certain time, such as 90 days. Many employers trying to draw in new employees have used these types of programs, called **employee referral programs**, very successfully. Employees who refer applicants are usually asked to fill out a referral form or card that may be handed in with the applicant's application form.

The idea behind this type of program is that if your current employees are good workers and are happy working for you, they are not likely to bring in someone who won't suit you or who won't fit into the work group. Bringing a total stranger into a group of workers can be very disruptive. Sometimes employees bring in relatives.

Among employers, there are two schools of thought about this: Some say that it is an absolute disaster, whereas others find that it works out well. It probably depends on the particular set of relatives. If a family fights all the time, you do not want them working for you. Some people point out that if one family member leaves or is terminated, the other will probably quit, too, and then you will have two jobs to fill. You have a similar problem if there is a family emergency; you will be short both employees.

Other internal recruiting methods include speaking with applicants who walk in, call in, or write in. These applicants should be asked to fill out an application form and should be interviewed when possible.

Date: Monday, September 25

Department: Food and Nutrition Services

Job Title: Food Service Worker

Job Code: 600026

Reports to: Operations Supervisor

Job Qualifications:

1. Six months experience in a health care facility.
2. High school graduate.
3. Courtesy and diplomacy in dealing with patients, hospital staff, fellow workers, and the department's management team.
4. Ability to consistently demonstrate the values of Sarasota Memorial Health Care System.
5. Communication skills, verbal and reading: required to read and understand written instructions, recipes, and labels.

Mental/Physical Demands:

1. Adaptability to routine work involving short-cycle repetitive duties under specific instructions.
2. Demonstration of good judgment consistently showing insight into problems.
3. Continuous physical activity involves standing, walking, bending, and stooping. Amount of weight lifted is routinely 25–30 pounds, and up to 50 pounds. Must be able to push carts weighing to 400 pounds.
4. Talking, hearing, and visual acuity essential.
5. Versatility required to adapt to frequently changing conditions in job duties covering a broad range of food service and production activities.
6. Finger and manual dexterity and motor coordination as required to manipulate kitchen utensils and food service supplies skillfully.

FIGURE 6.5: Job posting.

❋ EXTERNAL RECRUITING

external recruiting
Looking for job applicants outside the operation.

The remaining recruiting methods are all considered **external recruiting**, that is, seeking applicants from outside the operation. An advantage of bringing in outsiders is that they tend to bring in new ideas and a fresh perspective.

Today, hospitality companies use their own Web page as a recruiting tool. By having an icon for employment opportunities, they can save money by driving would-be applicants to their Web page, thus avoiding costly charges made by various *job search engines*.

In recent years, some hospitality companies have teamed with job search engines such as Monster.com and others to help find suitable applicants for their available positions.

Advertising

A job site on the Web is one common meeting place for job seekers and employers. It is also the best source for reaching large numbers of applicants. Many companies also use their own website or employee referrals.

Another way to screen is to include a specific instruction such as "Call Joe 9:00–11:00 A.M." The people who call Joe at 2:00 P.M. obviously do not follow written instructions, so if the job requires following written instructions, you can eliminate these callers (unless nobody calls between 9:00 and 11:00 and you are in a panic). Your company name and address will screen out people who do not want to work there for whatever reason.

When you are writing job advertisements, avoid terms that could be perceived as discriminatory, such as *busboy* or *hostess*. These terms indicate that the applicants should be male in the case of a busboy, or female in the case of a hostess. This is discriminatory, and therefore illegal, but you see it frequently in the newspapers. Also, avoid references to age, such as "young" or "recent high school graduate."

The number of applicants an ad pulls will vary greatly with the state of the economy. In good times, even an enticing ad may pull fewer responses than you would like. But when unemployment is high, even your most careful attempts to screen will not keep the numbers down. People who need a job are going to apply for it no matter what your ad says. You might have 50 applicants for one pot-washing job.

If you are going to advertise in the paper, it is well worth studying the ad pages to see what your competition is doing. Read all the ads with the mindset of a job seeker, and then write one that will top them all. Many ads mention incentives such as benefits, equal opportunity, job training, career growth, and other attractions. Usually, such ads are for large numbers of jobs (hotel openings, new units of chains, and so on) or for skilled labor or management jobs. If you are only looking for one pot-washer, you may not want to go all out in your ad, but if you want a competitive pot washer, run a good-looking, competitive ad. Some employers and job seekers use the services of companies such as CareerBuilder, Monster, indeed, snagajob, or usajobs to list their vacancy or résumé.

Today, most hospitality operators avoid expensive advertising in the major area paper; instead, they are running ads in places where potential workers will see them. Many cities have special area newspapers and shopping guides. Place your ads in those areas where your target employees live— people within commuting distance who may be candidates for your types of jobs.

For instance, if many of the potential employees in your area speak Spanish, consider running an ad in local Spanish-language newspapers. Other special places are the schools and colleges in your area. There are also websites that list job openings in the same ways as newspapers. Tommy Bahama in Sarasota, Florida, uses Craigslist, for example.

Many employers use their own website to advertise for jobs. A low-cost place to advertise is right in your operation. You can use any of the following to bring in applicants: placemats, indoor or outdoor signs (if done professionally), receipts, or table tents, to name just a few. Finally, you can advertise open jobs by posting notices in supermarkets, libraries, churches, synagogues, community centers, and health clubs. Some restaurants even put a sign in the window: "Now hiring smiling faces."

Employment Agencies

employment agencies
Organizations that try to place persons into jobs. Private employment agencies: privately owned agencies that normally charge a fee when an applicant is placed. Temporary agencies: agencies that place temporary employees into businesses and charge by the hour.

Employment agencies are a resource you should look into under certain circumstances. We look at three common types of agencies: private, temporary, and government. **Private employment agencies** normally charge a fee, which is not collected until they successfully place an applicant with you. In most cases, if this person does not stay with the company for a specified period of time, the agency must find a suitable replacement or return the fee. The fee is often 10 percent of the employee's first-year salary. These types of agencies most often handle management or high-skills jobs and should be used only if they specialize in your field.

Temporary agencies have grown in size and importance, and now a small number specialize in filling positions, including entry-level positions, for hotels, restaurants, and caterers. Temporary agencies charge by the hour for personnel who work anywhere from one day to as long as needed.

Using temporary employees is advantageous during peak business periods or other times when emergency fill-in personnel are needed. However, you can't expect a temporary employee to walk into your operation and go straight to work. You must be willing and able to spend time and money to orient and train these employees.

Another source of employees, at no cost, is the U.S. Employment Service, a federal and state system of employment offices called **job service centers**. Your local job service center will screen and provide applicants for entry-level jobs. The centers have many unemployed people on their books who are looking for jobs. It is a question of whether they are staffed enough to be able to sift through the people and send you suitable applicants who will not waste your time.

job service center
An office of the U.S. Employment Service.

Direct Recruiting

Direct recruiting, going where the job seekers are, is practiced primarily by large organizations seeking management talent or top-level culinary skills.

direct recruiting
On-the-scene recruiting where job seekers are, such as at schools and colleges.

Such organizations send recruiters to colleges that teach hospitality management or culinary skills to interview interested candidates. There are also certain situations in which direct recruiting is appropriate for entry-level and semiskilled personnel.

For example, when a hotel or restaurant closes, you might arrange to interview its employees. A large layoff at a local factory might be another such situation. It might be worthwhile to interview foodservice students in secondary or vocational schools. Some large cities hold job fairs in early summer to help high school students find summer work. This would be an appropriate place for direct recruiting. Summer employees, if they like the way they are treated, can also become part-time or occasional employees during the school year that follows.

One of the advantages of direct recruiting is that you might get better employees than you would by waiting for them to drop in or to answer your ad. Another advantage is the image-building possibilities of direct recruiting. You are not only hiring for the present; you are creating a good image of your company as a place of future employment. Some companies also have internal job fairs where managers are available to talk with employees about their jobs so they have a better idea of what it takes to be a manager. It shows the companies' willingness to promote from within.

Additional External Recruiting Sources

Organizations that are involved with minorities, women, veterans, disabled workers, immigrants, or other special groups will usually be very cooperative and eager to place their candidates. Examples of such organizations include the National Association for the Advancement of Colored People, the National Organization of Women, and the American Association for Retired Persons.

Since these organizations do not work only in hospitality, they may not be familiar with the demands of your jobs, and it is absolutely necessary that you be very clear and open and honest about what each job entails. Here again, detailed job descriptions and performance standards should be available. In addition, community organizations such as church groups, Girl Scouts, and Boy Scouts can be sources of employees.

It is a good idea to tell people with whom you do business when you are trying to fill a job. Many of the salespersons you deal with, for example, have wide contacts in the field, and they have good reason to help you out if you are a customer.

Sometimes friends and acquaintances in other fields know of someone who needs a job. Clergy or priests may be able to send people looking for work to you. Sometimes parents are looking for jobs for their children. Through individual contacts, you often reach people who are not yet actively looking for jobs but intend to start soon.

Many people say that one person's telling another that yours is a good place to work is the best advertising there is, and that it will provide you with a steady stream of applicants. Whether the stream of applicants appears or not, there is no guarantee that it will send you the people you want. You are more likely to get the type of people you are looking for through a systematic marketing plan to reach your target groups. But one thing is true: If yours is a good place to work, you will not need as many applicants because they will stay with you longer.

In the never-ending search for talent, some restaurant companies are considering podcasting as a way to attract young employees. Chris Russell recently launched JobsinPods.com, a website that allows employers to create online audio messages about their businesses for potential job applicants to download to their MP3 players or iPods.[12]

✱ EVALUATING YOUR RECRUITING

To determine which sources give you the best workers, you need to evaluate the results over a period of time. What is your successful rate of hire from each source? What is the cost, not only the cash paid out for ads but the hire ratio to numbers interviewed from each source? Interviewing is time-consuming, and if interviewing people from a certain source is just an exercise in frustration, that is not a good source.

What is the tenure of people from each source: How long, on average, have they stayed? How many have stayed more than 30 days or three months? How good is their performance? If you find that you are getting poor workers from a particular source, you should drop that source. If you are getting good people from a certain source, stick with it.

You should also evaluate your own recruiting efforts. Are you staying competitive? Do you explain the job clearly and completely and honestly, or do you oversell the job? Do you project a good image for your enterprise, or do you oversell the company? If you oversell, your mistakes will come back to haunt you.

> Before applying for jobs it is advisable to prepare a résumé. Please visit the book companion site to access two templates: a sample résumé template and a sample cover letter template. It is best practice when applying to any job to provide both a résumé and a cover letter.

Selecting the Right Person

LEARNING OBJECTIVE: List the best ways to find qualified employees, assessing the pros and cons of each method.

Let us suppose that you now have a number of applicants for a job you want to fill. Ten applicants for one job is considered by experts to be a good ratio, but that number will vary. Up to a point, the more you have to choose from, the better your chances are of finding someone who is right for the job. But even if you have only one applicant, you should go through the entire selection procedure. It may save you from a terrible mistake.

It is critical to select the right person for the open position. Companies such as Ritz-Carlton arrange for final applicants to complete a "talent" interview to determine if the candidate will fit with the Ritz-Carlton culture and be able to provide genuine caring service to guests. The most successful person is not always the most experienced person—natural talent plus a really positive attitude and desire to be a team player and to learn more every day will frequently be a better person for hospitality companies.

Other positive signs of a good candidate are things like—do they smile in the first few seconds, and what feeling do I get from them, and do they exhibit a passion for the hospitality business?[13] Some companies use current employees on a selection committee because they will be working with the new hire.

We all know that the hospitality industry has a high turnover rate, and much of this high turnover is due to poor selection. The cost of replacing employees is about $8,000 in a high-end hospitality business. This sounds like a lot, but by the time you add up all the costs involved with every stage of the process—position announcements, advertising, recruiting, selection, interviewing, testing, drug screening, talent interview, background checks, and job offers—you can see that this is no overstatement. For line employees in midmarket hospitality organizations, the typical cost of turnover is about $5,000 per position. However, the payoff is more than offset in reduced turnover that can occur with effective and efficient selection.[14]

So, if you want friendly, courteous service, you must hire friendly, courteous people. Hiring employees is like casting stars for a movie—if you do the job well, people will believe that the actor is actually the person they are portraying. Walt Disney World allows its best employees, known as star "cast members," to select future cast members.

Disney gives these star cast members three weeks of training in the selection process before they join the selection team. When screening, it is important to strike a balance. Extensive screening of potential and existing staff avoids breaking the law, failing to respect individuals' rights, and treating people in a discriminatory manner. Conversely, organizations could miss out on people who could be a highly valuable asset. Tread carefully.[15]

Assuming that you have already established job specifications and have done some preliminary screening through your ads or on the phone, the selection procedure from here on has five elements:

1. The application form
2. The interview and evaluation
3. Testing
4. The reference check
5. Making the choice

According to Jim Sullivan, people like to work with those who like them and are like them but, when hiring key employees, there are two qualities to look for: *judgment* and *honesty*. Almost everything else can be bought by the yard. Remember, there are two kinds of people who never succeed: those who cannot do what they are told and those who cannot do anything unless they are told.[16]

✻ APPLICATION FORM

An application form is a fact-finding sheet for each applicant. It is a standard form (see Figure 6.6) that asks relevant and job-related questions such as name, address, and phone number, type of job wanted, work history, education, references, and how the applicant heard about the job. As explained in Chapter 5, questions that can be viewed as discriminatory are not allowed (refer to Figure 5.2). You should instruct applicants to complete everything, especially the work history, including places and dates of employment and names of supervisors.

Before you interview an applicant, you should familiarize yourself with the material on the application and jot down questions. What about gaps in employment? Unanswered questions? The way applicants fill out applications can also be very revealing. Do they follow instructions? Can they read and write? Do they understand the questions? Are they neat or messy? Is their handwriting legible? Did they complete everything? Such things may relate to the job requirements. Did they sign the application form—because if they didn't, and you later find out that they had been convicted of a crime, they can always say, "I didn't sign the application form."

✻ THE INTERVIEW

The first essential for a good interview is a quiet place free of distractions and interruptions, and the first task is to put the candidate at ease. You can tell how they feel by looking for nonverbal clues: a worried look on his or her face, a tensed posture. It is important to remember that people get nervous about interviews. If you can make them feel comfortable and unthreatened, they are more likely to open up and be themselves, and this is what you are after. Listen attentively; this calls for your best listening skills. Remember that you want to impress them favorably on behalf of your organization. A careless mistake in the beginning can ruin the entire interview.

Prepare a list of questions based on the job description—this underlines the importance of a good job description. The best interview questions employers use start with *how, what,* and *why*. When employers use those words, they give the interviewee a chance to explain what they have done and why they did it.[17] With lower-level jobs, it is best to follow a preplanned pattern for the interview, so that you cover the same territory with every applicant. You can start off with general information about the job and the company. The interview involves a two-way exchange of information: you want to know about the applicant, and the applicant wants to know about the job.

APPLICATION FOR EMPLOYMENT
PLEASE PRINT ALL INFORMATION

Date

Month	Day	Year

Equal Opportunity Employer

The Company will not discriminate against an applicant or employee because of race, sex, age, religious creed, political affiliation, national origin, sexual preference, disability, or any veteran status.

Last Name	First	Middle Initial	Social Security Number

Present Address (Street & Number)	City	State	Zip Code	Home Phone Number
				( )

Address where you may be contacted if different from present address	Alternate Phone Number
	( )

Are you 16 years of age or older? ☐ YES ☐ NO U.S. Citizen or Resident Alien? ☐ YES ☐ NO If no, indicate type of Visa:

JOB INTEREST

Position you are applying for:	Type of position you eventually desire:

Available for: ☐ Full-Time ☐ Part-Time ☐ Per Diem ☐ Day Shift ☐ Evening Shift ☐ Night Shift ☐ Weekends ☐ Other

When would you be available to begin work?

Have you previously been employed by us? ☐ Yes ☐ No If yes, when	Previous Position(s)

Have you previously submitted an application to us? ☐ Yes ☐ No If yes, when

How were you referred to the Company? ☐ Employment Agency ☐ Your Own Initiative ☐ Advertisement—Publication ☐ Employee Referral—Name

EDUCATION

School	Name and Address	Circle Highest Year Completed	Type of Degree	Major Subject
High School Last Attended		1 2 3 4		
College, University, or Technical School		1 2 3 4		
College, University, or Technical School		1 2 3 4		
Other (Specify)				

FIGURE 6.6: Application for employment.

patterned interview
A highly structured interview in which the interviewer uses a predetermined list of questions to ask each applicant.

Some employers use a highly structured type of interview known as a **patterned interview**, in which the interviewer asks each applicant a predetermined list of questions. It is important to ask the same questions of all candidates. There may also be additional questions on the interviewer's form that are not asked of the applicant but are provided to help the interviewer interpret the applicant's responses. The training required for a patterned interview is minimal compared to other methods, and the standardized questions help to avoid possible charges of discrimination.

You are after two kinds of information about the applicant: hard data on skills and experience and personal qualities important to the job. As you go over the application in the interview, fill in all details that the applicant left unanswered and ask questions about gaps of 30 days or more on the employment record.

PREVIOUS EMPLOYMENT—BEGIN WITH PRESENT OR MOST RECENT POSITION

1. Employer

Employed _____ to _____

Address (include Street, City, and Zip Code)

May we contact?   ☐ Yes   ☐ No

Telephone Number ( )

Starting Position   Salary

Last Position   Salary

Name and Title of Last Supervisor

Telephone Number ( )

Brief Description of Duties:

Reason for Leaving:

Disadvantages of Last Position:

2. Employer

Employed _____ to _____

Address (include Street, City, and Zip Code)

May we contact?   ☐ Yes   ☐ No

Telephone Number ( )

Starting Position   Salary

Last Position   Salary

Name and Title of Last Supervisor

Telephone Number ( )

Brief Description of Duties:

Reason for Leaving:

Disadvantages of Last Position:

3. Employer

Employed _____ to _____

Address (include Street, City, and Zip Code)

May we contact?   ☐ Yes   ☐ No

Telephone Number ( )

Starting Position   Salary

Last Position   Salary

Name and Title of Last Supervisor

Telephone Number ( )

Brief Description of Duties:

Reason for Leaving:

Disadvantages of Last Position:

FIGURE 6.6: (continued)

Often, people will not list jobs on which they had problems. If they have something to hide, they will hide it, and these are exactly the things you need to find out. Don't hesitate to probe if you are not satisfied with either the applicant or his or her answers to your questions. Take care to avoid questions that could be considered discriminatory.

Getting people to talk may be agonizing the first time you interview. The best method is to avoid questions that have yes-or-no answers. Ask: "What did you do at . . . ?" "What did you like best about . . . ?" "Tell me why. . . " One owner always asked server applicants about the funniest thing that ever happened to them on the job. He would not hire people who said that nothing funny had ever happened to them because he believed that they could not deal with people effectively if they couldn't see the funny side of things. You should talk only about 20 percent of the time, with the candidate filling in the remaining 80 percent.

IF MORE THAN THREE PREVIOUS EMPLOYERS, PLEASE LIST OTHERS HERE

Employment Dates		Company and Address	Position or Type of Work	Salary or Wage	Reason for Leaving
From	To				

Please indicate if you were employed under a different name from the one shown on the first page of this application in any of your previous positions.

Employer	Name Used

U.S. MILITARY RECORD (If related to the job you are applying for)

Branch of Service _____

Active Duty _____ From _____ To _____

Nature of Duties _____

_____

_____

CONVICTIONS/COURT RECORD

Have you been convicted of a crime within the last 7 years?

_____ Yes _____ No   The existence of a record of convictions for criminal offenses is not considered an automatic bar to employment.

Date of Conviction _____   Describe circumstances: _____

_____

_____

ACKNOWLEDGMENT

I understand that this employment application and any other Company documents are not contracts of employment and that any individual who is hired may voluntarily leave employment upon proper notice and may be terminated by the Company at any time and for any reason. I understand that no employee of the Company has the authority to make any agreement to the contrary and I acknowledge that any oral or written statements to the contrary are hereby expressly disavowed and should not be relied upon by any prospective employee.

I hereby grant permission for the authorities of the Company, or its agents, to investigate my references, and I release the Company and all previous employers, educational institutions, persons, and law enforcement agencies from any and all liability resulting from such an investigation. Upon my termination, I authorize the release of information in connection with my employment.

I certify that the statements made on this application are true and correct, and thereby grant the Company permission to verify the information contained herein.

I understand that giving false information or the failure to give complete information requested herein shall constitute grounds, among others, for rejection of my application or my dismissal in the event of my employment by the Company.

DATE:_____   SIGNATURE OF APPLICANT: _____

FIGURE 6.6: *(continued)*

Other good work-related questions to ask are: Tell me about the strengths you bring to this job. This position requires good organizational skills; tell me about how organized you are. You can ask how the applicant would handle a situation that relates to your industry:

- A guest was unhappy with the room.
- A guest has eaten most of an entrée and then says, "I don't like it."
- A theme-park visitor says, "Why do I have to stand in line for two hours, just to go on one ride?"
- A patient says, "You forgot my dinner" to the hospitality server in a hospital room.

Getting applicants to talk takes practice.
Pressmaster/Shutterstock

Avoid asking about what the applicant likes to do for fun. What if the answer is, "On Wednesday night, I have Bible study," and he or she doesn't get the job? The applicant could claim that you discriminated on religious grounds. If it doesn't pertain to the job, don't ask it.[18]

Ask if you may take notes (but not on the application form because it goes in the applicant's file and can later be used as evidence in a legal case, and your comments may come back to haunt you). You can do this during the interview if it does not inhibit the applicant; otherwise, do it immediately after, lest you forget. Avoid writing down subjective opinions or impressions; instead, write down specific job-related facts and direct observations.

Be objective, factual, and clear. Evaluate the applicant immediately on your list of specifications for the job, using a rating system that is meaningful to you, such as a point system or a descriptive ranking: (1) exceptional, hire immediately; (2) well qualified; (3) qualified with reservations; (4) not qualified. Some large companies have evaluation forms or systems to use.

Look at the applicants from the perspective of what they can do and what they will do. **Can-do factors** include the applicant's job knowledge, past experience, and education—in other words, whether the applicant can perform the job. **Will-do factors** examine an applicant's willingness, desire, and attitude toward performing the job. You want the person whom you hire to be both technically capable to do the job (or be trainable) and willing to do the job. Without one or the other, you are creating a problem situation and possibly a problem employee.

One thing that happens frequently during interviews is that applicants are giving you the answers they think you want to hear and projecting the image they think you are looking for, and they may not be like that at all in real life. Yet often, they let their guard down when the interview is just about over and reveal their true selves in the last few minutes. If you are aware of this, perhaps you can exchange the first four minutes with the last few in making your evaluations.

It is very easy, in that first four minutes, to be influenced by one or two characteristics and extend them into an overall impression of a person. This is known as the **halo effect or overgeneralization**. You may be so impressed with someone who is articulate and well dressed that you jump to the conclusion that this applicant will make a great bartender. The first day on the job, this impressive person has drunk half a bottle of bourbon two hours into the shift.

Another form of overgeneralization is to assume that all applicants from a certain school or all people your pot-washer knows personally and says are okay are going to be good workers. This is not necessarily so; it is a generalization about personality rather than knowledge or skill.

can-do factors
An applicant's or employee's job knowledge, skills, and abilities.

will-do factors
An applicant's or employee's willingness, desire, and attitude toward performing a job.

halo effect or overgeneralization
The tendency to extend the perception of a single outstanding personality trait to a perception of the entire personality.

Another thing that happens easily is to let *expectations* blind you to reality. If someone has sent you an applicant with a glowing recommendation, you will tend to see that applicant in those terms, whether or not they are accurate.

Still another thing that is easy to do is to see some facet of yourself in someone else and to assume that this person is exactly like you. You discover that this person grew up in your old neighborhood, went to the same school that you did, had some of the same teachers, and knows people you know.

projection
In interviewing and evaluation, putting your own qualities on the applicant based on a perceived similarity.

A spark is kindled and you think, "Hallelujah, this person has got to be great!" This reaction is known as **projection**—you project your own qualities onto that person. Furthermore, you are so excited about finding someone exactly like you (you think) that you may even forget what that has to do with the job you are interviewing this person for.

What it all comes down to is that in interviewing and evaluating, you need to stick closely to the personal qualities needed *on the job* and to be on guard against your subjective reactions and judgments. Do not make snap judgments and do not set standards that are higher than necessary. Not all positions require people who are enthusiastic, articulate, or well educated, so don't be turned off by a quiet school dropout who can't put six words together to make a sentence.

When it comes to telling applicants about a job, you should be open and honest and completely frank. If they will have to work Sundays and holidays, tell them so. One supervisor told an applicant she would work a five-day week. The applicant assumed that it was Monday through Friday, and that was fine. But when she reported for work and they told her it was Wednesday through Sunday, she quit then and there. She felt that the supervisor had cheated, and from that point on the trust was gone.

truth in hiring
Telling an applicant the entire story about a job, including the drawbacks.

Be frank about days and hours, overtime, pay and tips, uniforms, meals, and all the rest, so the new employee will start the job with no unpleasant surprises. You might call this **truth in hiring**. Sure, you want to sell your jobs, but overselling will catch up with you.

Explain your pay scale and your promotion policy: "This is what you start at, this is what you can make with overtime, this is what you can realistically expect in tips, this is what you will take home, this is as high as you can go in this job, these are the jobs you can eventually work up to, these are your chances of that happening."

Give them a chance to ask questions, and then end the interview. Tell them when you will make your decision and ask them to call you in the next day or two if they have not heard from you. Altogether, it should take you 20–30 minutes to interview an applicant for an entry-level job and up to 60 minutes for a supervisory position. Tips for interviewing are summarized in Figure 6.7.

During the interview, clarify the important aspects of the job. For instance: "This job requires that you work Tuesday to Saturday; are you able to do that?" or "This position requires you to work evenings and weekends; are you able to do that?" Or, "This position requires you to lift up to 50 pounds. Can you do that?"

✻ TESTING

Some companies use tests as an additional method of evaluating applicants. Sometimes tests are given before the interview to screen out candidates. Sometimes they are given after interviews to the small group of candidates still in the running, to add objective data to subjective evaluations. Various kinds of tests are used:

1. Skills tests measure specific skills.
2. Aptitude tests are intended to measure ability to learn a particular job or skill. Manual dexterity tests are a form of aptitude test and measure manipulative ability.
3. Psychological tests are designed to measure personality traits; large companies often use them in hiring management personnel.
4. Medical examinations measure physical fitness.

Except for medical examinations and skills tests, most hospitality enterprises do not use tests for nonmanagement jobs. There are several reasons for this. One is the time it takes to give tests

1. Be nonjudgmental during the entire interview process. Do not jump to conclusions. A poor interviewer reaches a decision in the first 5 minutes.
2. Recognize your personal biases and try not to let them influence you. Be objective. Do not look for clones of yourself. Do not let an applicant's age, gender, attractiveness, or verbal fluency influence your opinions.
3. Spend most of your time listening attentively. Allow the candidate to do at least 70 to 80 percent of the talking. Listen to each answer before deciding on the next question. Do not interrupt.
4. Make notes so that vital information is not forgotten.
5. Repeat or paraphrase the applicant's statements to make sure that you understand the applicant and perhaps get more information, or you may repeat the last few words the applicant just said with a questioning inflection. Also, summarize the applicant's statements periodically to clarify points and to bring information together. A summary statement may begin with, "Let's state the major points up to now. . . ." In this manner, the applicant can confirm or clarify what has been discussed.
6. Another technique to get a quiet applicant to talk and show interest is to ask open-ended questions (questions without a yes or no answer) and use pauses. Pauses allow the applicant to sense that more information is desired and hopefully, the interviewee will feel compelled to fill the silence.
7. Use body language to show interest and elicit information. Use direct eye contact, nod, smile, and lean forward slightly.
8. Do not be bashful about probing for more information when it is needed.
9. Instead of asking about an applicant's "weaknesses," refer to areas of improvement.
10. Paint a realistic picture of the job. Be honest.
11. Always be sincere, respectful, courteous, friendly, and treat all applicants in the same way.
12. Allow the applicant to ask questions of you.

FIGURE 6.7: Tips for Interviewing.

and score them. Another is that many of the tests available have little relevance to the requirements of nonmanagement jobs. A third is that many tests, having been constructed for populations of a certain background and education, discriminate against applicants who do not have that background and education. It is illegal to use such tests either in hiring or in promotion.

To be usable, a test must be valid, reliable, and relevant to the job. To be valid, it must actually measure what it is designed to measure. To be reliable, it must be consistent in its measurement, that is, give the same result each time a given person takes it. To be relevant, it must relate to the specific job for which it is given. The user of any test must determine that it meets these criteria and must use it properly as its publisher designed it to be used.

All in all, the complications of testing, the risks of discrimination, and the possibilities for error at the hands of an untrained user make most tests more trouble than they are worth. Skills tests and specific aptitude tests such as manual dexterity tests are the exceptions. Your best bet, and the one most closely geared to your job needs, is a set of skills tests derived from your performance standards.

They must be adapted somewhat since the applicant will not know all the ins and outs of your special house procedures, but this can be done. It will give you an objective measure of an applicant's ability to perform on the job and an indication of how much additional training is needed.

Psychological tests are used to test for honesty and even broader qualities such as integrity. These tests are based on the assumption that honest and dishonest people have different values and see the world differently. Some employers use honesty tests in the hopes of providing a secure workplace for their employees. Using honesty tests properly requires some work.

First, some states and localities do not allow such testing, so check the regulations. Next, you need to examine independent reviews and validity tests (provided that the instrument actually tests what it is supposed to test) of the instrument you want to use. Even if you find a good instrument, and it is legal in your location, don't forget that testing also requires money and time and that the results are not a substitute for any of the other selection steps you take, such as interviewing or making reference checks.

A medical examination can be required only after a job offer has been made to the applicant. When a job offer is made prior to the medical exam, it is considered a conditional job offer because if the applicant does not pass the medical exam, the job offer is normally revoked.

The **Employee Polygraph Protection Act of 1988** prohibits the use of lie detectors in the screening of job applicants. Although lie detectors have been used in the past in some states, they are now illegal to use in the employment process.

> **Employee Polygraph Protection Act of 1988**
> A federal law that prohibits the use of lie detectors in the screening of job applicants.

❋ REFERENCE CHECK

You have now narrowed your choice to two or three people. So, why is it important to check references? Well, for starters, it is an important part of the selection process; they are more likely to help ensure successful hires by screening for a good fit for the organization/department. Conduct reference checks in compliance with all federal and state laws and regulations. The American with Disabilities Act prohibits asking non-job-related information from previous employers or other sources. Avoiding questions regarding marital status, religion, age, race, health-related issues, childcare, transportation, worker's compensation claims, and any other non-job–related questions can help avoid charges of *negligent hiring*.[19]

The reference check is the final step before hiring. It is a way to weed out applicants who have falsified or stretched their credentials or who in other jobs have been unsatisfactory. Reference information can be thought of in two ways: substance and style. *Substance* concerns the factual information given to you by the applicant. *Style* concerns how the person did in previous jobs, how he or she got along with others and how well he or she worked under pressure.

When requesting a reference check, prepare specific job-related questions and do not ask questions that are not permitted during the interview.[20] First, verify the substance of the application, such as dates of employment, job title, salary, and so on. You may wonder why applicants would falsify information on an application, but they do. If your job requires a particular educational degree or certification, ask applicants to supply a copy of the appropriate document. Otherwise, get the applicant's written permission to obtain a transcript.

Once you have confirmed that the person is who they say they are on paper, you can start checking previous work references. Often, former employers will only reveal neutral information such as job title, dates of employment, and salary because of fear of being charged with libel, slander, or defamation of character by the former employee.

Although there is nothing wrong with providing objective documented information, such as an attendance problem, past employers are often reluctant to discuss this sort of concern or even answer the one question you really need an answer to: "Would you rehire?" To reduce any possible liability, you should ask applicants to sign a release on the application form (Figure 6.7) that gives you permission to contact references and holds all references blameless for anything they say.

Because it is fast, checking references by phone is very common. Be sure to document your calls on a form. Ask to speak to the employee's former supervisor. Always identify yourself and your company, and explain that you are doing a reference check. Start by asking for neutral information such as salary and job title and work your way up to more telling information.

Despite the importance of checking references, few people in the hospitality industry bother with a reference check. It may be habit or tradition, or it may be fear and desperation: fear of finding out there is a reason not to hire and desperation to fill the job. It may just be too time-consuming or you may think that your gut feeling or intuition says it all. But it is really a serious mistake to neglect the reference check and thus run the risk of hiring a problem worker.

When calling for a reference check, talk to human resources, not the department supervisor—who might be a friend of the applicant or who might want to be rid of the applicant and therefore give him or her a good reference regardless. Do get background checks: these will include a credit check—you don't want someone with credit problems working in a cash-handling situation.

Do also get a criminal background check—you don't want to give a sex offender access to guest rooms.

❋ MAKING THE CHOICE

Choosing a new employee is your decision and your responsibility. Making the choice may mean choosing between two or three possibilities or looking further for the right person for the job. When making the hiring choice, avoid making any of these common mistakes:

1. Don't jump to hire someone who simply reminds you of yourself. Also try not to fall prey to the halo effect. Look at the big picture!

2. Many problems in hiring come about when you hire too quickly. Use the time involved in the selection process to go through each step thoughtfully. Aim to hire the best candidate, not simply the first reasonably qualified applicant who comes forth.

3. Don't rush to hire the applicant who interviewed the best. Although the interview process can certainly tell you a lot about an applicant, the applicant with the best interviewing skills (which can be learned and practiced by anyone) is not necessarily the best person for the job. Also, keep in mind that during an interview, some applicants will use their charismatic personalities and ability to tell you what you want to hear to get top consideration for the job in question.

4. Don't hire someone just because your "gut feeling," or intuition, says that this applicant is the best. Intuition is fine to use, but always combine it with the other tools of the trade, such as reference checking and testing.

5. Don't hire someone just because the person comes highly recommended. Perhaps an applicant comes highly recommended as a breakfast cook, but you are looking for an experienced pizza maker. It's fine to listen to a recommendation for an applicant, but as usual, that's only part of the story.

Every time you hire someone, even when you feel confident about your choice, there is the chance that you have made a mistake. You will not know this, however, until your new people have been with you awhile and you can see how they do the work, whether they follow instructions and learn your ways easily and willingly, how they relate to the customers and the other workers, whether they come in on time, and all the other things that make good workers. To give yourself the chance to make this evaluation, it is wise to set a probationary period, making it clear that employment is not permanent until the end of the period.

If you see that some of your new people are not going to work out, let them go and start over. Do not let them continue beyond the end of the probation period. It is hard to face the hiring process all over again, but it is better than struggling with an incompetent employee. It can be as hard to fire as it is to hire—but that's another story.

❋ HIRING CHECKLIST

The following is a checklist for hiring in the hospitality industry:

1. Review, update, or create a job description/specification
2. Gain permission to hire
3. Determine required and preferred qualifications
4. Gain agreement on the posting announcement and create computer screening questions
5. Post job announcement
6. Select interviewers
7. Decide on applicants to interview
8. Decide on telephone/Skype interview questions
9. Arrange interviews and questions

10. Conduct interviews
11. Determine top candidates
12. Do background checks (if necessary)
13. Make offer

❋ MAKING THE OFFER

Offers for all jobs should be made in writing. The offer letter typically is sent, or given, to the new hire after an offer has been made and accepted over the phone. When you are making an offer, be sure to include all the conditions that were discussed with the applicant. The following points should appear in the offer letter, as appropriate:

- Department
- Position title
- Supervisor
- Location
- Rate of pay
- Schedule of shift, days off
- When jobs start, where to report, whom to report to
- Clothing and equipment needed
- Meal arrangements
- Parking
- Orientation/training arrangements
- Brief description of benefits
- Probationary period
- Appointment time or whom to call for an appointment concerning filling out additional personnel forms (such as the I-9 form)

Negligent Hiring

LEARNING OBJECTIVE: Describe safeguards against negligent hiring.

Fear of negligent hiring and retention litigation is a hiring manager's worst nightmare and the most compelling reason to conduct in-depth criminal records searches of job applicants. A multi-level jurisdictional criminal records search is the greatest protection an employer has against a negligent hiring lawsuit. Additionally, companies such as Professional Screening and Information can conduct a background investigation. They have an online facilitator who, once the individual's information and payment have been received, can issue a security code so the company can view the results of the background check.[21]

Could your employer be sued if a guest were injured by a hostile employee who had a violent background that would have been uncovered if a proper reference check had been done? Yes, your employer could be sued for **negligent hiring**. In the past 10 years, lawsuits for negligent hiring have been on the rise. If a violent or hostile employee injures a guest or employee, the injured party may sue the employer and will probably win if he or she can prove that the employer did not take reasonable and appropriate precautions to avoid hiring or retaining the employee.

As a leader, you have the responsibility of taking reasonable and appropriate safeguards when hiring employees to make sure that they are not the type to harm guests or other workers. Such safeguards include conducting a reasonable investigation into an applicant's background and, especially, inquiring further about suspicious factors such as short residency periods or gaps in employment.

negligent hiring
The failure of an employer to take reasonable and appropriate safeguards when hiring employees to make sure that they are not the type to harm guests or other workers.

You also have a responsibility to counsel or discipline your employees when they become abusive, violent, or show any other deviant behavior. Follow up on complaints your employees and customers may make about another employee's negative behaviors. Use your employer's policies to dismiss dangerous or unfit employees after appropriate warnings. For hospitality companies, a well-oiled human resource team trained to screen for such hidden characteristics (prejudice) and identify people who will fit into a corporate culture with zero tolerance for prejudice is of the utmost importance.[22]

Orientation

orientation
A new worker's introduction to a job.

LEARNING OBJECTIVE: Explain the two primary purposes of employee orientation.

Orientation introduces each new employee to the job and workplace as soon as he or she reports for work. It is not uncommon in the hospitality industry for people to be put to work without any orientation at all. You don't even know what door to come in and out of and where the restrooms are, and on payday everyone else gets paid and you don't, and you wonder if you have been fired and didn't even know it.

The primary purpose of orientation is to tell new staff members (1) what they want and need to know, and (2) what the company wants them to know. As with any training, it takes time—the new person's time and the supervisor's time—anywhere from 30 minutes to most of the day.

Nevertheless, it is worth the time needed to do it and to do it well. It can reduce employee anxiety and confusion, ease the adjustment, and tip the balance between leaving and staying during the first critical days. In addition, it provides an excellent opportunity to create positive employee attitudes toward the company and the job.

Therefore, you have two goals for an orientation: (1) communicating information and (2) creating a positive response to the company and the job. Let us look at the second one first because it makes the first one easier and because it is more likely to be overlooked.

✳ CREATING A POSITIVE RESPONSE

If you do not have an orientation for each new employee, somebody else will—your other workers. Their orientation will be quite different from yours, and it may have a negative impact. They want to give a new person the inside story, the lowdown, and it will include everybody's pet gripes and negative feelings about the company and warnings to watch out for this and that, and your new worker will begin to have an uneasy feeling that this is not such a good place to work. People are always more ready to believe their coworkers, their peer group, than their boss, so it is important for you to make your impact first. Then, in the days that follow, you must live up to what you have told them in your orientation or their coworkers might undermine the impression you have made.

You want to create an image of the company as a good place to work. You also want to foster certain feelings in your new people: that they are needed and wanted, that they and their jobs are important to the company. You want to create the beginnings of a sense of belonging, of fitting in. You want to reduce their anxieties and promote a feeling of confidence and security about the company and the job and their ability to do it.

You do all this not only through what you say but how you say it and even more through your own attitude. You speak as one human being to another; you do not talk down from a power position. You assume that each is a person worthy of your concern and attention who can and will work well for you. You do not lay down the law; you inform. You treat orientation as a way of filling their need to know rather than *your* need to have them follow the rules (although it is that, too). You accentuate the positive.

If you can make a favorable impact, reduce anxieties, and create positive attitudes and feelings, new employees will probably stay through the critical first seven days. It will be much easier for you to train them, and they will become productive much more quickly.

❋ COMMUNICATING THE NECESSARY INFORMATION

Employees want to know about their pay rate, overtime, days and hours of work, where the restrooms are, where to park, where to go in and out, where the phone is and whether they can make or receive calls, where their workstation is, to whom they report, break times, meals, and whether their brother can come to the Christmas party. The company wants them to know all this plus all the rules and regulations they must follow; company policy on holidays, sick days, benefits, and so on; uniform and grooming codes; how to use the time clock; emergency procedures; key control; withholding of taxes; and explanation of paycheck and deductions. They must also

---

INTRODUCTION TO THE COMPANY

_____ Welcome.

_____ Describe company briefly, including history, operation (type of menu, service, hours of operation, etc.) and goals (be sure to mention the importance of quality service).

_____ Show how company is structured or organized.

POLICIES AND PROCEDURES

_____ Explain dress code and who furnishes uniforms.

_____ Describe where to park.

_____ How to sign in and out and when.

_____ Assign locker and explain its use.

_____ Review amount of sick time, holiday time, personal time, and vacation time as applicable.

_____ Review benefits.

_____ Explain how to call in if unable to come to work.

_____ Explain procedure to request time off.

_____ Review salary and when and where to pick up check, as well as who can pick up the employee's paycheck. If applicable, explain policy on overtime and reporting of tips.

_____ Discuss rules on personal telephone use.

_____ Explain smoking policy.

_____ Explain meal policy, including when and where food can be eaten.

_____ Review disciplinary guidelines.

_____ Explain guest relations policy.

_____ Review teamwork policy.

_____ Explain property removal policy.

_____ Explain responsible service of alcohol, if applicable.

_____ Explain Equal Employment Opportunity policy.

_____ Discuss promotional and transfer opportunities.

_____ Explain professional conduct policy.

_____ Explain guidelines for safe food handling, safety in the kitchen, and what to do in case of a fire.

_____ Explain notice requirement if leaving your job.

THE NEW JOB

_____ Review job description and standards of performance.

_____ Review daily work schedule including break times.

_____ Review hours of work and days off. Show where schedule is posted.

_____ Explain how and when employee will be evaluated.

_____ Explain probationary period.

_____ Explain training program, including its length.

_____ Describe growth opportunities.

_____ Give tour of operation and introduce to other managers and coworkers.

FIGURE 6.8: Sample orientation checklist.

fill out the necessary forms and get their name tags, and they should have a tour of the facility and be introduced to the people they will work with.

It is a lot to give all at once. It is best to give it one-on-one rather than waiting until you have several new people and giving a group lecture. A lecture is too formal, and waiting several days may be too late.

You can have it all printed in a booklet, commonly called an employee handbook. But you cannot hand people a book of rules and expect them to read and absorb it. It will really turn them off if you ask them first thing to read a little booklet about things they cannot do. *Tell them*. Give them the booklet to take home.

An orientation checklist, shown in Figure 6.8, is an excellent tool for telling your employees what they need to know. It lists sample topics covered during an orientation program, such as how to request a day off. These topics are grouped into three categories: "Introduction to the Company," "Policies and Procedures," and "The New Job." One benefit of using such a checklist is that it ensures consistency among managers and supervisors who are conducting orientation and makes it unlikely that any topic will be forgotten.

Similarly, you cannot expect new employees to soak up everything you say. As you are aware, communication is a two-way process, and you can send message after message but you cannot control the receiving end. They will listen selectively, picking out what interests them. Try to give each item an importance for them (e.g., "You can get any entrée under $5 free," "The employee parking lot is the only place that isn't crowded," "The cook will poison your lunch if you come in through the kitchen"). Give reasons ("The money withheld goes to the government"). Phrase things positively ("You may smoke on breaks in designated areas outside the building" rather than "Smoking is forbidden on the job").

Watch your workers carefully to make sure that you are understood, and repeat as necessary. Encourage questions. ("Can I clarify anything?") Be sure you cover everything (use a checklist). Even so, you will need to repeat some things during the next few days.

Taking the trouble to start new employees off on the right foot will make things easier as you begin their training for the job. They will feel more positive, less anxious, and more receptive to the new work environment.

employee handbook
A written document given to employees that tells them what they need to know about company policies and procedures.

CASE STUDY: The One That Got Away

Dennis is dining-room manager in the coffee shop of a large hotel. He is about to interview Donna, a drop-in applicant who is filling out an application form. Donna seems friendly and outgoing, and Dennis thinks she'd make a great replacement for Rosa, a server with an attendance problem and a difficult home life. Dennis is on duty as host for the lunch-serving period. He is seating a party of guests when Donna brings him her application. "Enjoy your lunch!" he says to the guests as he hands them the menus. Then he hurries over to ask Eleanor, a server who sometimes acts as hostess, to sub for him for a few minutes, and seats Donna at a table near the entrance. He can keep an eye on things while he interviews her. He glances at the application. A year as waitress at Alfred's Restaurant—good! A high school graduate taking a couple of courses at the community college—good! The application is filled out neatly and carefully—good! He looks up to compliment her but sees Eleanor waving at him. "Excuse me, I'll be right back," he says to Donna. He deals quickly with a customer who wants to get a recipe. Donna is fiddling with a spoon and looks up soberly when he comes back. "I'm sorry," he says. "Now, where were we? Oh yes, I was going to tell you—" Another waitress presents herself at the table. "Listen, Dennis," she says, "tell Eleanor to get off my back. I'm not taking orders from her, she's not my boss." "Look, Dolores, I'll talk to you in a minute. The guest at Table 9 is signaling you. Go tend to her." Donna has a fixed smile on her face. "I really think you'd like it here," says Dennis, "there's never a dull moment. Now tell me about your job at Alfred's." After getting a chance to discuss Donna's work experience, he sees that Eleanor is gesturing that he is wanted on the phone, so he excuses himself. "Yes, of course, I'll take care of it," he says to his boss, and rushes back to Donna, who is sitting with hands folded, looking straight ahead. "Now tell me about yourself." "Well . . . what would you like to know?" She smiles politely. "Are you married?" Dennis asks abruptly. "Yes." Not so good, he thinks to himself. "Any kids?" "A baby boy." Worse! Upon seeing Dennis's facial expression, she looks at him levelly and says, "My mother takes care of him." "Would you—oh damn!" Eleanor is gesturing madly and a customer looking like he has very bad news is heading his way. He rises hastily. Donna rises, too. "I have to go," she says. "I'll call you," Dennis says over his shoulder before facing a furious man with a long string of complaints. The day goes on like this, one thing after another. The next morning he thinks about Donna again. Never mind about the baby: He decides to hire her on a probationary basis. When he finally finds time to call her, she tells him she has taken a job at the hotel across the street and promptly slams down the phone.

Case Study Questions

1. Dennis has made a number of mistakes in this interview. Identify as many as you can and discuss their adverse effects.
2. What did he find out about Donna during the interview?
3. What did he tell her about the job? What did she learn about the job in other ways?
4. On what basis did Dennis decide to hire her? Is it a good basis for making a hiring decision?
5. Do you think Donna would have decided to work for Dennis if he had gone about the interview differently?

KEY POINTS

1. *Labor market* refers to the supply of employees looking for jobs and the jobs available in a given area.
2. Many hospitality jobs require hard physical labor. The days and hours of work vary, but many employees work part-time hours, including evenings and weekends.
3. Possible sources of employees include those already working in your operation, people looking for their first job, women, immigrants, retired people, moonlighters, the unemployed, the disabled, and people who just want to get away from what they have been doing.
4. You will find it helpful to know something about the labor market in your own area, such things as prevailing wages for various jobs, unemployment rates for various types of workers, demographics, and the kinds of companies you are competing with for workers.
5. To determine labor needs, you must define the qualifications for each job in a document called a job specification. Job qualifications include knowledge, skills and abilities, work experience, and education and training.
6. When forecasting staff needs, look at your schedules and consider the amount of time it takes to replace an employee and get the new employee trained. Anticipate openings using a staff forecast form, shown in Figure 6.3.
7. Figure 6.2 shows recommended ways to ask questions of job applicants to avoid charges of discrimination.
8. Recruiting should be appropriate, competitive, constant, and use a multifaceted approach.
9. Recruiting is either internal or external. Examples include employee referral programs, direct recruiting, advertising, employment agencies, community organizations, personal contacts, and word of mouth.
10. The selection process includes the application form, the interview and evaluation, testing, the reference check, and making the choice.
11. Tips for interviewing are given in Figure 6.8.
12. To be usable, a test must be valid, reliable, and relevant to the job.
13. When hiring, you must make good-faith efforts to safeguard employees and guests from harmful people. If the proper checks are not made, your employer could be charged with negligent hiring.
14. Orientation tells new staff members what they want to know and what the company wants them to know.

KEY TERMS

can-do factors
demographics
direct recruiting
employee handbook
Employee Polygraph Protection Act of 1988
employee referral programs
employment requisition form
external recruiting
halo effect (overgeneralization)
internal recruiting
job posting
job service centers
job specification

labor market
negligent hiring
orientation
patterned interview
private employment agencies
projection
promoting from within
recruiting
scheduling
temporary agencies
truth in hiring
will-do factors

 # REVIEW QUESTIONS

Answer each question in complete sentences. Read each question carefully and make sure that you answer all parts of the question. Organize your answer using more than one paragraph when appropriate.

1. Describe the labor market in the area in which you live. What jobs are available? Are there many jobs advertised in the classified section of the newspapers? Is it hard to get a job because of a large number of applicants?
2. Describe five sources of potential employees.
3. List the job qualifications detailed in a job specification.
4. Which of the following questions are okay to ask applicants?
   ■ Do you own a car?
   ■ Do you own a home?
   ■ In this job, you will be lifting boxes up to 50 pounds. Can you lift 50 pounds?
   ■ Are you healthy?
   ■ Can you supply a photograph?
   ■ If you came from Greece, are you a Greek citizen?
   ■ Are you married?
   ■ What professional organizations do you belong to?
   ■ What ages are your children?
   ■ What clubs do you belong to?
   ■ Are you 40-something?
   ■ Do you have any disabilities?
   ■ Are you able to perform the job I have just described?
   ■ Do you have any outside activities that would keep you from observing the required days and hours of work?
5. What is negligent hiring? How can you avoid it?
6. Describe two methods of internal recruiting and three methods of external recruiting.
7. Discuss three methods you might use to evaluate your recruiting efforts.
8. List seven dos and seven don'ts for interviewing.
9. Why is checking references so important? Why is it so difficult to check references?

ACTIVITIES AND APPLICATIONS

1. **Discussion Questions**
   - What recruiting methods would be most appropriate to the situation in your area?
   - Which is better in your opinion: to hire experienced workers or to train people? Defend your opinion. Are there other alternatives?
   - How can you guard against your own subjectivity in an interview?
   - How could performance standards be used in recruiting and selection?
   - Do you think you have ever been discriminated against while trying to get a job? If so, describe.
   - Describe various experiences you have had when taking a job interview. Which interviewers struck you as being good? What did poor interviewers do or forget to do?

2. **Role-Play: Interviewing**
   Using the job description for a server that was presented in Chapter 5, work in groups of four to develop a series of interview questions for a part-time server position (Thursday through Saturday evenings) in an Italian restaurant serving pizza, pasta, and other Italian meals. When completed, have two students role-play an interview, with the two extra students acting as observers. When the first role-play is done, the observers will act as interviewer and interviewee. The role of the observers is to look for questions that are illegal and also to judge the ability of the interviewer to do a good job.

3. **Group Activity: Job Specifications**
   Working in groups of four, each group decides on a job classification, such as cook or housekeeper, for which they will write a job specification. Use the format in Figure 6.1.

ENDNOTES

1 https://www.google.com/search?q=us+unemployment+rate&ie=utf-8&oe=utf-8&aq=t&rls=org.mozilla:en-US:official&client=firefox-a&channel=sb. Retrieved November 20, 2014.

2 www.restaurant.org/Restaurant/media/Restaurant/SiteImages/News%20and%20Research/forecast/2014/EmploymentGrowth.jpg. Retrieved November 20, 2014.

3 Jim Sullivan, "Word to the Wise: Stop Hiring or Retaining People who Make Your Job as a Restaurant Operator Harder," *Nation's Restaurant News*, vol. 41, no. 10 (May 7, 2007), p. 22.

4 Matt Owens, Director of Human Resources, Suso Beach Resort, personal communication, October 16, 2014.

5 www.shrm.org/about/foundation/products/pages/recruitingandattractingtalent.aspx. Retrieved November 21, 2014.

6 Ibid.

7 https://www.recruiter.com/i/top-15-online-recruiting-resources-the-complete-guide/. Retrieved November 21, 2014.

8 Mary Bolz Chapman. "Do Green Efforts Aid?" *Chain Leader* (February 1, 2009). www.slideshare.net/gayu_vijay/green-recruiting. Retrieved November 21, 2014.

9 http://humanresources.about.com/od/glossaryj/g/job_specification.htm. Retrieved November 21, 2014.

10 Garry Colpitts, CEC, Culinary Manager, Culinary Innovation Laboratory, University of South Florida, Sarasota-Manatee, personal communication, November 12, 2014.

11 Leslie A. Weatherly, Selection Tests, SHRM Research. www.shrm.org/research/briefly_published/Employee%20Testing%20Series%20Part. Retrieved February 14, 2008.

12 http://jobsinpods.com. Retrieved November 22, 2014.

13 Charlotte Jordan, retired human resources director, Ritz-Carlton, personal communication, October 25, 2014.

14 Chris Chapman, HR Director, Sheraton Hotels, New York, personal communication, September 14, 2014.

15 http://connection.ebscohost.com/c/articles/12895024/looks-good-paper. Retrieved November 23, 2014.

16 Sullivan, p. 22.

17 Getting Hired Practice Questions. Hospitality Jobs Online. Retrieved November 23, 2014.

18 Katerina Ameral, presentation to USF HR class, November 4, 2014.

19 www.mtu.edu/equity/hiring/pdfs/checkingreferences.pdf. Retrieved November 23, 2014.

20 Ibid.

21 Robert Capwell, "Due Diligence in Screening: No Room for Shortcuts," Society for Human Resource Management, Staffing Management Library—Reference Checking, January 2008. https://www.psibackgroundcheck.com/Background-check-service.shtml. Retrieved November 23, 2014.

22 Ellen Koteff, "Lethal Weapon: Use Zero Tolerance Policies to Eradicate Prejudice in the Workplace," *Nation's Restaurant News*, vol. 40, no. 50 (December 11, 2006), p. 25.

Chapter 7

Performance Effectiveness

A s organizations strive for 100 percent effectiveness, they look to *performance management* as a complete system to not only set goals but also measure and evaluate the results. Performance management may be defined as translating goals into results.[1] Performance management may therefore encompass goal setting, employee selection and placement, performance appraisal, compensation, training and development, and career management; in other words, performance management focuses not only on individual employees, but also on teams, programs, processes, and the organization as a whole.[2]

Hospitality organizations, like other businesses, establish a vision, mission, and then *goals* and *strategies*, which are the tactics for how to meet the mission (discussed in earlier chapters). Given that a hospitality business will likely have guest satisfaction as part of the mission statement, a goal would be to increase guest satisfaction from 87 to 94 percent by December 31.

One of the most useful tools for sorting out this kind of confusion is a job description that incorporates performance standards. Once you start to grasp these concepts and learn how to use them, they will become some of the most useful devices in your entire supervisory repertoire.

In this chapter, we explain how to develop job descriptions and performance standards and examine their use in standardizing routine jobs. We then discuss performance appraisal.

After completion of this chapter, you should be able to:

- List the basic operating principles of a performance improvement plan.
- Explain why some performance standard systems succeed and others fail.
- Analyze the purposes and benefits of performance reviews.
- Describe the five key factors of a successful performance standard system.
- Compare the value of a daily employee evaluation to a formal, periodic employee review.
- Recall the factors that a supervisor must consider when performing employee evaluations.
- Name the appropriate actions that a supervisor must take after an employee's appraisal review. Discuss the potential legal issues surrounding employee evaluations.

Performance Improvement

LEARNING OBJECTIVE: List the basic operating principles of a performance improvement plan.

Performance improvement operates on the premise that to be effective, individuals and teams require the following:[3]

Direction about What the Organization Wants to Accomplish
- Consistent direction from the leadership
- Management processes that allow for efficient decision making

Clear Expectations
- A clear understanding of what is expected of them
- An explanation of the criteria that will be used to judge the adequacy of their work

The Equipment to Do the Job

- Training and information required for the job
- Performance support tools to help them remember how to do the job right and effectively

Information and Incentives to Keep Them on Track

- Feedback on how well they are doing and coaching on how to improve
- Rewards and incentives that support the behaviors that produce the results

If done properly and holistically, performance management can lead to excellent relations between managers and employees.

This chapter addresses these and other important performance issues, beginning with performance standards.

Performance Standards

LEARNING OBJECTIVE: Describe how performance standards can be used to develop a system of managing employees and their work.

performance standards
Codes of conduct that describe the what and how of a job, and explain what an employee is to do, how it is to be done, and to what extent.

unit of work
Any one of several work sequences that together form the content of a given job.

tasks
In job analysis, a procedural step in a unit of work.

Performance standards form the heart of the job description, and they describe the whats, how-tos, and how-wells of a job. Each performance standard states three things about each unit of the job: (1) what the employee is to do, (2) how it is to be done, and (3) to what extent it is to be done (how much, how well, how soon)

Traditionally, job descriptions have simply listed the duties and responsibilities (what the employee is to do) for each job. Although this approach is better than no approach, a job description using performance standards is much more useful, as is discussed shortly.

Here is an example of a performance standard for one unit of a server's job at a certain restaurant: "The server will take food and beverage orders for up to five tables with 100 percent accuracy, using standard house procedures." Figure 7.1 breaks this standard down to give you the structure of a performance standard. The "what" of the standard is the **unit of work**. The **tasks** become the "hows" that make up the standard procedure. When you add a performance goal for each unit, you set a performance standard: how much, how many, how good, how fast, how soon, how accurate—whatever it is that is important for establishing how well that unit of work should be done in your operation.

Supporting materials explaining or illustrating the specifics of the how (in this case, "standard house procedures") are necessary to complete each performance standard. They explain the action to be taken in order to reach the goal or standard.

Ken Blanchard, renowned management author, gives a great example of Jim, whose work performance is only "so-so" but who is the star of the Wednesday night bowling league. The story goes . . . imagine that you are Jim as he is about to bowl . . . and a curtain comes down so that he can't see where to bowl the ball. Sounds familiar? There are numerous associates who are not clear as to what they are aiming at—in other words, they have no clear goals.

So Jim knocks down six pins, but he doesn't know that because no one tells him—also sounds familiar. Well, thousands of associates carry on working without any feedback on how well they did. Suppose that a supervisor stood there and said, "Jim, you missed four pins." We would know how Jim would feel. But guess how many times this actually happens to associates? Wouldn't it be better to have clear goals—to knock down all 10 pins—and have immediate feedback and encouragement?[4]

Job descriptions are used often in recruiting, evaluating applicants, and training. They are also useful in assigning work, evaluating performance, and deciding on disciplinary action. In the next section we discuss the uses and benefits of performance standards in more detail.

ANATOMY OF A PERFORMANCE STANDARD

Job classification: Server (waiter/waitress)

Unit of work: Takes food and beverage orders

Performance standard:

The server will take food and beverage orders for up to five tables with 100 percent accuracy using standard house procedures.

Breakdown:

The server will:

What: *take food and beverage orders*
How: *using standard house procedures*
To what standard: *for up to five tables with 100 percent accuracy*

FIGURE 7.1: Anatomy of a performance standard.

What a Good Performance Standard System Can Do

LEARNING OBJECTIVE: Explain why some performance standard systems succeed and others fail.

If you develop a full set of performance standards for each job classification that you supervise, you have the basis for a management system for your people and the work they do. You can use them to describe the jobs, to define the day's work for each job, to train employees to meet standards, to evaluate employees' performance, and to give them feedback on how they are doing.

You can use performance standards as a basis for rewarding achievement and selecting people for promotion. You can use them as diagnostic tools to pinpoint ineffective performance and as a basis for corrective action. You can also use them in disciplining workers as a means of demonstrating incompetence. They provide the framework for a complete system of people management. This system operates successfully in many areas of supervisory responsibility.

Intelligent and consistent use of a performance standard system reduces or eliminates those five major reasons for low productivity and high turnover. Employees are told clearly what to do. They are taught how to do it. They know how well they are doing because there is an objective standard of measurement. The supervisor helps and supports them with additional training or coaching when standards are not being met. All this makes for much better relationships between workers and the supervisor.

Performance standards improve individual performance. When people are not given explicit instructions but are left to work out their own ways of getting their work done, they usually choose the easiest methods they can find. If this meets your standards, well and good, but often it does not. People also begin to find certain parts of their job more to their liking than other parts and will slack off on the parts they like least. Procedures and standards put all these things into the right perspective.

Once employees know what to do and how to do it, they can concentrate on improving their skills. Improved skills and knowledge, coupled with goals to be met, encourage people to work more independently. If a reward system is related to achievement—as it should be—people will respond with better and better work. Better and better work means better productivity, better guest service, more sales, and higher profits. Who could complain about that?

Morale benefits greatly. People feel secure when they know what to do and how to do it, and when their work is judged on the basis of job content and job performance. If they have participated in developing the objectives, they have a sense of pride and a commitment to seeing that the objectives work. Participation also contributes to their sense of belonging and their loyalty to the company.

A performance standard system can reduce conflict and misunderstanding. Everybody knows who is responsible for what. They know what parts of the job are most important. They know the level of performance the boss expects in each job. This reduces the likelihood that one person is doing less than another, who is being paid the same wage—often a cause for discontent and conflict.

Well-defined standards can eliminate problems caused by the overlapping of functions. In a restaurant, for example, the functions of busing and serving overlap. Who resets the tables after customers leave? The busperson and the server might try to do it and run into each other, or each of them might think it is the other's job and it does not get done, and the table is out of service while people are waiting to be seated.

At a hotel front desk, who will do the report for the housekeeper, the night auditor or the early morning desk clerk, and will there be two reports or none? In a hospital, who picks up all the dirty trays, the kitchen personnel or the aides on the floor, or do they sit in the patients' rooms all day? These gaps and overlaps will be eliminated in a performance standard system because the responsibility for performing the tasks is spelled out.

A critical area for standard procedures to be in place is for alcohol liability. Under the third-party liability of the Dram Shop legislation, if a guest becomes or is intoxicated at your establishment and goes out and causes an accident, then the server, manager, and owner are liable. Creating a responsible alcoholic beverage service program is, in itself, a powerful lawsuit defense. This legislation can cause serious problems for supervisors when other members of a party purchase drinks and then give them to the one who is or is close to being intoxicated.

Performance standards help ensure quality and consistency.
Goodluz/Shutterstock

✳ IN RECRUITING AND HIRING

The typical job description spells out in general terms the content of the job, the duties, and perhaps the kind of experience or skills desired. Performance standards, by contrast, clearly define the jobs and the duties, the methods of performing the duties, and the competencies required. This will help you as a supervisor to find the right people and to explain the jobs to prospective employees. It will also help in planning and forecasting personnel needs because you will know exactly what you can expect from each trained employee. If you are looking for experienced people, performance standards are helpful for testing skills.

When you select a new employee, you have a ready-made definition of the day's work for the job. You and your new worker can start off on the right foot with a clear understanding of what is to be done in return for a paycheck. It is a results-oriented approach to defining the job.

We discussed recruiting and selection more fully in Chapter 6.

✳ IN TRAINING AND IN EVALUATING PERFORMANCE

A complete set of performance standards gives you the blueprint for a training program. Each standard sets the competency goals for on-the-job performance toward which the training is guided. The training that forms the heart of a successful performance standard system begins with a written procedure.

A complete performance standard system should include periodic evaluations of each worker's performance, with feedback to the workers on how they are doing. Realistic and well-developed standards of performance form a solid basis for objective evaluation.

After evaluation, the supervisor is responsible for helping those who are working below standard to improve their performance. An evaluation system based on performance standards can pinpoint specific deficiencies needing corrective training. It is a positive approach; the focus is on the work, not the person; it does not put the person down. The problem is addressed and corrected, and everyone benefits. A performance standard evaluation system can also help you to identify superior workers by the way they meet or exceed the standards set. Such people merit your attention as candidates for development and promotion.

✳ IN YOUR JOB AND IN YOUR CAREER

A performance standard system will simplify your job as supervisor. Once it is in place and running, you will spend less of your time supervising because your people will be working more independently and things will run more smoothly of their own accord. You will have fewer misunderstandings, fewer mistakes in orders, fewer broken dishes, and fewer irate customers. You will have more time to spend in planning, training, thinking, observing, and improving product and method instead of managing on a crash-and-crisis basis.

After experiencing the standard-setting process, you will have a much better conceptual grasp of your own department, your own area of command, and everything that goes on. You will be able to better coordinate the various aspects of the work you are responsible for, be able to see how things can be better organized, and be able to run a tight ship. It will be a growth experience for you, and it will make you a better manager. The experience will stand you in good stead as you pursue your career, and so will the improved results in your department.

Setting Up a Performance Standard System

LEARNING OBJECTIVE: Analyze the purposes and benefits of performance reviews.

Developing a complete performance standard system is not something you can do overnight. There are a number of steps to the process, and there are certain essentials for success that must be included in the planning and operation.

Three essentials for successful operation must be built into the system from the beginning. The first is *employee participation*. The people who are currently working in a given job category should work with you as you analyze that job, set the standards for performance, develop the standard procedures, and determine a fair day's work. Employee input is very important to you. Often, employees know the job better than you do, particularly the procedural steps involved.

The give-and-take of discussion will often produce better results than one person working alone. In many cases your workers will set higher standards of performance than they would have accepted if you alone had set them. In the end, there must be mutual agreement between supervisor and workers on the procedures and the standards and the fair day's work, although the supervisor always has the final say.

Helping to hammer out the *whats, hows,* and *how-wells* will inevitably commit workers to the goals. They will work much harder for something they have helped to develop than for something handed down by the boss. The experience will make them feel recognized, needed, and important, as well as helping to build that sense of belonging that is so necessary to morale.

The second essential for a successful system is *active supervisory leadership and assistance throughout.* As supervisor, you will make the final decisions on the work units to be included and their relative importance. You will determine how much leeway to give your people in working out the procedures and standards of performance.

As the leader, you will be in charge at all times. But you will all work together as much as possible in identifying the units, specifying the methods and procedures, and setting the performance standards. Under your leadership, performance standards will represent a joint acceptance of the work to be done and responsibility for achieving it.

In training and on the job, the supervisor's leadership continues. Now your role is the supportive one of facilitating the learning of skills, giving feedback, and providing additional training as necessary. Frequent evaluations, whether formal appraisals or a "Hey, you're doing fine" must be an integral part of the system. If the supervisor neglects this aspect of the system, the entire system will soon deteriorate.

The third essential is a *built-in reward system* of some sort, with the rewards linked to how well each worker meets the performance standards. People who do not want to work hard must understand that the better shifts, promotions, and other rewards will go to those whose work meets or exceeds the standards set. There is a definite order of steps to be taken in developing a performance standard system. Figure 7.2 is a flowchart depicting the entire process. The next several pages will follow in detail each progression on the chart. As you read, you will find it helpful to refer back to the flowchart to see the relationship of each step to the whole process.

❋ DEFINING THE PURPOSE AND ANALYZING THE JOB

The first step is to define the purpose for which the standards are to be used. Our purpose here is to develop a system for one job classification that can be used to define a day's work, set standards, develop training plans, and evaluate on-the-job performance. A performance standard is to be developed for each unit of the work.

Once you have defined your purpose, your next step is to analyze the job and break it down into units. First, your employees can help to identify all the different work units they perform. When your list of units is complete, you and your crew should list in order of performance all the tasks or steps to be taken in completing that unit of work.

Once you have agreed on a list of tasks or procedural steps for each unit, you have the data for the first two parts of each performance standard that you are going to write. The unit is the *what* of the objective. The tasks become the *hows* that make up the standard procedures. When you add a standard for each unit, you have a complete performance standard.

The supervisor and the people working at the job should set the standards of performance together, as already discussed. Although the supervisor has the final say in the matter, it is critical to have the workers' input on the standard and their agreement that it is fair. If they don't think that it is fair, they will stop cooperating and your entire system will fail. They will let you succeed only to the degree that they want you to succeed.

Sometimes it is appropriate to define three **levels of performance**: an optimistic level, a realistic level, and a minimum level. An **optimistic level** is your secret dream of how a fantastic crew would do the work. A **realistic level** is your estimate of what constitutes a competent job and the

levels of performance
Employee performance measured against a performance standard—optimistic-level performance = superior, near perfection; realistic-level performance = competent; minimum-level performance = marginal—below which a worker should be terminated.

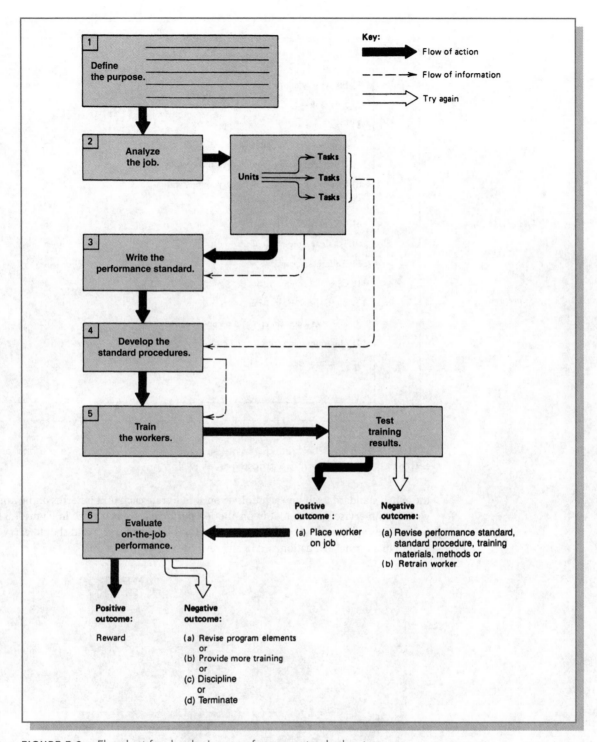

Key:
→ Flow of action
---→ Flow of information
⇨ Try again

1 Define the purpose.

2 Analyze the job.

Units → Tasks
→ Tasks
→ Tasks

3 Write the performance standard.

4 Develop the standard procedures.

5 Train the workers.

Test training results.

Positive outcome:
(a) Place worker on job

Negative outcome:
(a) Revise performance standard, standard procedure, training materials, methods or
(b) Retrain worker

6 Evaluate on-the-job performance.

Positive outcome:
Reward

Negative outcome:
(a) Revise program elements
or
(b) Provide more training
or
(c) Discipline
or
(d) Terminate

FIGURE 7.2: Flowchart for developing a performance standard system.

way that good steady workers are doing it now. A **minimum level** is rock bottom—if people did any less, you would fire them.

It is best to write your performance standards for a realistic level. A minimum level simply sets the standard at what a worker can get away with—and some of them will. An optimistic level is appropriate for the high achiever who is not challenged by a goal that is too easy. Achievement on this high level must be rewarded if you want that kind of effort to continue. When you have determined all the elements of each performance standard in a given job—the *what*, the *how*, and

SERVER UNIT RATINGS

1) Stocks service station	4 points
2) Sets tables	4
3) Greets guests	8
4) Explains menu to guests	8
5) Takes food and beverage orders and completes guest check	8
6) Picks up order and completes plate preparation	4
7) Serves food	6
8) Recommends wines and serves them	8
9) Totals and presents checks	8
10) Performs side work	4
11) Operated equipment	4
12) Meets dress and grooming standards	8
13) Observes sanitation procedures and requirements	8
14) Maintains good customer relations	10
15) Maintains desired check average	8
	100

FIGURE 7.3: Server unit ratings. Point values represent the importance of good performance in each unit of work (page from a procedures manual).

the actual standard itself for each unit—you should rate each unit in terms of the importance that you, as a supervisor, attach to it in on-the-job performance, as is done in Figure 7.3 for the job of server. This value scale should be made very clear to your servers and should carry considerable weight in a formal evaluation and in any reward system you set up.

Written performance standards are crucial to setting up a performance standard system.
auremar/Shutterstock

You may want to ask your servers for their ideas about relative importance, but the final decisions are your responsibility alone. You are the one with the management point of view and the company goals in mind. It will help your people if you explain clearly just why you rate the units as you do. In Figure 7.3, their relative importance is shown by assigning a point value to each unit. The rewards go to those people with the highest total points on their evaluation score.

❋ WRITING THE PERFORMANCE STANDARDS

Now we are ready for step 3 on the flowchart (Figure 7.4), writing the performance standards for each unit of the job. First, let us review the essential features of a performance standard. It is a concise statement made up of three elements that together describe the way a unit of work is to be carried out in a given operation: (1) what is to be done? (2) how is it to be done? and (3) to what extent (quality, quantity, accuracy, speed) is it to be done?

You can use the form in Figure 7.4 and simply fill in the blanks as we go. Let us take the first unit on the server list (Figure 7.3) and go through the process step by step. The first unit is *stock service station*. You make this abbreviated description more precise by limiting the scope of the work sequence: "The server will stock the service station for one serving area for one meal. . . ." Notice two things here:

1. You must use an action verb: the server will *do* something, will perform—not "be able to" or "know how to" or "understand," but actually do something. (The other phrases are used in objectives written for training purposes. Here we are writing objectives for day-in, day-out, on-the-job performance.) Use Figure 7.5 for help choosing a verb.

2. You limit the action as clearly and precisely as possible—which service station, what for. Limiting the action in this way makes it easier to measure performance.

Next, you define the how: ". . . as described in the Service Station Procedures Sheet . . ." or ". . . following standard house procedures. . . ." The standard simply states how or where this information is spelled out.

Finally, you state the standard of performance: ". . . completely and correctly in 10 minutes or less. . . ." That is, everything must be put in its assigned place within 10 minutes and nothing must be missing.

Now you can put together the whole performance standard: "The server will stock the service station for one serving area for one meal as described in the Service Station Procedures Sheet. The server will stock the station completely and correctly in 10 minutes or less."

Here are the requirements for the finished product: a good, useful, workable performance standard.

1. *The statement must be specific, clear, complete, and accurate.* It must tell the worker exactly what you want. Instructions cannot be vague so that it is not misinterpreted or misread. If it is not specific, clear, complete, and accurate, it can be more confusing than anything else.

2. *The standard of performance must be measurable or observable.* "Good" and "well" are not measurable or observable; they are subjective judgments. "Correct" and "accurate" are measurable *if there is something to measure by*—a set of instructions, a diagram, mathematical accuracy. The waiter delivers the order correctly if he serves the customer what the customer ordered. The bartender measures accurately if she pours 10 two-ounce martinis and 20 ounces are gone from the gin bottle. There are ways to measure these performances. There *must* be a measurable or observable way for the supervisor to tell whether a person is meeting the performance standard.

3. *The standard must be attainable.* It must be within the physical and mental capabilities of the workers and the conditions of the job. If nobody can meet it, expectations are unrealistic and you should reexamine the objective.

4. *The standard must conform to company policies, company goals, and applicable legal and moral constraints.* It must not require or imply any action that is legally or morally wrong (such as selling liquor to minors or misrepresenting ingredients or portion sizes).

WRITING A JOB DESCRIPTION

Job classification:

Unit of work:

What must be done?   (state the performance)

The employee will:

How is it to be done?   (the standard procedures are
where they are spelled out)

. . . according to . . . using . . . as shown in . . .

To what standard?   (how you measure it or what
you must be able to observe):

FIGURE 7.4:   Sample form to use in writing a job description.

5. *Certain kinds of standards must have a time limit set for achievement.* This applies to training objectives and performance improvement objectives, to be discussed shortly.

Performance standards are a specialized and demanding form of communication, and writing them may be the most difficult part of developing a performance standard system. But it is precisely this process that requires you to make things clear in your own mind. If you have problems

Anticipate	Evaluate	Operate
Calculate	Explain	Organize
Classify	Give examples	Pace
Construct	Identify	Predict
Define	Interpret	Recognize
Demonstrate	Label	See
Describe	Lead	Show
Diagram	List	Smell
Discuss	Locate	Solve
Display	Make	Use
Distinguish	Measure	Work
Estimate	Name	

FIGURE 7.5: Action verbs for performance standards.

with writing performance standards, don't worry—even experts do. But try it anyway. If it forces you to figure out just what you as a supervisor expect of your workers, you will have learned a tremendous lesson.

❋ DEVELOPING STANDARD PROCEDURES

The fourth step on the flowchart (Figure 7.6) is to develop standard procedures. Standard procedures complete each package and in many ways are the heart of the matter. The procedures state what a person must do to achieve the results—they give the instructions for the action. They tell the worker exactly how things are supposed to be done in your establishment. Spelling out, step by step, each task of each unit in a given job develops them. There might be many tasks involved—5, 10, 20, 100—whatever is necessary to describe precisely how to carry out that unit of the job.

The standard procedures have two functions. The first is to standardize the procedures you want your people to follow; the second is to provide a basis for training. You can use various means of presenting the how-to materials that make up each standard procedure: individual procedure sheets, pages in a procedural manual, diagrams, filmstrips, videotapes, slides, and photographs. It depends on what will be easiest to understand and what will best meet the requirements of the individual standard.

For stocking the service station, you might use a list of items and quantities along with a diagram showing how they are placed. For opening a wine bottle, you might have a DVD or a series of slides or pictures showing each step. For dress and grooming you would have a list of rules (Figure 7.6). The important thing is to have them in some form of accessible record so that they can be referred to in cases of doubt or disagreement and so that the trainers can train workers correctly. Show-and-tell is not enough in a performance standard system.

Two areas of caution: Don't get carried away with unnecessary detail (you don't need to specify that the menu must be presented right side up), and don't make rigid rules when there is a choice of how things can be done (there are many acceptable ways to greet a customer). You do not want your people to feel that they have tied themselves into a straitjacket in helping you to develop these procedure sheets. In fact, this entire process should free them to work more creatively.

You simply specify what must be done in a certain way and include everything that is likely to be done wrong when there are no established procedures. The rest should be left up to the person on the job as long as the work is done and the standards are met.

APPEARANCE AND GROOMING CHECKLIST

Immaculate cleanliness is required.
1. Personal hygine: Clean body: take a daily shower or bath, and use deodorant.
2. Personal hygine: Clean teeth and breath: brush your teeth often, use mouthwash after meals if brushing is impossible.
3. Clean hair: shampoo at least twice a week.
4. Clean hands and nails at all times: wash frequently.
5. Clean clothing, hose, and underwear daily.
6. Clean shoes: polish well, in good repair.

Grooming must be neat and in good taste.
7. Hair neatly styled.
8. Nails clean, clipped short, no polish.
9. Minimal makeup.
10. No strong scents from perfume, cologne, after-shave, soap, or hair spray.
11. No jewelry except wedding ring and/or service award pins.

Uniforms will be issued by hotel. You must care for them.
12. Clean, wrinkle free, in good repair at start of shift.
13. Hose required.
14. Closed-toe shoes, low heels.

Health and Posture
15. Report any sickness, cuts, burns, boils, and abrasions to your Supervisor immediately.
16. Stand straight, walk tall, and look confident!
17. Smile, you want to be noticed!

FIGURE 7.6: Standard procedures for the unit on appearance and grooming: 16 tasks must be completed to meet the objective (page from a procedures manual).

❋ TRAINING ASSOCIATES TO MEET THE PERFORMANCE STANDARDS

training objective
A trainer's goal: a statement, in performance standard terms, of the behavior that shows when training is complete.

Training is the fifth step in developing a performance standard system. A training program should have its own **training objective** for each standard. Each training objective will have a time limit added within which the worker must reach a required performance standard.

For example: "After 1 hour of training and practice, the trainee will be able to stock the service station for one serving area for one meal completely and correctly in 15 minutes or less following standard house procedures." You will notice that the performance time limit is changed from the previous example because this is a training goal and not an on-the-job goal. In training, the procedures form the basis of the training plans and the training itself.

At the end of the training period, the results of the training will be tested. In a new performance standard program this is a test of both the worker and the various elements of the program. If the results are positive, you can put the worker right into the job (or that part of the job for which training is complete). If the results are negative, you have to consider where the problem lies. Is it the worker? The standard? The procedures? The training itself? Something calls for corrective action.

❋ EVALUATING ON-THE-JOB PERFORMANCE

The final step in developing a performance standard system is to evaluate worker performance on the job using the performance standards that apply to that job. This first evaluation is a test of both the workers and the system so far. If a worker meets all the standards, the outcome is positive

Courtesy of Patrick Accord

Regardless of the industry, leadership is a key constant: key because the quality of leadership affects the bottom line as well as most of the operations required to reach that point; constant because you can be assured that there will always be someone looking over your shoulder. There are certain qualities and characteristics that define a good supervisor in any industry. My personal experience, based on work as a kitchen/chef, coupled with other experiences as an employee under supervision, has taught me that certain traits common to the most effective supervisors cross industry boundaries.

While supervising a kitchen, I learned quickly that personal communication is one of the main ingredients needed by an effective supervisor. Actually, I think I learned this while working under supervisors who weren't able to communicate effectively. There is nothing more disheartening for an employee than when he or she is blamed for something that was altered or lost in the channels of communication.

To minimize the chance that information is lost in the chain of command, shorten the chain. The most effective supervisors and managers that I've witnessed are the ones who are in the field with employees, talking with them and keeping lines of communication intact. There is no need to ask Cook 1 what Cook 2 has been doing when you can just ask Cook 2 himself.

Through hard work and communication, a supervisor can then begin to foster a team atmosphere. A supervisor shapes a group's atmosphere or company culture through his work ethic, attitude, and set regulations. The supervisor must set the standard for hard work, do it in an affable manner, and set rules for the benefit of the entire unit.

These rules might include a certain dress code or even a dress-down day. Functioning as a group in the workplace, so that the work can be accomplished efficiently, is the key to profitability. Hopefully, a supervisor who begins to foster a good atmosphere will be able to retain employees. Quality in almost any industry is synonymous with consistency. Two out of the four employees left during my first week of supervising the kitchen. Even though they had served notice and we started training in preparation, their loss set our consistency off for almost a month. Customers were complaining that their favorite recipes over the last several years tasted different and they weren't going to give us another chance. I stress the importance of consistency in everything that a supervisor does.

Obviously, good supervision is multidimensional and much more involved than the limited discussion here indicates. But I believe that the basic tenant of open lines of communication leads to good relationships, a good working environment, and retention of employees. Quality is consistency, and as we keep employee turnover at a minimum and build a valued and trusted staff, our jobs as supervisors become that much easier and more profitable.

and a reward is in order. A positive outcome is also an indication that your standards and procedures are suitable and workable.

If a worker rates below standard in one or more areas, you again have to diagnose the trouble. Is the standard too high? Are the procedures confusing, misleading, or impossible to carry out? Or is it the worker? If it is the worker, what corrective training does he or she need? If the worker is far below standard in everything, is there hope for improvement, or should the worker be terminated?

Implementing a Performance Standard System

LEARNING OBJECTIVE: Describe the five key factors of a successful performance standard system.

Once you have fine-tuned your system, you have a permanent set of instruments for describing jobs, defining a fair day's work for each job, training workers to your standards, evaluating performance, and rewarding achievement. How well can you expect it to work? It depends on five key factors.

❋ KEYS TO A SUCCESSFUL PERFORMANCE STANDARD SYSTEM

The first key to making your system work is the *workers' cooperation* in the developmental stage and their agreement to the standards of performance. On the one hand, if they have participated fully in developing them, they will participate fully in carrying them out. If, on the other hand, the development sessions were full of wrangling, bargaining, and manipulation, and in the end you more or less forced your people to agree to your decisions, they will find ways to sabotage the system. They will also be resentful and uncooperative if they are required to put in time and work in addition to their regular duties and hours without extra compensation or reward.

The second key to success is to *put the system to work slowly* over a period of time, one job at a time. It cannot be done in a day or a week or a month. A performance standard system is a total management system, and it takes a great deal of time to develop it and put it in place. It takes a long time to develop good standards, to standardize the procedures, to translate the standards and procedures into training programs, and to train your people to meet the performance standards. It takes total commitment to the system, and if you do not have that commitment it will never work for you.

The third key to success is *an award or incentive system.* This is something you work out alone, since you are the only one who knows what you have to offer. It could be money: a bonus, a prize, a pay raise, a promotion. But it does not have to be money; it could be a better shift, an extra day off, a better serving area, a bottle of champagne, or a certificate of merit displayed for all to see. Whatever it is, it is important that all your people understand what the rewards are for and how they are allotted, that they feel the system is fair, and that you practice it consistently.

The fourth key to success is to *recognize your employees' potential* and use it as fully as you can within the limits of your authority. That means empowering your staff, investing a chef's time to educate his or her staff, acknowledging the staff's good points and contributions to the organization. Performance standards tend to uncover talent that has been hidden under day-after-day drudgery. Numerous surveys have shown that many people in the hospitality industry are truly underemployed.

If they are encouraged to become more productive, to take more responsibility, to learn new skills, you will get a higher return. Human assets are the most underutilized assets in the hospitality industry today. A performance standard system gives you new ways to capitalize on them. Better products, better service, more customers—who knows how far you can go?

The fifth key to continued success is to *review your system periodically,* evaluating and updating and modifying if your ways of doing things have changed. For example, allow your staff to have input in updating the system. You may have changed your menu or your wine list. Have you also changed the list of what wine goes with what food? You may have put in some new pieces of equipment. Have you adapted your procedures to include training the workers to use them properly?

If you do not keep your materials up to date, if you begin to let them slide, you may begin to let other things slide, too—the training, the evaluations, the reward system. It will run by itself for a time, but not indefinitely. It works best when everyone is actively involved in maintaining it.

❋ HOW A PERFORMANCE STANDARD SYSTEM CAN FAIL

Performance standards do not work everywhere. Good, clear, accurate, understandable standards are often hard to write unless you or one of your workers is good at putting words together. (This may be one of those hidden talents that the process uncovers.) *If the standards are not clearly stated and clearly communicated to everyone, they can cause confusion instead of getting rid of it.* The objectives are communications tools, and if they do not communicate well—if the people do not understand them—the program will never get off the ground.

The supervisor can cause the system to fail in several ways. The worst thing that you can do is to change standards without telling your people. You just do not change the rules of the game while you are playing it, especially without telling anyone. You can make changes—often you have to—but you have to keep your people informed, especially when such critical matters as evaluations and rewards are at stake.

Another way in which the supervisor can bring about the failure of the system is to neglect its various follow-up elements. It is especially important to help your people attain and maintain the performance standards you and they have set—to correct underperformance through additional instruction and training, and to do this in a positive, supportive way rather than criticizing or scolding. You must help, and you must maintain a helping attitude.

If you neglect the follow-up elements—if you do not help underperformers, if you fail to carry out a consistent reward system, if you do not recognize superior achievement and creativity, if you do not analyze individual failures and learn from them—all these things can make a system die of neglect. *Similarly, it will die if your people find no challenge or reward in the system—if the goals are too low to stimulate effort, if the supervisor is hovering all the time "evaluating," or if for some reason the system has not succeeded in putting people on their own.*

What it often comes down to is that if the supervisor believes in the system and wants to make it work, it will, bringing all its benefits with it. If the supervisor is half-hearted, you will have a half-baked system that will fail of its own dead weight.

Sometimes a supervisor can become so preoccupied with maintaining the system that the system will take over and become a straitjacket that prevents healthy change in response to new ideas and changing circumstances. This happens at times in large organizations where the dead weight of routine and paperwork stifles vitality and creativity. It can also happen with a rigid, high-control supervisor whose management style leans heavily on enforcing rules and regulations. A performance standard system should not lock people in; it should change and improve in response to changes in the work and the needs of the workers.

Sometimes the system is administered in a negative way: "You didn't meet your objectives." "You won't get a raise." "You're gonna be fired if you don't meet these standards." People can experience it as a whip or a club rather than as a challenge, and that is the end of its usefulness. This is not the fault of the system, however, but of the way in which it is administered. Truly, the supervisor is the key to success.

At the start of this chapter, we identified two parts, one to design a performance rating system and the second to design and conduct the performance appraisal. The remainder of the chapter looks at the second part, designing and conducting the performance appraisal.

Essentials of Performance Evaluation

LEARNING OBJECTIVE: Compare the value of a daily employee evaluation to a formal, periodic employee review.

performance evaluation, performance appraisal, performance review
Periodic review and assessment of an employee's performance during a given period.

In management terms, the phrase **performance evaluation** refers to a periodic review and assessment of each employee's performance during a given period: three months, six months, a year, or a certain number of hours worked. The assessment is recorded, usually on a company rating form, and is then discussed with the employee in an interview that answers the perennial question, "How am I doing?" and explores the possibilities for improvement.

Other terms used for this process are **performance appraisal** and **performance review**. We use these terms to distinguish the system of periodic evaluation from the informal performance evaluation that is a daily part of the supervisor's job.

Performance reviews are not always used for hourly workers in the hospitality industry. This is partly because supervisors are busy, partly because many workers do not stay long enough to be evaluated, and partly because many operations are under the immediate direction of the owner. But the practice is increasing, especially in chain operations. It is part of their general thrust toward maintaining consistency of product and service, improving quality and productivity, and developing the human resources of the organization.

A performance review does not substitute in any way for the informal evaluations you make in checking on work in progress. Where things happen so fast, where so many people are involved and so much is at stake in customer satisfaction, you cannot just train your people, turn them loose, and evaluate their performance six months later. You must be on the scene every day to see how they are doing, who is not doing well, and how you can help those who are not measuring up.

This is an informal blend of evaluation and on-the-job coaching and support to maintain or improve performance right now and to let people know when they are doing a good job. Performance reviews every six months or so cannot substitute for it. Feedback must be immediate to be effective.

In fact, if you had to choose between periodic reviews and daily evaluations, the daily evaluations would win hands down. But it isn't a choice; one complements the other.

✻ PURPOSE AND BENEFITS

If you are evaluating people every day, why do you need a performance review? There are five good reasons or purposes:

1. *In your day-to-day evaluations you tend to concentrate on the people who need to improve,* the people you have trouble with, the squeaky wheels who drive you crazy. You may also watch the outstanding performers because they make you look good and because you are interested in keeping them happy and in developing them. But you seldom pay attention to the middle-of-the-road people. They come in every day, they are never late, they do their work, they don't cause any problems, but they never get any recognition because they do not stand out in any way. Yet they really are the backbone of the entire operation and they ought to be recognized. Everybody who is performing satisfactorily should be recognized. In a performance review you evaluate everybody, so you will notice these people and give them the recognition they deserve.

2. *Looking back over a period of time gives you a different perspective.* You can see how people have improved. You will also look at how they do the entire job and not just the parts they do poorly or very well. You evaluate their total performance.

3. *A performance review is for the record.* It is made in writing, and other people—the personnel department, your own supervisor, someone in another department looking for a person to fill another job—may use it. It may be used as data in a disciplinary action or in defending a discrimination case. It may be and should be used as a basis for recognition and reward.

4. A performance review requires you to get together with each worker to discuss the results. *It lets people know how well they are doing.* You may forget to tell them day by day, but you cannot escape it in a scheduled review. And if you know you will have to do ratings and interviews at evaluation time, you may pay more attention to people's performances day by day.

5. *A performance review not only looks backward, it looks ahead.* It is an opportunity to plan how the coming period can be used to improve performance and solve work problems. It is a chance for setting improvement goals, and if you involve the worker in the goal setting, it increases that person's commitment to improve. The improvement goals then become a subject for review at the next appraisal, giving the entire procedure meaningful continuity.

Performance reviews have many uses beyond their primary concern with evaluating and improving performance. *One is to act as the basis for an employee's salary increase.* This type of salary increase is called a **merit raise** and is based on the employee's level of performance. For example, an employee who gets an outstanding evaluation may receive a 6 percent increase, the employee who gets a satisfactory evaluation may receive a 4 percent increase, and the employee with an unsatisfactory evaluation may not receive any increase. In one survey of U.S. businesses, 75 percent of respondents reported using appraisals to determine an employee's raise.

Another use is to identify workers with potential for advancement: people you can develop to take over some of your responsibilities, people you might groom to take your place someday or recommend for a better-paying job in another department. As you know, managers have an obligation to develop their people, and a performance review is one tool for identifying people capable of doing more than they are doing now.

merit raise
A raise given to an employee based on how well the employee has done his or her job.

Other managers may use your performance reviews. Since they are a matter of record, others may use them to look for people to fill vacancies in their departments. If they are going to be used this way, it is important that you make your evaluations as accurate and objective as you can. (It is important anyway—more on this subject later.) If someone has been promoted on the basis of your inaccurate evaluation and the promotion does not work out, you may be in hot water.

Your boss, to rate your own performance as a supervisor, may use your performance reviews. If the records show that most of your workers are poor performers, this might indicate that you are not a very good supervisor.

Performance reviews can *provide feedback on your hiring and training procedures.* When workers turn out to lack skills they should have been trained in, your training procedures might be inadequate. Workers you hired who rate poorly in every respect reflect on your hiring practices. Both indicate areas for improvement on your part. (Good selection and training programs are discussed in Chapters 4 and 8.)

Workers who rate poorly across the board are of special concern and are possible candidates for termination. Performance evaluations can help to identify such workers. If they do not respond to attempts to coach and retrain, their performance evaluations should document inadequacies to support termination and help protect your employer from discrimination charges.

Finally, *performance reviews provide the occasion for supervisors to get feedback from employees about how they feel about their job, the company, and the way they are treated.* Supervisors who are skilled interviewers and have good relationships and open communication with their people may be able to elicit this kind of response.

It takes a genuine interest plus specific questions such as, "How can I help you to be more effective at your job?" or "Are there problems about the work that I can help you solve?" Many people will hesitate to express anything negative for fear that it will influence the boss to give them a lower rating, but questions with a positive, helpful thrust can open up some problem areas.

When carried out conscientiously and when there is constant communication between reviews, performance reviews have many benefits. They help to maintain performance standards. By telling workers how they are doing, they can remove uncertainty and improve morale. By spotlighting areas for improvement, they can focus the efforts of both worker and supervisor to bring improvement about.

They can increase motivation to perform well. They provide the opportunity for improving communication and relationships between supervisor and worker. They can identify workers with unused potential and workers who ought to be terminated. They can give feedback on supervisory performance and uncover problems that are getting in the way of the work.

All these things have great potential for improving productivity, the work climate, and person-to-person relations. And they all benefit the customer, the company image as an employer, and the bottom line.

❋ STEPS IN THE PROCESS

A performance review is a two-part process: making the evaluation and sharing it with the worker. There should also be a preparation phase in which both supervisors and workers become familiar with the process, and there should be follow-up to put the findings to work on the job. In all, there are four steps: (1) preparing for evaluation, (2) making the evaluation, (3) sharing it with the worker, and (4) providing follow-up.

Companies that use performance review systems usually give supervisors some initial training. They are told why the evaluation is important and what it will be used for: promotions, raises, further employee training, whatever objectives the company has. They are given instruction in how to use the form, how to evaluate performance fairly and objectively, and how to conduct an appraisal interview. This initiation may take the form of a briefing by the supervisor's boss, or it may be part of a companywide training program. It depends on the company.

When carried out conscientiously and when there is constant communication between reviews, performance reviews have many benefits.
CandyBox Images/Shutterstock

The people being evaluated should also be prepared. They should know from the beginning that performance reviews are part of the job. Good times to mention it are in the employment interview, in orientation, and especially during training, when you can point out that they will be evaluated at review time on what they are being trained to do now. Showing people the evaluation form at this point can reinforce interest in training and spark the desire to perform well.

People must also know in advance when performance reviews will take place, and they must understand the basis for evaluation. They should be assured that they will see the completed evaluation, that they and the supervisor will discuss it together, and that they will have a chance to challenge ratings they consider unfair.

Making the Evaluation

LEARNING OBJECTIVE: Recall the factors that a supervisor must consider when performing employee evaluations.

evaluation form
A form on which employee performance during a given period is rated.

performance dimensions or categories
The dimensions of job performance chosen to be evaluated, such as attendance and guest relations.

The performance evaluation is typically formalized in an evaluation form that the supervisor fills out. There are probably as many different forms as there are companies that do performance reviews, but all have certain elements in common. Figures 7.7 and 7.8 are sample evaluation forms.

✳ PERFORMANCE DIMENSIONS

An evaluation form typically lists the performance dimensions or categories on which each worker is to be rated. Examples include the quality and quantity of the work itself, attendance, appearance, work habits such as neatness or safety, and customer relations. The dimensions of job performance chosen for an evaluation form should be related to the job being evaluated and clearly defined in objective and observable terms, as in a performance standard.

Many evaluation forms go beyond specific performance to include such personal qualities as attitude, dependability, initiative, adaptability, loyalty, and cooperation. Such terms immediately invite personal opinion; in fact, it is hard to evaluate personal qualities in any other way. The words mean different things to different people, the qualities are not in themselves measurable, and they do not lend themselves to objective standards. Yet some of these qualities may be important in job performance. Some evaluation forms solve the problem by defining the qualities in observable, job-related terms. For instance, dependability can be defined as "comes to work on time."

Some qualities that are pleasant to have in people who work for you are not really relevant to doing a certain job. A dish machine operator does not have to be "adaptable" to run the dish machine. A "cooperative" bartender may cooperate with customers by serving them free drinks. "Initiative" may lead people to mix in areas where they have no authority or competence or to depart from standard procedures (change recipes or portion size).

Sometimes the personal qualities found on company evaluation forms are included because they are important in assessing potential for advancement. But where you are concerned only with evaluation of the performance of routine duties, it is not appropriate to include such qualities in an overall evaluation on which rewards may be based.

People who polish silver or wash lettuce should not be penalized for lacking initiative. In such cases, the question can be answered "NA," not applicable. Concern with promotions should not be allowed to distort an evaluation system intended primarily for other purposes.

❄ PERFORMANCE STANDARDS

A quality evaluation form defines each performance dimension in measurable or observable terms by using performance standards (Figure 7.8). There should be standards, measurable or observable, wherever possible to make evaluation more objective.

Unfortunately, subjective evaluations are not legally defensible if an employee ever takes you to court for matters such as employment discrimination. To be legally defensible, your evaluation of job performance should be based on measurable and objective performance standards that are communicated to employees in advance.

On the face of it, an evaluation based on performance standards may look intricate and difficult to carry out. But supervisors who have used performance standards in training and in informal day-by-day evaluations find them to be a very simple way to rate performance. Usually, they don't have to test people; they recognize performance levels from experience.

Probably not many organizations have such a system, and not all jobs lend themselves to this kind of evaluation. It is best suited to jobs where the work is repetitive and many people are doing the same job, a situation very common in hospitality operations.

❄ PERFORMANCE RATINGS

rating scale
A system, usually a scale, for evaluating actual performance in relation to expected performance or the performance of others.

Many evaluation forms use a **rating scale** ranging from outstanding to unsatisfactory performance. A common scale includes ratings of outstanding, above average, average, needs improvement, and poor. In the case of performance standards, you can simply check off that the employee either meets or does not meet the standard. In some systems, there is also a category for "exceeds standard" (Figure 7.9).

The major problem with ratings such as outstanding or excellent is figuring out what they mean in performance terms. What constitutes excellent? What is the difference between fair and poor? If there is no definition, the ratings will be entirely subjective and may vary greatly from one supervisor to another. Where raises and promotions are involved, the results are not always fair to everyone. And nothing bugs employees more than seeing an employee who puts in half the amount of work they do receive the same raise as everyone else.

Some forms take pains to describe what is excellent performance, what is average, and so on. The more precise these descriptions are, the fairer and more objective the ratings will be. In some cases, point values are assigned to each performance dimension (Figure 7.8), indicating its relative importance to the job as a whole. These point values add up to 100 percent, the total job. A different set of point values is used to weight each level of performance (3 points for superior, 2 for competent, 1 for minimum, 0 for below minimum).

After evaluating each item, you multiply the point value by the performance level. Then you add up the products to give you an overall performance rating. This will provide a score for each

UNIVERSITY OF MAINE SYSTEM

Hourly Employee Performance Review

Employee Name and ID	Job Title	Department	Position Review Date
Assessment Period From To	Date of Assessment Meeting	Supervisor	Type of Assessment Probationary Interim Annual

PERFORMANCE RATING CATEGORIES

Outstanding*	Exceptional performance, exceeds all performance expectations for this factor, contributes significantly to organizational effectiveness and efficiency
Commendable	Performs beyond normal requirements and compe tence
Effective	Fulfills the normal job requirements with some strong points
Needs Improvement**	Fails to meet one or more of the significant performance expectations for this factor , performance must improve to be acceptable
Unsatisfactory**	Performance must improve substantially to be acceptable

*   Ratings of Outstanding should be accompanied by a comment or example.

** Ratings of Need Improvement or Unsatisfactory require a comment or example, and develop of a Performance Improvement Plan (PIP) is recommended to help the employee achieve satisfactory performance. Information about developing a Performance Improvement Plan is available at your campus Human Resources office.

PERFORMANCE FACTORS	PERFORMANCE EXPECTATIONS: COMMENTS OR EXAMPLES (attached extra sheets if needed)	RATING
QUALITY OF WORK Competence, accuracy, neatness, thoroughness, safety		Outstanding Commendable Effective Needs Improvement Unsatisfactory
QUANTITY OF WORK Use of time, volume of work accomplished, ability to meet schedules, productivity		Outstanding Commendable Effective Needs Improvement Unsatisfactory
JOB KNOWLEDGE Degree of technical knowledge, understanding of job procedures, methods , use of tools and technology		Outstanding Commendable Effective Needs Improvement Unsatisfactory
WORKING RELATIONSHIPS Cooperation, dependability and ability to work with supervisor, co-workers, students and customers		Outstanding Commendable Effective Needs Improvement Unsatisfactory
SUPERVISORY SKILLS-Indicate whether employees supervised are regular, temporary and/or students Training and directing employees, delegating, evaluating employees, planning and organizing employees' work		Not Applicable Outstanding Commendable Effective Needs Improvement Unsatisfactory
ORGANIZATION AND COMMUNICATION SKILLS Planning and organizing own work, problem solving, decision making, following directions, communication skills, judgment, adaptability to change		Outstanding Commendable Effective Needs Improvement Unsatisfactory
ATTENDANCE -Do not consider approved leaves.		Outstanding Commendable Effective Needs Improvement Unsatisfactory

FIGURE 7.7: An hourly position performance appraisal review form.

Courtesy of the University of Maine.

1. **Specific Achievements**

2. **Performance Goals For The Past Year**

Goals and Accomplishments	Were Goals Achieved?	Reason if Not Achieved

3. **Performance Goals For The Next Year**

Goals	Steps to Achieve Goals	Target Date

4. **Training and Development Recommendations** (may be directly related to the employee's current job or to help the employee develop skills for advancement)

Overall Assessment: The employee's overall job performance for the assessment period is **Satisfactory** ☐ **Unsatisfactory** ☐
If unsatisfactory is selected, please enclose documentation and a PIP (Performance Improvement Plan) or a PIP that is already in progress.

Supervisor's Name/Supervisor's Title	Signature	Date

Supervisor's Comments? (optional)

Employee's Comments? (optional)

For Employees: In what ways can your supervisor help in your job performance and/or career development?

I have reviewed this performance review with my supervisor and have received a copy. I understand that a copy will be placed in my personnel file and that if I do not sign and return the review within seven (7) days, an unsigned copy will go in the personnel file. My signature does not necessarily indicate agreement with the assessment. I understand that I may attach a written response or comments to this assessment.

Employee's Name	Signature	Date
Department Head's Name/Title	Signature	Date

FIGURE 7.7: *(continued)*

Performance Standards	Exceeds	Meets	Does Not Meet
12. Meets at all times the Dress Code requirements. Comments:	———	———	———
13. Uses at all times the sanitation procedures specified for serving personnel in the Sanitation Manual; maintains work area to score 90 percent or higher on the Sanitation Checklist. Comments:	———	———	———
14. Maintains a "Good" or higher rating on the Customer Relations Checklist; maintains a customer complaint ratio of less than 1 per 200 customers served. Comments:			
15. Maintains a check average of not less than $7 per person at lunch and $15 per person at dinner. Comments:	———	———	———
16. Is absent from work less than 12 days in a year. Comments:	———	———	———
17. Is late to work less than 12 times in a year. Comments:	———	———	———
18. Can always be found in work area during work hours or supervisor knows where he or she is. Comments:	———	———	———
19. Attends or makes up all required meetings and training. Comments:	———	———	———
20. Supervisor receives positive feedback from peers with minimal complaints. Comments:	———	———	———

OVERALL RATING:
Outstanding Performance: 75–100 points (must meet or exceed all standards)
Good Performance: 50–74 points
Marginal Performance, Reevaluate in 60 Days: Below 50 points

EVALUATOR'S COMMENTS: _____
_____
_____

EMPLOYEE'S COMMENTS: Please comment freely on this evaluation.
_____
_____
_____

EMPLOYEE'S OBJECTIVES: What would you like to accomplish in the next 12 months? _____
_____
_____

EMPLOYEE'S OBJECTIVES FOR THE NEXT 12 MONTHS:
(Plan should be specific, realistic, measurable, and include target dates.)
_____
_____
_____

SIGNATURES:

_____ _____ _____
Employee   Evaluator   Reviewer
Date: _____

SERVER PERFORMANCE EVALUATION

Name: _____
Position: _____
Date of Hire: _____ Yearly or 60-Day Evaluation: _____
Department: _____

Please use COMMENT section whenever "Exceeds" or "Does Not Meet" is checked. POINTS: Exceeds—5, Meets—3, Does not meet—0.

Performance Standards	Exceeds	Meets	Does Not Meet
1. Stocks the service station for one serving area for one meal completely and correctly, as specified on the Service Station Procedures Sheet, in 10 minutes or less. Comments:	———	———	———
2. Sets or resets a table properly, as shown on the Table Setting Layout Sheet, in not more than 3 minutes. Comments:	———	———	———
3. Greets guests cordially within 5 minutes after they are seated and takes their order if time permits; if too busy, informs them that he or she will be back as soon as possible. Comments:	———	———	———
4. Explains menu to customers: accurately describes the day's specials and, if asked, accurately answers any questions on portion size, ingredients, taste, and preparation method. Comments:	———	———	———

Performance Standards	Exceeds	Meets	Does Not Meet
5. Takes food, wine, and beverage orders accurately and legibly for a table of up to six guests according to Guest Check Procedures; prices and totals check with 100 percent accuracy. Comments:	———	———	———
6. Picks up order and completes plate preparation according to Plate Preparation Procedure. Comments:	———	———	———
7. Serves a complete meal to all persons at each table in an assigned station in not more than 1 hour per table using the Tray Service Procedures. Comments:	———	———	———
8. If asked, recommends wines appropriate to menu items selected, according to the What Wine Goes with What Food Sheet; opens and serves wines correctly as shown on the Wine Service Sheet. Comments:	———	———	———
9. Accepts and processes payment with 100 percent accuracy as specified on the Check Payment Procedures Sheet. Comments:	———	———	———
10. Performs side work correctly according to the Side Work Assignments Sheet and as requested. Comments:	———	———	———
11. Operates all equipment in assigned area according to the Safety Manual. Comments:	———	———	———

FIGURE 7.8: Performance appraisal form based on performance standards.

Performance (abbreviated here)	Point Value		Performance Level*		Overall Evaluation**
1. Stocks service station	4	×	2	=	8
2. Sets/resets a table properly	4	×	2	=	8
3. Greets guests	8	×	3	=	24
4. Explains menu	8	×	3	=	24
5. Takes orders	8	×	3	=	24
6. Picks up and completes order	4	×	2	=	8
7. Serves meal	6	×	3	=	18
8. Recommends and serves wines	8	×	1	=	8
9. Totals and presents check	8	×	3	=	24
10. Performs side work	4	×	0	=	0
11. Operates equipment	4	×	2	=	8
12. Meets dress and grooming standards	8	×	3	=	24
13. Observes sanitation procedures	8	×	2	=	16
14. Maintains good customer relations	10	×	3	=	30
15. Maintains check average	8	×	3	=	24
	100				248

*Superior = 3	**Overall rating of 300 = outstanding: highest reward
Competent = 2	250–300 = superior: middle reward
Minimal = 1	200–250 = competent: minimum reward
Below minimum = 0	100–200 = improvement needed
	100 = marginal
	below 100 = hopelessly inadequate

FIGURE 7.9: Performance dimensions rated using point values.

person based entirely on performance; raises can then be based on point scores. This system allows you to rate performance quality in different jobs by the same standard—a great advantage.

Perhaps the most valuable feature of this rating method is that it pinpoints that part of the job the employee is not doing well and indicates how important that part is to the whole. It gives you a focus for your discussions with the worker in the appraisal interview, and it shows clearly where improvement must take place.

The evaluation form shown in Figure 7.7 achieves a good balance in what it asks of a supervisor. It is simple, yet it requires a fair amount of thought. The required ratings provide both a means of assessing excellence for reward purposes and a way of determining where improvement is needed. The evaluation form features another element often used in the overall review process: improvement objectives. Each evaluation considers how well past objectives have been met and sets new objectives for the upcoming period. This tends to emphasize the ongoing character of the performance review process rather than a report card image.

No evaluation form solves all the problems of fairness and objectivity. Probably those that come closest are designed exclusively for hourly workers, for specific jobs, and for evaluating performance rather than promotability. Some experts suggest that a single form cannot fulfill all the different purposes for which performance reviews are used, and that questions needed for making decisions on promotion and pay be eliminated where reviews are used primarily for feedback, improvement, and problem solving.

The form you encounter as a supervisor will probably be one developed by your company. Whatever its format and its questions, its usefulness will depend on how carefully you fill it out. You can make any form into a useful instrument if you complete it thoughtfully and honestly for each person you supervise.

If your company doesn't have an evaluation system, you can develop your own forms, tailored to the jobs you supervise. If they evaluate performance rather than people and are as objective as you can make them, they will serve all the basic purposes of performance review on the supervisory level: feedback, improvement, incentive, reward, and open communication between you and your people.

❋ PITFALLS IN RATING EMPLOYEE PERFORMANCE

Whatever form or system you use, evaluating performance consists of putting on paper your ratings of each person's work over the period since hiring or since the last performance review. It is based on your day-to-day observations plus relevant records such as attendance records. No matter how well you think you know an employee's work, the process demands thought and reflection and a concentrated effort to be fair and objective.

There can be many pitfalls on the way to objectivity. *We have noted how the form itself* may in some cases encourage subjective judgments. Another pitfall is the *halo effect* in the selection interview. Something outstanding, either positive or negative, may color your judgment of the rest of a person's performance.

Kevin may sell more wine than anyone else, so you might miss that he comes in late every day. Sharon broke a whole tray of glasses her first day on the job, so you don't even notice that she has not broken anything since and that her check average has risen steadily.

Letting your feelings about a person bias your judgment is another easy mistake. If you don't like someone, you see their mistakes and forget about their achievements. If you like someone, you reverse the process.

Comparing one person with another is another trap. If John were as good as Paul, if he were even half as good as Paul . . . But Paul really has absolutely nothing to do with John. You have to compare John's performance with the job standards, not John with Paul.

Sometimes supervisors' feelings about the entire evaluation process will affect their ratings. They may be impatient with the time that evaluations take and the cost of taking people away from work for interviews. Even supervisors who believe in evaluation and practice it informally all the time often resent putting pencil to paper (and the interviews, too) as an intrusion into their busy days.

Some supervisors do not take evaluations seriously and simply go through the motions. Some are really not familiar with the details of their employees' work. Some simply hate paperwork and feel that daily informal evaluation and feedback are enough.

Some supervisors let concern about the consequences influence their ratings. They may fear losing good workers through promotions if they rate them high. They may fear worker anger and reprisals if they rate them low. They may not want to be held responsible or take the consequences of being honest, so they rate everybody average. That way they do not have to make decisions or face the anger of people they have rated negatively, and nobody is going to argue.

Procrastination is another pitfall. Some people postpone ratings until the last minute on grounds of the press of "more important" work. Then the day before evaluations are due they work overtime and rush through the evaluation forms of 45 people in 45 minutes. Obviously, this is not going to be a thoughtful, objective job.

Another pitfall is the *temptation to give ratings for the effect they will have.* If you want to encourage a worker, you might give her higher ratings than her performance warrants. If you want to get rid of somebody, you might rate him low—or recommend him for promotion. If you want to impress your boss, you might rate everybody high to show what a good supervisor you are.

If you are a *perfectionist*, and few employees measure up to your standards, you might rate everyone poorly. Another pitfall occurs when you *rate employees on their most recent performance* because you kept insufficient documentation of their past performance. This often results in vague, general statements based only on recent observations. This can upset the employee, especially if earlier incidents of outstanding performance are forgotten.

Sooner or later, false ratings will catch up with you. Unfair or wishy-washy evaluations are likely to backfire. You are not going to make a good impression with your superiors, and you will lose the respect of your workers. Such evaluations tend to sabotage the entire evaluation system, the value of which lies in accuracy and fairness.

The defense against such pitfalls and cop-outs lies in the supervisor's own attitude. You can never eliminate subjectivity entirely, even by measuring everything. But you can be aware of your own blind spots and prejudices, and you can go over your ratings a second time to make sure that they represent your best efforts. You can make the effort needed to do a good job. You can also do the following:

- Evaluate the performance, not the employee. Be objective. Avoid subjective statements.

- Give specific examples of performance to back up ratings. Use your supervisor's log or other documentation to keep a continual record of past performance so that doing evaluations is easier and more accurate.

- Where there is substandard performance, ask why. Use the rule of finger, which means looking closely at yourself before blaming the employee. Perhaps the employee was not given enough training or the appropriate tools to do the job.

- Think fairness and consistency when evaluating performance. Ask yourself, "If this were my review, how would I react?"

- Get input from others who have a working relationship with the employee. Write down ideas to discuss with the employee on how to improve performance.

- If you set out to be honest and fair, you probably will be. If you keep in mind what evaluations are for and how they can help your workers, you will tend to drop personal feelings and ulterior motives and to see things as they are.

❋ EMPLOYEE SELF-APPRAISAL

employee self-appraisal
A procedure by which employees evaluate their own performance, usually as part of a performance appraisal process.

As part of some performance appraisal systems, some employees are asked to fill out the appraisal and evaluate themselves. **Employee self-appraisal** is surprisingly accurate. Many employees tend to underrate themselves, particularly the better employees; less effective employees may overrate themselves. If given the chance to participate, and the manager takes the self-appraisal seriously, the employee gets the message that his or her opinion matters.

This may result in less employee defensiveness and a more constructive performance appraisal. It may also improve motivation and job performance. Self-appraisal also helps put employees at ease as they know what will be discussed during the appraisal. Employees may also tell you about skills they have or tasks they have accomplished that you have forgotten. Self-appraisal is particularly justified when an employee works largely without supervision.

The Appraisal Interview

appraisal interview evaluation interview, appraisal review
An interview in which a supervisor and an employee discuss the supervisor's evaluation of the employee's performance.

LEARNING OBJECTIVE: Identify appraisal interview best practices for a hospitality supervisor.

The **appraisal interview (evaluation interview, appraisal review)** is a private face-to-face session between you and an employee. In it you tell the worker how you have evaluated his or her performance and why, and discuss how future performance can be improved. The way you do this with each person can determine the success or failure of the entire performance review.

❋ PLANNING THE INTERVIEW

Each interview should take place in a quiet, technology-free area, with no distractions. Schedule your interviews in advance, and allow enough time to cover the ground at a comfortable pace. A sense of rush or hurry will inhibit the person being appraised. If you encourage people to feel that this is a time with you that belongs to them alone, they are likely to be receptive and cooperative.

It is important to review your written evaluation shortly before the interview and to plan how you will communicate it to the employee for best effect. Indeed, in some companies you may give a copy of the evaluation to the employee ahead of time. Your major goal for the interview is to establish and maintain a calm and positive climate of communication and problem solving rather than a negative climate of criticism or reprimand. Although you might have negative things to report, you can address them positively as things that can be improved in the future rather than dwelling on things that were wrong in the past. If you plan carefully how you will approach each point, you can maintain your positive climate, or at the very least stay calm if you are dealing with a hostile employee.

You will remember that in communications, the message gets through when the receiver wants to receive it. Successful communication is as much a matter of feeling as of logic, so if you can keep a good feeling between you and the worker, you have pretty much got it made. Your own frame of mind as you approach the interview should be that any performance problems the worker has are your problems, too, and that together you can solve them.

✳ CONDUCTING THE INTERVIEW

Usually, a bit of small talk is a good way to start off an interview—a cordial greeting by name and some informal remarks. You want to establish rapport; you want to avoid the impression of sitting in judgment, talking down, or laying down the law. You want to be person-to-person in the way you come across. You want your workers to know you are there to help them do their jobs well, not to criticize them. Criticism diminishes self-esteem, and people who have a good self-image are likely to perform better than people who don't.

Employees who are facing their first appraisal interview are often worried about it. Even though they have been told all about it before, it might seem to them like a day of judgment or like getting a school report card. *It is important to make sure that they understand the evaluation process*: the basis for evaluation, its purpose, how it will be used, and how it affects them. Stress the interview as useful feedback on performance and an opportunity for mutual problem solving. Conveying your willingness to help goes a long way toward solving problems ("How can I help you to do a good job?").

After explaining the purpose of the interview, *it is often useful to ask people to rate their own performance on the categories listed on the form*. If you have established clear standards, they usually know pretty well how they measure up. Often, they are harder on themselves than you have been on them. The two of you together can then compare the two evaluations and discuss the points on which you disagree. Stress the positive things about their work, and approach negative evaluations as opportunities to improve their skills with your support.

Encourage them to comment on your judgments. Let them disagree freely with you if they feel you are unfair. You could be wrong: You may not know the whole story or you may have made a subjective judgment that was inaccurate. Do not be afraid to change your evaluation if you discover that you were wrong.

Get them to do as much of the talking as possible. Ask questions that make them think, discuss, and explain. Encourage their questions. Take the time to let them air discontent and vent feelings. Let them tell you about problems they have and get them to suggest solutions.

Be a good listener. Don't interrupt; hear them out. Maintain eye contact. If the people being reviewed feel that you are not seeing their side, if they begin to feel defensive, you have lost their cooperation. An evaluation that is perceived as unfair will probably turn a complacent or cooperative employee into a hostile one.

Although you encourage them to do most of the talking, do not relinquish control of the interview. Bring the subject around to improvement goals and *work with them on seeing objectives for improvement*. Many evaluation systems make goal setting a requirement of the appraisal interview.

The employee, with the supervisor's help and guidance, sets goals and objectives with specific performance standards to be achieved between now and the next appraisal. These goals are recorded (as in Figure 7.6) and become an important part of the next evaluation. It is best to concentrate on two or three goals at most rather than on the whole range of possibilities. Goals should be measurable and attainable.

Be a good listener. Do not interrupt; hear them out. Maintain eye contact.
wavebreakmedia/Shutterstock

If you can get people to set their own improvement goals, they will usually be highly motivated to achieve them because they themselves have made the commitment. You should make it clear that you will support them with further training and coaching as needed to meet their goals.

It is a good idea to *summarize the interview* or ask the worker to summarize it and to make sure that you both have the same understanding of what the employee is to do now. Have the employee read the entire evaluation and sign it, explaining that signing it does not indicate agreement and that he or she has the right to add comments. Discuss your reward system openly and fully and explain what is or is not forthcoming for the person being interviewed. Make sure that the employee receives a copy of the completed evaluation form.

End the interview on a positive note—congratulations if they are in order, an expression of hope and support for the future if they are not. Your people should leave their interviews feeling that you care how they are doing and will support their efforts to improve, and that the future is worth working for.

❋ COMMON MISTAKES IN APPRAISAL INTERVIEWS

A poorly handled appraisal interview can undermine the entire evaluation process, engender ill feelings and antagonisms, cause good people to leave, and turn competent workers into marginal performers and cynics. Interviewing is a human relations skill that requires training and practice.

If you have established good relations with your people, the appraisal review should not present any problems. It is simply another form of communication about their work—a chance to focus on their problems, reinforce acceptable behavior, and help them improve—no big deal for either of you. If you are new to supervision or if you are a hard-driving, high-control type of person, you may have difficulty at first in carrying out the human relations approach recommended for a productive appraisal review. But you will find it worth the effort. Here are some major mistakes to avoid.

If you take an *authoritarian approach* (this is what you have done well, this is how I want you to improve, this is what you will get if you do, this is what will happen if you don't), it will often antagonize employees rather than produce the improvements you want. It may work with the employee who thinks you are right or with dependent types of people who are too insecure to disagree.

But people who think your evaluations are unfair in any way and do not have a chance to present their point of view may not even listen to your message, and they probably won't cooperate if the message does come through. *You* cannot improve their work; only they can improve their work, and few will improve for a carrot-and-stick approach unless they desperately need the carrot or truly fear the stick. They will leave, or they will remain and become hostile and discontented. Discontented people complain about you to each other, morale declines, and problems multiply. Improvement does not take place.

The *tell-and-sell approach* is a mild version of the authoritarian approach: The supervisor tells the worker the results of the evaluation and tries to persuade the employee to improve. It is a presentation based on logic alone, rather like a lecture. It seems to be a natural approach for someone who has not developed sensitivity in handling people.

The assumption is that the worker will follow the logic, see the light, and respond to persuasion with the appropriate promise to improve. No account is taken of the feelings of the people being evaluated, and the supervisor has no awareness of how the message is being read as the interview proceeds. There is also the assumption that the supervisor's evaluation is valid in every respect, so there is no need for the worker to take part in discussing it.

The result for the people being evaluated is at best like getting a report card; it is a one-sided verdict handed down from the top, and it leaves them out of the process. Usually, they sit silent and say nothing because the format does not invite them to speak. If they do challenge some part of the evaluation, the supervisor brushes aside the challenge and doubles the persuasion (being sure that there are no mistakes or perhaps being afraid to admit them). The supervisor wins the encounter but loses the worker's willingness to improve. The results are likely to be the same as in the hard-line authoritarian approach.

Certain mistakes in interviewing technique can destroy the value of the interview:

- *Criticizing and dwelling on past mistakes* usually make people feel bad and may also make them defensive, especially if they feel you are referring to them rather than to their work. Once they become defensive, communication ceases. The best way to avoid such mistakes is to talk in terms of the work, not the person, and in terms of the future rather than the past, emphasizing the help and support available for improvement.

- *Failing to listen, interrupting, and arguing* make the other person defensive, frustrated, and sometimes angry. Avoiding these mistakes requires you to be aware of yourself as well as of the other person and to realize continuously what you are doing and the effect it is having. It takes a conscious effort on your part to maintain a cooperative, problem-solving, worker-focused interview.

- *Losing control of the interview* is a serious mistake. There are several ways this can happen. One is to let a discussion turn into an emotional argument. This puts you on the same level as the worker: You have lost control of yourself and have abdicated your position as the boss. Another way of giving up control is to let the worker sidetrack the interview on a single issue so that you do not have time for everything you need to cover. You can recoup by suggesting a separate meeting on that issue and move on according to plan. Still another way of losing control is to allow yourself to be manipulated into reducing the standards for one person (such as overlooking poor performance because you feel sorry for someone or in exchange for some benefit to yourself). Although you may think you have bought future improvement or loyalty in this way, you have actually given away power and lost respect.

Your first appraisal interviews may not be easy. Many supervisors have trouble telling people negative things about their work in a positive, constructive way. As with so many other management skills, nobody can teach you how; it is something you just have to learn by experience, and if you are lucky you will learn it under the skillful coaching of a good supervisor.

A good interview comes from preparing yourself, from practicing interviewing for other purposes (hiring, problem solving), from knowing how to listen, from knowing the worker and the job, from staying positive, and from keeping tuned in to the interviewee, whose feelings about what you are saying can make or break the interview. It is probably one of the best learning experiences you will have in your entire career in the industry.

Follow-Up

LEARNING OBJECTIVE: Name the appropriate actions that a supervisor must take after an employee's appraisal review.

The evaluation and the appraisal review have let employees know how they are doing and have pointed the way toward improved performance. If you have done the reviews well, they have fostered momentum for improvement in responsive employees. You have become aware of where people need your help and support and probably also of where your efforts will be wasted on people who will not change or are unable to meet the demands of their job. So the appraisal review has marked the end of one phase and the beginning of a new one. How do you follow up?

The first thing you do is to see that people receive the rewards they have coming to them. You must make good on rewards promised, such as raises in pay, better shifts, better stations, and so on. If there is some problem about arranging these things, devise several alternative rewards and discuss them promptly with the people concerned. Never let people think you have forgotten them.

For people you have discovered need more training, arrange to provide it for them. For people you feel will improve themselves, follow their progress discreetly without hovering or breathing down their necks. Coaching is in order here, day-to-day counseling as needed.

Remember that in the appraisal review you emphasized your help and support. It doesn't take much time, just touching base frequently to let people know you will come through for them, frequent words of praise for achievement, readiness to discuss problems. Put them on their own as much as possible, but do not neglect them.

There will be some people who you are sure will make no attempt to improve, who will continue to get by with minimum performance. Reassess them in your mind: Was your appraisal fair? Did you handle the interview well? Is there some mistake that you are making in handling them? Are you hostile or merely indifferent? Are *they* able to do better, or is minimal performance really their best work?

Would they do better in a different job? Is their performance so poor on key aspects of the job (customer relations, absenteeism, sanitation, quality standards) that discipline is in order? Should they be retrained? Should they be terminated (hopelessly unwilling or unable to do the job)?

If employees are complacent or indifferent, you might as well give up trying to make them improve unless you can find a way to motivate them. If employees are hostile, you should try to figure out how to turn them around or at least arrive at an armed truce so that they will do the work and get their pay without disrupting the entire department. We have more to say on motivation and discipline in later chapters.

There are two important facets of follow-up. One is actually carrying it out. If, after you have done your reviews, you let the process drop until the next appraisal date, you will let all its potential benefits slip through your fingers. The other important facet is using all you have discovered about your people and yourself to improve your working relationship with each person you supervise. It can be a constantly expanding and self-feeding process, and it will pay off in the morale of your people and in your development as a leader.

Legal Aspects of Performance Evaluation

LEARNING OBJECTIVE: Discuss the potential legal issues surrounding employee evaluations.

Four major equal employment opportunity laws affect the process of performance evaluation: Title VII of the Civil Rights Act of 1964, the Equal Pay Act, the Age Discrimination in Employment Act, and the Americans with Disabilities Act (see Chapter 5). Knowing how to avoid violations of these laws in the evaluation process can save time and money as well as create goodwill with your employees and a positive public image. Following are six ways to ensure fair and legal evaluations:

1. Evaluation of performance should be based on standards or factors obtained from a job analysis of the skills, tasks, and knowledge required to perform the job.

2. Performance standards should be observable, objective, and measurable.

3. Keep a positive rapport during your discussions with the employee. This helps tremendously to avoid complaints of being unfair and, possibly, charges of discrimination.

4. Do not enter into discussions that focus on qualities of employees based on their membership in a group protected by EEO laws. If employees refer to their membership, it is best not to respond. For example, suppose that Jack, who is 60, says: "At my age, it gets harder to see the small details. I guess that explains my trouble with this." It would be appropriate for you to focus on how to ensure that Jack is able to see well enough to perform his job. Do not talk about his age or write anything about it on the review.

5. Employee performance should be documented more frequently than once a year at appraisal time. An employee should not be surprised at performance appraisal time.

6. If an employee disagrees with his or her evaluation, he or she should be able to appeal.

 ## CASE STUDY: The First Appraisal Interview

Sandy is sitting outside her boss's office awaiting her first appraisal interview. She is nervous, but confident. She has improved tremendously since she dropped that whole tray of dinners when she first came to work three months ago.

Her boss has stopped coming around and telling her not to do this and that, so she thinks that she's doing all right (although, of course, you never know). She gets along very well with the guests, and, in fact, sometimes people ask to be seated at her tables. Her tips are higher than almost anyone else's, and that must count for something.

The door opens and the boss motions her to come in and sit down. "Good morning, Sally," he says. "We're a little bit rushed for time, so I'll just go through this evaluation form with you—er—Sandy. Read it over, won't you? Then we'll talk."

Sandy glances through the ratings: average, average, average, needs improvement. Well, she has to admit she still has trouble opening wine bottles and sometimes breaks the cork. Average, average, average.

She sighs, hands the form back to her boss, sits back in her chair, folds her hands tightly, and looks down at them.

"Well, what do you think, Sandy? Do you agree? We need to make a plan for your improvement on wine service. I know you sometimes ask Charlie to open your bottles and that's not really what good customer service is all about. Why don't you get Charlie to give you some tips on what you're doing wrong? Then maybe next time you'll get a better rating. Now, do you have any comments or questions?"

"What's average?" Sandy asks.

"Well, I guess it means no better and no worse than anyone else. Actually, it means you're doing okay; you're just not as good as people like Ruth and Charlie. But you certainly don't need to worry about losing your job or anything like that—you're all set here! Anything else?"

"Well," Sandy begins, gathering up her courage, "I thought I was really above average in customer service—people ask for me and they tip me a lot, so I must be—"

"But don't forget the time you dropped the dishes, Sandy! I do think you're doing very well indeed now, but we're talking about the whole evaluation period! Now, if you'll just sign this. . . ."

Case Study Questions

1. What do you think of the boss's ratings and his defense of them?

2. How do you think Sandy feels? Will she be motivated to improve? Is it enough to know you are not going to lose your job?

3. List the mistakes the boss makes in his interview. How could he have handled things better?

4. What do you think of the boss's improvement plan? How will Charlie feel about it?

5. If the boss's supervisor could have heard this interview, what would have been the supervisor's opinion of it? What responsibility does the boss's boss have for the way that interviews are handled? What means could be set up for evaluating supervisors on their interviews?

KEY POINTS

1. If you develop a full set of performance standards for each job that you supervise, you have the basis for a management system for your people and the work they do. You can use them in recruiting, training, and evaluation. You can use them with employees to reduce conflict and misunderstanding. Everyone knows who is responsible for what.

2. Three essentials to setting up a successful performance standard system are worker participation, active supervisory leadership and assistance throughout, and a built-in reward system.

3. Figure 7.1 depicts how to develop a performance standard system: define the purpose, analyze the job, write the performance standard, train the workers, and evaluate on-the-job performance.

4. Figure 7.4 shows a sample form to be used when writing performance standards.

5. Performance standards must be specific, clear, complete, accurate, measurable or observable, attainable, and in conformance with company policies and legal and moral constraints.

6. The first key to making your system work is the workers' cooperation and agreement in the developmental stage. The second key to success is to put the system to work slowly over a period of time, one job at a time. Other keys to success include having an award or incentive system, recognizing your workers' potential, and reviewing the system periodically.

7. A performance standard system can fail if the standards are not clearly stated and communicated to everyone, if the supervisor does not follow up properly, if the supervisor does not provide enough challenge or reward, or if the system is administered in a confining or negative manner.

8. Performance evaluation refers to the periodic review and assessment of each employee's performance during a given period, such as a year. This is in addition to the informal performance evaluation that is a daily part of a supervisor's job.

9. When carried out conscientiously and when there is constant communication between reviews, performance reviews have many benefits. They help to maintain performance standards. By telling employees how they are doing, they can remove uncertainty and improve morale. By spotlighting areas for improvement, they can focus the efforts of both worker and supervisor to bring about improvement.

10. Performance reviews can increase motivation to perform well and provide the opportunity for improving communication and relationships. They can identify workers with unused potential and workers who ought to be terminated. They can give feedback on supervisory performance and uncover problems that are getting in the way of the work.

11. The performance review process includes these four steps: preparing for evaluation, making the evaluation, sharing it with the worker, and providing follow-up.

12. An evaluation form typically lists the performance dimensions or categories on which each worker is to be rated. The performance dimensions should be related to the job being evaluated and defined clearly in objective and observable terms, as in a performance standard.

13. A rating scale is used for each performance dimension, such as outstanding to unsatisfactory. The more precise the descriptions for each rating, the more objective the ratings will be.

14. No evaluation form solves all the problems of fairness and objectivity. Probably those that come closest are designed exclusively for hourly workers, for specific jobs, and for evaluating performance rather than promotability.

15. Some pitfalls when rating employee performance include the halo effect, letting your feelings about a person bias your judgment, comparing one person with another, letting your feelings about the evaluation process affect rating, procrastination, giving ratings for the effect they'll have, and being too lax or too much of a perfectionist.

16. Evaluate the performance, not the employee.
17. Employee self-appraisals are especially justified when an employee works largely without supervision.
18. Plan a quiet location for appraisal interviews, review your written evaluation shortly beforehand, and plan how you will communicate it to the employee for best effect.
19. When conducting the appraisal interview, start with a bit of small talk. Make sure that the employee understands the evaluation process, ask the employee to rate his or her own performance first, encourage employees to comment on your judgments, get the employee to do most of the talking. Work with them on setting improvement objectives, summarize the interview, and end on a positive note.
20. Common mistakes in appraisal interviews include taking an authoritarian or tell-and-sell approach, criticizing, dwelling on past mistakes, failing to listen, and losing control of the interview.
21. Follow-up after performance appraisals is crucial. If you let the process drop until the next appraisal date, you will let all its potential benefits slip through your fingers.
22. Equal employment opportunity laws apply to performance evaluation, so the evaluation process needs to be nondiscriminatory.

KEY TERMS

appraisal interview (evaluation interview appraisal review)

employee self-appraisal

evaluation form

levels of performance

merit raise

minimum level

optimistic level

performance appraisal (performance evaluation performance review)

performance dimensions or categories

performance standards

rating scale

realistic level

training objectives

tasks

unit of work

REVIEW QUESTIONS

Answer each question in complete sentences. Read each question carefully and make sure that you answer all parts of the question. Organize your answer using more than one paragraph when appropriate.

1. How is a job different from a position?
2. How is a job description different from a job specification?
3. How is a job evaluation different from job analysis?
4. List four possible units of work for a baker.
5. Describe common parts of a job description.
6. What is a job description used for?
7. Briefly describe how you would set up a performance standard system in a restaurateur hotel. How would you implement it? What would you do to make sure that it was successful?
8. Think of a unit of work performed by a hospitality worker, such as taking reservations or setting up a salad bar, and write a possible performance standard for that unit. Did your performance standard meet the five standards mentioned in the book?
9. Describe the major points in each of the four steps in a performance review: preparing for evaluation, making the evaluation, sharing it with the employee, and providing follow-up.
10. Which equal employment opportunity laws affect the process of performance evaluation?
11. What can you do to ensure legal evaluations?

ACTIVITIES AND APPLICATIONS

1. **Discussion Questions**
   - Describe the goals of a performance standard system as you see them. How must the performance standards be used to attain these goals?
   - What elements of scientific management are contained in a performance standard system? Of human relations theory? Of participative management? Explain each answer.
   - What managerial style would go best with a performance standard system? Why do you think so?
   - In your opinion, what are the chief values in using a performance standard system? What are the chief drawbacks? Do you think it is worthwhile or even possible to develop such a system in an industry with such high turnover?
   - What is the relationship between ongoing day-by-day evaluations and periodic performance reviews? Is either one valid without the other? Defend your answer.
   - Explain the following statement: "Performance appraisals are about as beloved as IRS audits." As a supervisor, what can you do to reduce employee fear and anxiety about performance appraisals?
   - Do you think periodic performance reviews are worth the time and trouble they take? Why or why not?
   - What type of evaluation form do you think is most suitable for hourly employees in hospitality operations? Consider performance dimensions, performance ratings, length, and ease of completion. Explain your choice.
   - In your opinion, which part of the appraisal review is most important: the written evaluation, the interview, or the follow-up? Explain.
   - Explain the following statement: "A poorly handled appraisal interview can undermine the entire evaluation process." Give examples of poor handling and their effects.
   - When you allow employees to do a self-appraisal, do you think they can rate their own performance accurately? What are the advantages and disadvantages of having an employee do a self-appraisal?

2. **Group Activity: Critique of Job Descriptions**
   Using a variety of job descriptions from student workplaces or other hospitality operations, critique them using this chapter as a guide. What works well with each job description? What could be improved, and how would you improve it? In your personal experience, do hospitality operations really rely on job descriptions to recruit, hire, train, and evaluate?

3. **Role-Play: The Appraisal Interview**
   In a group of four, complete one of the performance evaluation forms in this chapter for a fictitious person. While two students role-play the supervisor and employee in an appraisal interview, the other two students can be observers. When the role-play is done, the observers can take a turn doing the appraisal interview.

4. **Writing Performance Standards**
   The following standards represent first attempts by a housekeeping supervisor to write some of the standards for the job of housekeeper in a motel. They are for on-the-job performance. You are to evaluate them according to the criteria listed here and correct any that do not meet the criteria. Then discuss each one as it was and as you have rewritten it. (This is not intended to be a complete list for the job.)

   ### Drafts of Standards

   1. Before starting work, the cleaner will load the cleaning cart correctly according to the Cleaning Cart Diagram and Supply Sheet.
   2. The cleaner will make all beds using the procedures shown on the Bed-Making Procedures Sheet.

3. The cleaner will scrub the tub, shower, basin, floor, and toilet according to the Bathroom Cleaning Procedures Sheet, using the cleaning supplies and utensils specified on that sheet. The result must score 90 percent or higher on the Cleanliness Checklist for each room cleaned.

4. The cleaner will vacuum carpeting according to the instructions on the Vacuum Cleaning Procedures Sheet.

5. The cleaner will operate all cleaning equipment correctly and safely.

6. The cleaner must be able to clean 15 rooms per day in an average time of 25 minutes per room.

Criteria for Evaluation

■ Specific, clear, and complete: states what, how, and how well

■ Measurable or observable standard of performance

■ Attainable, possible

■ Correct verb type for on-the-job performance

 # ENDNOTES

1. www.opm.gov/services-for-agencies/performance-management/. Retrieved November 25, 2014.
2. Ibid.
3. Judith Hale, *Performance-Based Management: What Every Manager Should Do to Get Results* (San Francisco: Pfeiffer, 2004), p. 1; http://performance-based-management.com/?p=57. Retrieved December 18, 2014.
4. From an address to the National Restaurant Association by Ken Blanchard, May 14, 2001.

Chapter 8

Motivation

Susan just started working a month ago at the front desk of an airport hotel. So far, she is not very happy with the job. To begin with, she has trouble finding a parking spot every afternoon when she comes to work, although she was promised that there were plenty. When she reports to work, she is lucky if she can find her boss, who is often away from the work area, to question him about her training program, which is going very slowly. Often she wonders if anyone would notice if she just took off out the front door and did not come back. Randy, a cook in a downtown restaurant, loves where he works. Although his pay and benefits are good, there are many other reasons why he loves his job. He feels like part of a quality team at work, management always keeps him informed of what's going on, hourly employees frequently get promoted when there are open positions, the kitchen is comfortable to work in and he has just the equipment he needs, everyone is on a first-name basis, he gets bonuses based on the number of guests served, and the restaurant owners give him time off to go to college and pay his tuition.

Employees want to be treated as individuals first and as employees second. They want a lot more out of work than just a paycheck. They want, for example, respect, trust, rewards, and interesting work.

Today's employees tend to have higher expectations from their jobs, and they have less tolerance for mismanagement, frustrating coworkers, or poor working conditions. They expect companies to invest in them—in coaching and job development—so they can grow professionally and personally. If they do not feel valued personally, they will leave. If the economy is good, they will leave quickly. To retain good workers—and even to transform mediocre workers into good workers—supervisors must know how to motivate employees.

After completion of this chapter, you should be able to:

- Describe common employee expectations of their leaders.
- Explain the hospitality supervisor's role in motivating his or her employees.
- Classify the different theories of motivation, analyzing the advantages and disadvantages of each theory.
- Explain the challenges that hospitality supervisors may face in motivating employees.
- Identify ways to build a positive work climate.
- Explain why individual strategies of motivation are essential.
- Describe the most effective ways of creating an attractive job environment.
- Discuss why the leader is the key to a positive work environment.

Employee Expectations and Needs

LEARNING OBJECTIVE: Describe common employee expectations of their leaders.

When you become a leader, you will have certain expectations of your employees. You will expect them to do the work they have been hired to do—to produce the products and services to the quality standards set by the enterprise that is paying you both. You sometimes wonder whether their performance will meet your expectations, and you have some plans for improving productivity.

But you might not realize that what these people expect from you and how *you* meet *their* expectations has as much to do with their performance as your expectations of them. If you handle their expectations well, if they recognize your authority willingly, you will have a positive relationship going for you—one on which you can build a successful operation. Let us look at some categories of things workers typically expect and need from the boss.

Delighted, motivated employees are productive, loyal employees. Today's cooks are not only interested in cuisine and what they expect to learn from the chef, they are also interested in paying off student loans, health insurance coverage, 401K plans, and a balanced personal life.[1]

✳ YOUR EXPERIENCE AND TECHNICAL SKILLS

Employees expect you to be qualified to lead. First, they want you to have worked in the area in which you are leading: a hotel, a hospital kitchen, a restaurant, whatever it is. Coming into a restaurant from a hospital kitchen might discount your experience, and you will have to prove yourself. Coming into a big hotel from a job in a budget motel, you will also have to prove yourself. Your associates want to feel that you understand the operation well and appreciate the work they are doing. They want to feel that they and their jobs are in good hands—that you are truly capable of leading them and their work.

In some circumstances, being a college graduate will make you distrusted. Your associates might assume that you think you know it all and that you will look down on them. They might think that *they* know it all and that you have not paid your dues by coming up the hard way.

In other places, if you are *not* a college graduate and other leaders have college degrees—in a hospital setting, perhaps—you will have to work harder to establish yourself with your associates. If they are satisfied with what you have done on other jobs and how you are doing on this one, they will each decide at some point that you are qualified to lead here. But it will take time and tact and determination on your part.

Second, they want you to not only be experienced but also technically competent. Every employee who works with you expects you to be able to do his or her particular job. This can become a sort of game. Employees will question you, they will check your knowledge, and they will make you prove you know what you are doing: "Why doesn't the bread rise?" "Why doesn't the sauce thicken?" There might be instances when they will have sabotaged that recipe just to see if you know what is wrong. They may unplug the slicer and tell you it is broken, and you will start checking the machine and the fuses before you catch on. You are going to have to prove your right to supervise. You don't have to know how to do every job *as well as* each person doing that job—but you do have to know how to do it.

✳ THE WAY YOU BEHAVE AS A LEADER

Nearly everyone wants a leader who will take stands and make decisions, who will stay in charge no matter how difficult the situation is, who is out there handling whatever emergency comes up. Hardly anyone respects a boss who evades issues and responsibilities, shifts blame, hides behind the mistakes of others, or avoids making decisions that will be unpopular even though they are necessary. In the foodservice industry they will likely "size you up" and in some cases they may play mind games with you, such as immature behavior or actions, or letting on that they are not paying attention. One who motivates leads and one who leads creates a culture of belonging in the kitchen and restaurant.

Many people expect authority and direction from the boss. These people want you to tell them what to do; they do not know how to handle too much independence. Some of them will want you to supervise every single thing they do—"Is that okay?" "Is this the way you want it?" Others just want you to define the job, tell them what you want done, and let them go at it—"Hey, get off my back and leave me alone." It is very important not to have favorites as this will cause a rife among staff, and the environment in your kitchen will deteriorate quickly, which will then affect your overall business.

Sometimes you will have an employee who is totally opposed to authority, who will reject everything you say simply because you are the boss; this one will give you a hard time. When you get to know each person's special needs and expectations, you can adjust your style of directing them accordingly—your style, but not what you require of them. You must do what is correct, not what pleases them.

Your people expect you to act like a leader toward them, not like one of the gang. They want you to be friendly, but they expect you to maintain an objective, work-oriented relationship with each person. They do not want you to be everyone's pal, nor will they like it if you have special friends among the workers.

If you do socialize off the job with some of the people you supervise, you are running certain risks. Can you go out and party with them, form close friendships, and then come back and supervise them on the job without playing favorites or making other workers jealous? Maybe you can. But can your worker friends handle this closeness, this double relationship? Will they think they are special and that they can get away with things? These are friendships to approach with caution or to avoid altogether.

Your people expect you to treat them fairly and equally, without favoritism. The fairness that people expect is fairness as *they* see it, not necessarily as you see it. If there is someone on your staff that you don't like, it is going to be difficult, if not impossible, for you to treat this person without bias. If there is someone else who you like a lot, you will tend to favor that person. Will employees think you are being fair when there is a difference in the way you instruct, discipline, and deal with these two people? They might think that you are playing favorites or are really putting somebody down. You must always think of how these things look to the other associates and how it will affect their acceptance of you. Sometimes they might be right and you are not aware of it.

Fairness includes honesty with your associates and with the company. Your people expect you to evaluate their work honestly, to follow company rules, to put in your time, to fulfill your promises, and to carry out your threats. One of the worst mistakes you can make is to promise something you cannot deliver, whether it is a threat or a reward. People will not respect the authority of a leader who does this. If you do not come through for your workers, they will not come through for you.

Consider Joe Clark, operator of a Chick-Fil-A restaurant. "The people who work for me are my guests," says Clark. He greets them when they arrive, and says goodbye when they leave, and talks with them in between. "The most fun I have is right before we open, when everyone is doing food prep," he says. "It's an opportunity for us to catch up with each other." Some of Clark's employees have been with him for 8, 9, even 10 years. "They've found job satisfaction right here."[2]

✺ COMMUNICATION BETWEEN LEADER AND EMPLOYEES

Your workers expect several things from you in the way of communication. First, *they expect information.* They expect you to define their jobs and to give them directions in a way they can comprehend. Probably 90 percent of the people who work for you want to do a good job, but it is up to you to make it clear to them what the job is and how it should be done. It often takes a little extra time to make sure that each associate has grasped the full meaning of what you have said. But if you expect them to do a good job, they expect you to take the time necessary to tell them exactly what a good job is.

Telling them what to do and how to do it should include the necessary skills training. In the foodservice industry, it is typical to skip this training or to ask another associate to train the new person while the two of them are on the job. It is not uncommon to hire people to bus tables, put them to work without training, yell at them for doing everything wrong, and then fire them for breaking so many dishes; that is, unless they quit first. Lack of clear direction is a major reason for the high rate of employee turnover in this industry. The leader does not meet the associate's expectations.

The second type of communication that people want from the boss is *feedback on their performance.* The most important thing a worker wants to know is, "How am I doing? Am I getting along all right?" Yet this expectation, this need, is usually met only when the worker is *not* doing all right. We tear into them when they are doing things wrong, but we seldom take the time to tell them when they are doing a good job. A few seconds to fill that basic human need for

Clear communication between supervisors and associates is critical in a creative and positive work climate.

approval can make a world of difference in your associates' attitude toward you, and the work they do for you.

A third form of communication that employees expect from you is to have you *listen* when they tell you something. They can give you useful information about their jobs and your customers, and they can often make very valuable suggestions if you will take the time to listen—really listen—to what they have to say. But they do expect you to take that time and to take them seriously because they are offering you something of their own.

Two cardinal rules on suggestions from employees are:

1. Never steal one of their suggestions and use it as your own.
2. If you cannot use a suggestion, explain why you can't, and express your appreciation.

If you violate either of these rules, suddenly your associates will stop telling you anything. They will not even respond when you ask for their input. You have closed the door they expected to be open, and they are not going to open it again.

✳ UNWRITTEN RULES AND CUSTOMS

In most enterprises certain work customs become established over the years, and employees expect a new supervisor to observe them. They are not written down anywhere, they have just become entwined in the culture, and they are treasured by workers as inviolable rights, never to be tampered with, especially by newcomers. In many kitchens, for example, a new worker is always given the grungy jobs, such as vegetable prep or cleaning shrimp.

In a hotel, a new night cleaner will have to clean the lobby and the public restrooms. If the leader brings in somebody new and he or she isn't started off with the grungy jobs, that's just not right. If a new waiter is brought in and given the best station in the restaurant—the one with the best tips or the one closest to the kitchen—there's going to be a mutiny; that's just not done.

People will lay claim to the same chair day after day to eat their lunch, they will park their cars in the same place, and if you disrupt one of these things established by usage and custom, they will take it as a personal affront. If you want to make changes to the established customs, you will be wise to approach them cautiously and introduce changes gradually.

Another type of rule or custom, sometimes written down but more often unwritten, is the content of a job as seen by the person performing it. When people begin a new job, they quickly settle in their own minds what constitutes a day's work in that job and the obligations and expectations that go with it. If you as a leader go beyond your workers' expectations, if you ask them to do something extra or out of the ordinary, you have violated their concept of what they were hired to do, and they feel you are imposing on them, taking advantage. They will resent you, and they will resent the whole idea.

Suppose that you are a dishwasher and you finish early, and the leader is so pleased that she asks you to clean the walk-in. The next day you finish early again and the boss says, "This is terrific, today we are going to clean the garbage cans." "Hey, no," you say, "I was hired to wash dishes, not to clean walk-ins and garbage cans." And you are about ready to tell her off but you think better of it; you need the job. On the third day you have only 30 people for lunch instead of your usual 300, but how long does it take you to finish the dishes? All afternoon and 30 minutes of overtime, at least.

In sum, people expect the leader to observe what associates believe their jobs to be, whether they have been defined on paper by management or defined only in the associates' own minds. Rightly or wrongly, they resent being given more to do than they were hired to do, and they may refuse to do the extra work, or won't do it well, or will take overtime to do it.

One way to avoid this kind of resistance is to make clear when you hire people that you may ask them to vary their duties now and then when the work is slow or you are shorthanded or there is an emergency. An all-purpose phrase included in each job description—"other duties as assigned"—will establish the principle.

However, as a new supervisor you need to be aware of the way people perceive what you ask them to do. In our example, the worker who finishes early is rewarded with two unpleasant jobs totally unrelated to running the dishwashing machine. There is no immediate and urgent need and no warning that the worker might be expected to fill idle time with other tasks. We have more to say about defining job content in later chapters. A clear understanding is essential to a successful relationship between associate and leader.

✳ PERSON-TO-PERSON RELATIONSHIPS

Today's associates expect to be treated as human beings rather than as part of the machinery of production. They want the leader to know who they are and what they do on the job and how well they are doing it. They want to be respected and treated in the same way you want to be treated. If a leader or executive chef develops sensitivity toward his/her staff, they in return gain mutual respect back. Staff want to be treated as individuals, and they want to feel comfortable talking to the leader, whether it is about problems on the job or about hunting and/or fishing and/or the weather and the new baby at home. They want the leader's acceptance and approval, including tolerance for an occasional mistake or a bad day. They want recognition for a job well done. Whether they are aware of it or not, they want a sense of belonging on the job.

To your people, you personify the company. They don't know the owners, the stockholders, the general manager, and the top brass. To most hourly workers, you are the company—you are it. If they have a good working relationship with you, they will feel good about the company. If they feel good about the company, they can develop that sense of belonging there. And if they feel that they belong there, they are likely to stay.

Successful leaders develop a sensitivity to each person, to the person's individual needs and desires and fears and anxieties as well as talent and skills. They handle each person as much as possible in the way that best fills the associates' personal needs. If you can establish good relationships on this one-to-one level with all your workers, you can build the positive kind of work climate that is necessary for success.

Motivation

LEARNING OBJECTIVE: Explain the hospitality supervisor's role in motivating his or her employees.

motivation
The why of behavior; the energizer that makes people behave as they do.

The term **motivation** refers to what makes people tick: the needs and desires and fears and aspirations within people that make them behave as they do. Motivation is the energizer that makes people take action; it is the *why* of human behavior. In the workplace, motivation goes hand in hand with productivity. Highly motivated people usually work hard and do superior work. Poorly motivated people do what is necessary to get by without any hassles from the leader, even though they may be capable of doing more and better work. Unmotivated people usually do marginal or substandard work and often take up a good deal of the leader's time.

Sometimes, people are motivated by resentment and anger to make trouble for the supervisor, to beat the system, or to gain power for themselves. Such motivations are at cross-purposes with the goals of the operation and have a negative effect on productivity. Motivation, as we have noted many times, is a major concern of the leader. If employees are unmotivated they are more likely to bring down a good team. Hiring a good leader is critical and probably the most difficult thing to do for any hospitality business.

Leadership success is measured by the performance of the department as a whole, which is made up of the performance of individuals. Each person's performance can raise or lower overall productivity and leadership success. The big question is how to motivate poor performers to realize their potential and raise their productivity, and how to keep good performers from going stale in their jobs or leaving for a better opportunity.

Actually, you cannot motivate people to do good work. Motivation comes from within. The one thing you as a leader can do is to turn it on, to activate people's own motivations. To do this you must get to know your associates and find out what they respond to. It may be the work itself. It may be the way you lead. It may be the work environment. It may be their individual goals: money, recognition, achievement, or whatever. Figure 8.1 shows the relationship of key motivators: needs, desires, fears, and aspirations.

How do you find out what will turn people on? It isn't easy. There are many theories and few answers. What motivates one person might turn someone else off completely. Everybody is different. People do the same things for different reasons and different things for the same reasons. People's needs and desires and behaviors change from day to day and sometimes from minute to minute. You can never know directly why they behave as they do, and they might not know why either, or would not tell you if they could.

In sum, motivation is a complicated business, and motivating people to do their jobs well has no one simple answer. It takes something of an experimental approach; you try to find out what each person responds to, and if one thing doesn't work, maybe the next thing will. But it need not be just a trial-and-error process. You can get quite a bit of insight into human behavior from people who have spent their lives studying the subject, and you will find much in their theories that will help you to figure out how to motivate individual associates to do their best for you. Motivation is critical not only for employee retention, but also for providing the best possible dining

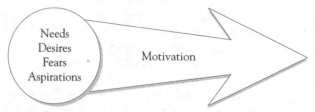

FIGURE 8.1: Needs, desires, fears, and aspirations lead to motivation.

experience. The National Restaurant Association recommends sharing the purpose of the business with employees and soliciting their involvement, such as asking for and rewarding good suggestions for making improvements. Another recommendation is to share responsibility with a "we're in this together" attitude.[3]

The one thing that you can seldom do is to develop a set of rules applying this or that theory to a certain person or particular situation on the job. For this reason, we give you the various theories first. Then we spend the balance of the chapter investigating ways of motivating people by using your broadened understanding of human nature along with a mixture of theory, sensitivity, and ingenuity.

❋ WHAT MAKES YOU WORK?

In *Why Work?* MacCoby suggests that there are five different character types of work. Each responds to different values. Recognizing the kind of person you are will help determine what intrinsically motivates you:[4]

1. *Expert:* motivated by mastery, control, autonomy. Example: craftsman—excellence in making things.
2. *Helper:* motivated by relatedness, caring for people. Example: institutional helper—skill in resolving conflict.
3. *Defender:* motivated by protection, dignity. Example: supporting the oppressed—power, self-esteem, survival.
4. *Innovator:* motivated by creating, experimenting. Example: gamesman—glory, competition.
5. *Self-developer:* Motivated by balancing competence, play, knowledge, and personal growth. Example: seeking harmony.

Today's best supervisors and managers, says MacCoby, are self-developers—a well-rounded person who seeks to balance personal growth, knowledge, and competence.[5]

In the workplace, motivation goes hand in hand with productivity.
racom/Shutterstock

Theories of Motivation

LEARNING OBJECTIVE: Classify the different theories of motivation, analyzing the advantages and disadvantages of each theory.

Whether they realize it or not, everyone has a theory of how to get people to perform on the job. Several are familiar to you, although you might not think of them as theories. Chapter 2 introduced some of them in terms of leadership; now we look at those theories and others in terms of motivation.

✳ MOTIVATION THROUGH FEAR

One of the oldest ways of motivating people to perform on the job is to use fear as the trigger for getting action. This method makes systematic use of coercion, threats, and punishment: "If you don't do your job and do it right, you won't get your raise," "I'll put you back on the night shift," "I'll fire you."

This approach to motivation is sometimes referred to as a "kick in the pants." It is still used surprisingly often, with little success. Yet people who use it believe that it is the only way to get results. They are typically autocratic, high-control, authoritarian bosses with Theory X beliefs about people, and they think other theories of motivation are baloney—you must be tough with people.

Motivation through fear seldom works for long. People who work in order to avoid punishment usually produce mediocre results at best, and fear actually reduces the ability to perform. At the same time, it arouses hostility, resentment, and the desire to get even. Absenteeism, tardiness, poor performance, and high turnover are typical under this type of supervision.

Fear will sometimes motivate people who have always been treated this way, and it can function as a last resort when all other methods have failed. But it will work only if the supervisor is perceived as being powerful enough to carry out the punishment. If the boss continually threatens punishment and never punishes, the threats have no power to motivate. In fact, not even fear works in this situation.

No one recommends motivation through fear except the people who practice it. On average, workers in the United States simply will not put up with that kind of leader unless they are desperate for a job.

✳ CARROT-AND-STICK METHOD

A second philosophy of motivation is to combine fear with incentive reward for good performance, punishment for bad. You may recognize this as carrot-and-stick motivation: the carrot dangled in front as a promised reward, the stick hitting the employee from behind as goad and punishment. It is another high-control method, one that requires constant application. Once the reward is achieved or the punishment administered, it no longer motivates performance, and another reward must be devised or punishment threatened or applied.

In effect, the leader is pushing and pulling employees through their jobs; they themselves feel no motivation to perform well. At the same time, employees come to feel that they have a continuing right to the rewards (such as higher wages, fringe benefits), and these get built into the system without further motivation. Meanwhile, the punishments and threats of punishment breed resentment and resistance.

✳ ECONOMIC PERSON THEORY

economic person theory
The belief that people work solely for money.

A third motivation theory maintains that money is the only thing that people work for. This classical view of job motivation was known as the **economic person theory**. Frederick Taylor was perhaps its most influential advocate. Taylor developed his scientific management theories on the cornerstone of incentive pay based on amount of work done. He firmly believed that he was offering workers what they wanted most, and that the way to motivate workers to increase their productivity was to relate wages directly to the amount of work produced. What he did not know was that the employees in his plant were far more strongly motivated by their loyalty to one another. In fact, for three years they united to block every effort he made to increase output despite the extra wages that they could have earned.

There is no doubt that money has always been and still is one of the most important reasons that people work. For some people it may be the most important reason. That paycheck feeds and clothes and houses them; it can give them security, status, a feeling of personal worth. For people who have been at the poverty level, it can be the difference between being hungry and being well fed or between welfare and self-support with self-respect. For teenagers, it can mean the difference between owning a car and being without transportation. For most people on their first job, whether it is an hourly job or an entry-level management job, money is often the primary motivator.

But the amount of money in the paycheck does not guarantee performance on the job. The paycheck buys people's time and enough effort to get by, but it does not buy quality, quantity, and commitment to doing one's work well. If people work for money, does it follow that they will work better for more—the more the pay, the better the performance?

There are certainly instances in which it works: the expectation of wage increases, bonuses, tips, and rewards is likely to have this outcome. But money does not motivate performance once it is paid; the incentive comes from the expectation of more to come.

Furthermore, people do not work for money alone. A number of research studies have shown that, for most people, money as a motivator on the job has less importance than achievement, recognition, responsibility, and interesting work. In sum, money is only one of the resources you have for motivating people, and it does not necessarily have a direct relationship to productivity.

❋ HUMAN RELATIONS THEORY

social person theory
The idea that fulfillment of social needs is more important than money in motivating people.

After the Hawthorne experiments uncovered the human factors affecting productivity, the **social person theory** succeeded the economic person in motivation theory. The human relations enthusiasts pushed their convictions that if people are treated as people, they will be more productive on the job. Make people feel secure, they said, treat them as individuals, make them feel they belong and have worth, develop person-to-person relationships with each one, let them participate in plans and decisions that affect them, and they will respond by giving their best to the organization.

Putting this theory to work brought about higher wages, better working conditions, pension plans, paid vacations, insurance plans, and other fringe benefits, making workers happier but not necessarily more productive. The question remained: What motivates people to work?

❋ MASLOW'S HIERARCHY OF NEEDS

An influential answer to this question was the motivation theory of psychologist Abraham Maslow. Human beings, he pointed out, are *wanting animals*, and they behave in ways that will satisfy their needs and wants. Their needs and desires are inexhaustible; as soon as one need is satisfied, another appears to take its place.

hierarchy of needs
A theory proposed by Maslow that places human needs in a hierarchy or pyramid. As one's needs at the bottom of the pyramid are met, higher-level needs become more important.

In *Motivation and Personality*, Maslow proposed a hierarchy of universal human needs representing the order in which these needs become motivators of human behavior.[6] This **hierarchy of needs** is represented by the pyramid in Figure 8.2.

At the bottom of the pyramid are people's most basic needs—the *physiological needs* related to *survival*, such as food and water. When these needs are not being met, every effort is directed toward meeting them. People who are truly hungry cannot think of anything but food. For many hospitality employees, this equates to salary or wages.

But when survival needs are being met, they no longer motivate behavior, and the next level of needs comes into play. These relate to *safety*; they include protection, security, stability, structure, order, and freedom from fear, anxiety, and chaos. For hospitality employees, this equates to benefits and pension plans.

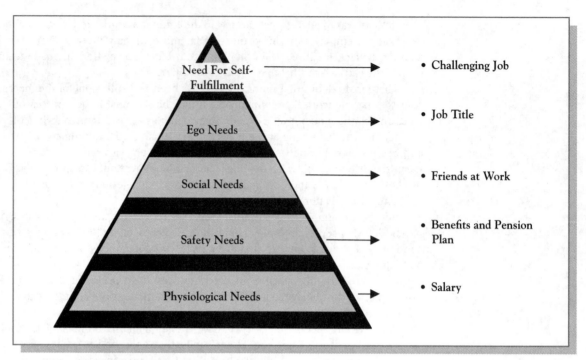

FIGURE 8.2:   Relating Maslow's hierarchy of needs to a work setting.

As these needs, in turn, are more or less satisfied, *social needs* become the predominant motivators. These include the need to be with others, to belong, to have friends, to love, and to be loved. For hospitality employees, this means socializing at work—and we do plenty of that, don't we?

Above these three groups of needs (sometimes called *primary needs*) is a higher level of needs centered on esteem. These are sometimes referred to as ego needs. One of them is the desire for *self-esteem* or *self-respect* and for the strength, achievement, mastery, competence, confidence, independence, and freedom that provide such self-esteem.

Another is the desire for the *esteem of others*: for status, fame and glory, dominance, recognition, attention, importance, dignity, and appreciation. The need for esteem gives rise in some people to the need for power as a way of commanding the esteem of others. Satisfaction of the need for self-esteem leads to feelings of self-confidence, strength, and worth. When these needs go unsatisfied, they produce feelings of inferiority, weakness, and helplessness. For hospitality employees, this equated to job title and perks.

At the top of the hierarchy is the need for *self-fulfillment*, or what Maslow called **self-actualization**. This includes the need to be doing what one is best fitted for, the desire to fulfill one's own potential. For the hospitality employee this equates to a challenging job where people can always learn more.

One or another of all these personal needs or various combinations of needs is what motivates people to do what they do. If a lower need goes unsatisfied, people will spend all their time and energy trying to fill it, and they will not experience the next level of needs until the lower needs are met. When a need is satisfied it is no longer a motivator, and the next level of needs becomes the predominant motivation.

Thus, motivation is an unending cycle of need and satisfaction, need and satisfaction. You have a need, you look for a solution, you take action to satisfy the need, and another need appears because human beings are wanting animals whose needs and desires are never completely satisfied. This continuing cycle explains why workers' needs evolve and change as their own situation changes.

Maslow's theory of motivation does not give you a tool you can use directly; you cannot sit down and analyze each person's needs and then know how to motivate that person. What it can do is to make you aware of how people differ in their needs and why they respond to certain things and not to others. It can help you understand why some of your associates behave as they do on the job.

self-actualization
The desire to fulfill one's own potential.

❋ THEORY Y AND MOTIVATION

Maslow's theories were the springboard for psychologist Douglas McGregor's Theory X and Theory Y, two opposing views of the way that supervisors and managers look at their workers. Theory X and Theory Y applied Maslow's theories directly to the problem of motivating workers on the job. McGregor made two particularly significant contributions with Theory Y. One was to revise the typical view of the way that people look at work: It is "as natural as play or rest" when it is satisfying a need. This is a flat reversal of the Theory X view of the worker, and it suggests a clear reason why people work willingly.

McGregor's second contribution to motivational theory was the idea that people's needs, especially their ego and self-actualization needs, can be made to operate on the job in harmony with the needs and goals of the organization. If, for example, people are given assignments in which they see the opportunity for achievement, responsibility, growth, and self-fulfillment, they will become committed to carrying them out. They will be self-directed and self-controlled, and external controls and the threat of punishment will be unnecessary. In other words, if you can give people work that will fill some basic need, their own motivation will take care of its performance. People will work harder and longer and better for the company if they are satisfying their own needs in the process.

❋ HERZBERG'S MOTIVATION-HYGIENE THEORY

dissatisfiers
A factor in a job environment that produces dissatisfaction, usually reducing motivation.

The work of another psychologist, Frederick Herzberg, explained why human relations methods failed to motivate performance and identified factors that truly motivate (see Figure 8.3).[7] Herzberg found that factors associated with the job environment (compensation, supervision, working conditions, company policy, etc.) create dissatisfaction and unhappiness on the job when they are inadequate; they become **dissatisfiers**.

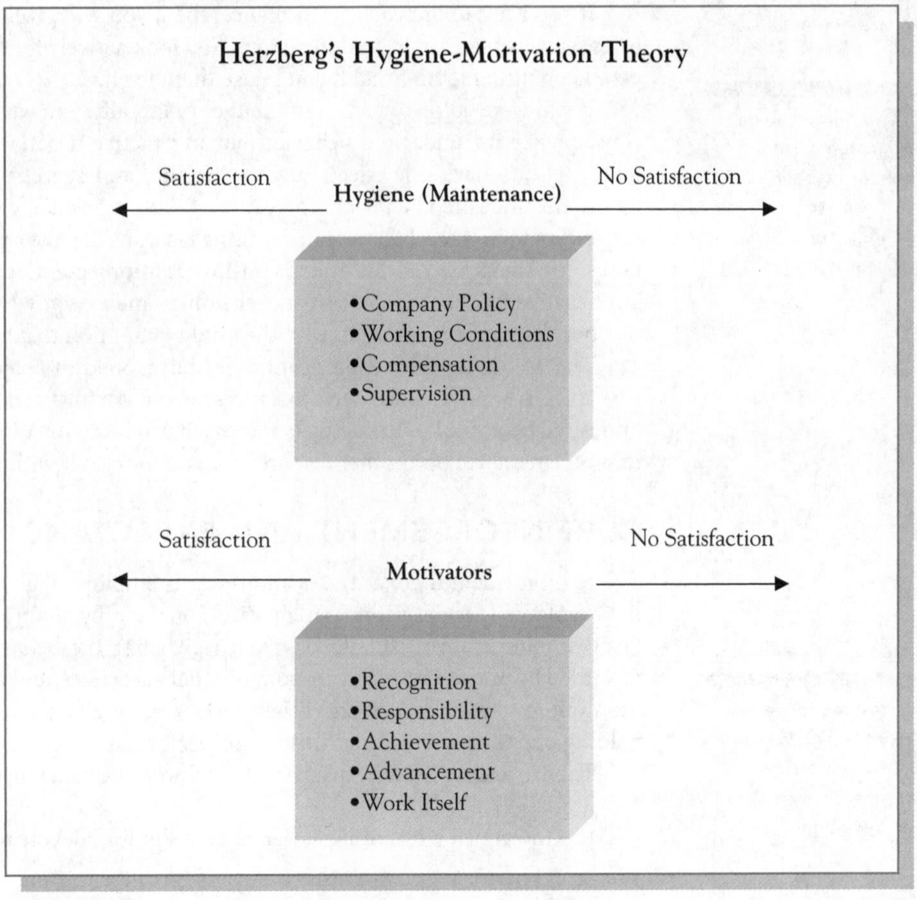

Herzberg's Hygiene-Motivation Theory

Satisfaction   **Hygiene (Maintenance)**   No Satisfaction

- Company Policy
- Working Conditions
- Compensation
- Supervision

Satisfaction   **Motivators**   No Satisfaction

- Recognition
- Responsibility
- Achievement
- Advancement
- Work Itself

FIGURE 8.3: An adaptation of Herzberg's hygiene-motivation theory, which was originally proposed by Herzberg, Mausner, and Snyderman in *The Motivation to Work* (New York: Wiley, 1958).

hygiene factors (maintenance factors) Factors in the job environment that produce job satisfaction or dissatisfaction but do not motivate performance.

However, removing the causes of dissatisfaction (the human relations approach) does not create satisfaction, and it therefore does not motivate performance. Herzberg called these environmental factors **hygiene factors**. They are also commonly called **maintenance factors**. For example, if you think you are underpaid, if you don't get along with your boss, if the kitchen isn't air-conditioned—these things can reduce motivation and cause absenteeism, poor work, and less work.

They are related to motivation only in the sense that they reduce it. Such factors must be maintained at satisfying levels to avoid negative motivation. But air-conditioning the kitchen or raising wages will not make the cooks work harder once the novelty wears off. In contrast, a second group of factors provides both motivation and job satisfaction.

motivators Whatever triggers a person's inner motivation to perform. In Herzberg's theory, motivators are factors within a job that provide satisfaction and that motivate a person to superior effort and performance.

These, Herzberg found, consist of opportunities in the job itself for achievement and growth—such factors as recognition, responsibility, achievement, advancement, the work itself. He called these factors **motivators**. If you give a cook who loves to invent new dishes a chance to develop a special menu item, you will see a motivator at work.

The answer to motivating employees, then, lies in the job itself. If it can be enriched to provide opportunity for achievement and growth, it will not only motivate the worker to perform well but will also tap unused potential and use personnel more effectively. We look at job enrichment in more detail later in the chapter.

❋ BEHAVIOR MODIFICATION

behavior modification Effecting behavioral change by providing positive reinforcement (reward, praise) for the behavior desired.

Behavior modification, a newer method for improving performance simply bypasses inner motivation and deals instead with behavior change. It takes off from the behaviorist's theory that *all behavior is a function of its consequences;* people behave as they do because of positive or negative consequences to them. If the consequences are positive, they will tend to repeat the behavior. If they are negative, they will tend not to.

If you want to improve performance, then, you will give **positive reinforcement** (attention, praise) whenever people do things right. You look actively for such behavior, and when you catch people doing something right, you praise them for it.

positive reinforcement Giving attention and praise for superior performance in order to encourage continued good results.

If you were going to carry out the theory literally, you would provide some form of negative consequence for undesired behavior, but in practice negative consequences (blaming, punishment) tend to have side effects such as hostility and aggressive behavior. However, you cannot ignore the undesired behavior. You can deal with it positively without threatening the person by suggesting the correct behavior in coaching fashion: "Let me show you how." But the really important side of behavior modification is positive reinforcement. It reverses the usual story of nothing but negative feedback ("The boss never notices me except when I do something wrong"), and it satisfies the need for attention with the kind of attention that builds self-worth.

The use of behavior modification has burgeoned in recent years, and it can sometimes be very effective. There have been instances where positive reinforcement has not only corrected undesired behavior but has actually increased productivity. Whatever its theoretical base, positive reinforcement can be another resource for you to try out with your people.

❋ REINFORCEMENT AND EXPECTANCY THEORY

negative reinforcement Withholding praise and rewards for inferior performance in order to encourage better results.

The reinforcement theory praises and rewards employees' good behavior. Undesired behavior is not reinforced. Supervisors can modify behavior by giving appropriate praise and rewards. Positive reinforcement should be given right after the behavior occurs. Good performance is rewarded by praise, preferably in front of other associates, and other incentives like bonuses, gifts, promotions, pay increases, and other perks can be given. **Negative reinforcement** is the withholding of praise and rewards for inferior performance.

The expectancy theory explains that employees are concerned about three important questions:[8]

1. How much effort, diligence, and care should I devote to my work?

2. If I perform well as a result of my effort, diligence, and care, will I obtain desired outcomes to satisfy my needs?

3. Does my employer provide work outcomes that satisfy my needs?

With the work expectancy theory it is vital that supervisors provide the training and coaching necessary so that associates will have the expectancy of achieving superior performance. If, however, the superior performance goes unrewarded or even if the reward does not match up to associates' expectations, then dissatisfaction will result. It is important to realize that not all employees want the same reward. The best approach is to find out what will motivate them and offer a selection of rewards.

Applying Theory to Reality: Limiting Factors

LEARNING OBJECTIVE: Explain the challenges that hospitality supervisors may face in motivating employees.

Now, what can you do with all this theory? There is a great deal in it that you can put to work if you can adapt it to your particular situation and to your individual workers. There are also circumstances that limit how far you can go. One limiting factor that immediately comes to mind is the *nature of many jobs* in the hospitality industry.

They are dull, unchallenging, repetitive, and boring. On the surface, at least, there does not seem to be much you can do to motivate the pot-washer, the security guard, the cleanup crew, the bed-maker, and the lightbulb-changer to keep them working up to standard and to keep them from leaving for another job.

Even among the less routine jobs there is little you can change to make the work itself more interesting and challenging. The great majority of jobs are made up of things that must be done in the same way day after day. At the same time, many jobs depend to some extent on factors beyond your control: What people do each day and how much they do varies according to customer demand. Unless your workers happen to find this interesting and challenging (and some people do), it is difficult to structure such jobs to motivate people. But the situation is not hopeless. Later in the chapter we see what creative management can do for even the dullest jobs.

A second limiting factor is *company policy, administration, and management philosophy*. Everything you do must be in harmony with company goals (customer-oriented and cost-effective) and must meet company rules and regulations. Furthermore, you do not control wage rates, fringe benefits, promotion policies, controls, and other companywide systems and practices. If jobs are totally standardized by scientific management methods, you cannot tamper with job content and method at all unless you go through proper channels and procedures established by the company.

Positive reinforcements can encourage associates to improve their performance.
wavebreakmedia/Shutterstock

The style of leadership characteristic of the organization will greatly influence what you can and cannot do. If the philosophy of management is authoritarian and high control, you will have a hard time practicing another approach. In particular, your relationship with your own boss and your boss's management style will influence the nature and scope of what you can do to motivate your people.

A third factor, closely related to the second, is the *extent of your responsibility, authority, and resources*. You cannot exceed the limits of your own job. You may be limited in your authority to spend money, to make changes in job duties, and so on. Remember, too, that your boss is responsible and accountable for your results, and this goes all the way up the chain of command. If you are going to innovate extensively, you will need the blessing of your superiors. But maybe you can get it!

PROFILE Bob Haber,
Former Director of Human Resources, Grand Hyatt Tampa Bay

Courtesy of
Bob Haber

Early in my management career, I quickly realized how important it was to be able to motivate people and how this motivation would lead to the success of my department or the operation I was responsible for. I found this to be true no matter what type of profession or business I was in.

As a human resource professional, I believe that one of my most important responsibilities is to maintain and develop a motivated workforce. Associates who are motivated to perform are key to the successful operation of any business or company. This is especially true in the hospitality business. After all, true hospitality requires associates to perform in a genuine, caring manner that is perceived by the guest as such.

This genuine approach is largely derived from an associate's desire to want to be hospitable or enjoy what they do. So, how do you get associates to enjoy their work? One important part is the way you motivate or lead them in the workplace. Although everyone has different wants and needs to be motivated, there are several things you can do as a leader to motivate your associates to perform to their fullest.

Associates should be "bought in" to what you want the end result to be. They have to see what's in it for the company and them. Understanding the "why" of what is expected of them is important. People rarely perform well for no reason. Being part of the process requires their understanding of the desired end result and why it is important. Because of this, leaders must engage their associates. They have to communicate openly and directly. They have to give information freely and accept it from their associates. This is especially true with today's younger workforce who will not perform with the "because I said so" approach. In order to motivate associates, leaders must both communicate needed information regularly and listen to feedback on an ongoing basis.

The environment you provide as a leader is also a key part of an associate's motivation. Having someone who enjoys what they do and the surroundings they do it in will help maximize their contributions. I don't mean the physical surroundings; rather, the environment they are provided through dignity and respect.

Treating associates with respect greatly adds to their emotional or mental well-being, and as a result will help them be more productive for you and the company. You must also ensure that this mutual respect is shown from coworker to coworker. Get to know your associates and provide them with a comfortable workplace that makes them feel good and want to come to work each day.

Making it rewarding to work hard and perform will also add a lot of value to your motivation techniques. Both formal and informal praise, on an ongoing basis, are crucial. Associates want to know they are doing well, and what it takes to be good. There are many long-term or formal types of recognition that will help motivate your associates. Although this is important, positive reinforcement and praise that is timelier will motivate continued performance and behavior like no other.

This on-the-spot reward will best relate the performance with your desired outcome and motivate for continued good performance. Praise often and be genuine! Say thank you when a job is well done. Provide ongoing coaching when associates don't quite perform as expected. Give them the tools and opportunity to succeed through their own efforts. As I said, associates want to know what it takes to be good.

I like to measure our motivation efforts by seeing how much associates will rally around you and the company when the "going gets tough." How difficult is it to get someone to perform out of a normal routine, or to take on an unusually difficult task? This can be an excellent indicator of how well motivated they really are. In any case, proper motivation will only lead to good things and help maximize your associate's contributions.

Another limiting factor is the *kinds of people who work for you*. If they are only working there *until*, the job does not really motivate them; they are just putting in time. They do not put forth their energy and enthusiasm because work is not the central interest in their lives. They have something going on outside—family, studies—that takes care of most of their personal needs and interests, and they don't want to work any harder than they have to.

The large numbers of workers who are dependent personalities often pose a motivation problem—they want you to tell them what to do at every turn, until they sometimes seem like millstones around your neck. How do you shake them loose and put them on their own?

The *constant pressures* of the typical day in the life of the hospitality manager tend to fix attention on the immediate problems and the work itself. It is all too easy to become work-oriented rather than people-oriented, especially if you have been an hourly worker and are more at home managing work than managing people. This is a limitation that managers can deliberately strive to overcome once they see how motivating people can help to accomplish the work better.

Another limitation is *time*. You probably think your day is already too full, and it may well be. It takes time to get to know your people. It takes time to figure out ways of changing things that will make people more motivated.

It takes a lot of time to get changes through channels, if that is necessary. It takes time to get people used to changes in their jobs, and it usually takes time before you begin to see results. But the effective manager will make the time and will gain time in the end by making more effective use of people.

There are limitations in the theories themselves when it comes to applying them. The primary one is that there is no law of motivation or set of laws that you can apply as you can with scientific or mathematical formulas. This, of course, is true of everything having to do with human beings. Everyone is different, and their needs and desires and behavior respond in a kaleidoscope of change triggered by anything and everything—other people, the environment, the task, their memories, their expectations, *your* expectations, and what they ate for breakfast.

The theories themselves change. New experiments shed new light. The enthusiasms of the past give way to the fads of the future.

Who has the answers? What works? You have to translate the findings of others in terms of your individual workers and the jobs you supervise. These are judgments you make; there are no sure-fire answers. But there is plenty of guidance along the way.

Building a Positive Work Climate

LEARNING OBJECTIVE: Identify ways to build a positive work climate.

A positive work climate is one in which employees can and will work productively, in which they can do their best work and achieve their highest potential in their jobs. Meeting employee expectations and needs is one way to create a positive work climate. Before we take a look at others, let's discuss a similar concept: morale.

morale
Group spirit with respect to getting a job done.

Morale is a group spirit with respect to getting the job done. It can run the gamut from enthusiasm, confidence, cheerfulness, and dedication to discouragement, pessimism, indifference, and gloom. It is made up of individual attitudes toward the work that pass quickly from one person to another until you have a group mood that everyone shares. It may change from moment to moment. You see it when it is very high, and you notice it when it is very low; and if it is average, nobody says anything about it.

When people are unhappy in their jobs, they just plain don't feel good at work. They feel exhausted, they get sick easily and miss a lot of days, and eventually they give up because the job is not worth the stress and unhappiness. In an industry where many people are working "until" and do not have a sense of belonging, these kinds of feelings and behavior are contagious, and morale becomes a big problem. Absenteeism, low-quality work, and high employee turnover multiply production problems and cost money. It probably costs at least $1,000 every time you have to replace a busperson, for example.

High morale has just the opposite effects and is the best thing that can happen in an enterprise. To build a positive work climate, you need to focus on three areas: the individual, the job,

1. Write effective vision, mission, and goals statements, and ensure everyone knows them.
2. Actively listen to your employees.
3. Give a hand to your employees when appropriate.
4. Treat employees fairly and consistently.
5. Keep your employees informed.
6. Involve and empower your employees.
7. Use up-to-date and accurate job descriptions.
8. Orient, train, and coach your employees.
9. Formally evaluate employee performance at least twice yearly.
10. Praise and reward your employees.
11. Pay for performance.
12. Institute a profit-sharing or other gain-sharing program for employees.
13. Let your employees make as many of their own decisions as possible.
14. Cross-train employees, rotate their positions, and have a career ladder and promote from within.
15. Be able to perform the job you supervise.
16. Manage your time.
17. Be a good role model.
18. Establish competitive and equitable pay rates.
19. Offer a competitive benefit package suited to your employees.
20. Provide a pleasant, safe, and clean work environment.

FIGURE 8.4:   Twenty ways to build a positive work climate.
Source: Jay R. Schrock.

and the supervisor. Figure 8.4 lists 20 ways to build a positive work climate. Let's look at some of the most important ways that you can make work enjoyable.

✻ PURPOSE = MOTIVATION

You've got to have a purpose in life![9] Before you can motivate, you need a purpose. People yearn for purpose—for doing something that's important, that engages their full potential in a way that's meaningful beyond their bank accounts. And that makes your job as their supervisor a little harder. You may not have as much control over their compensation and benefits as you would like to have. But you do have control over how inspired they are and how connected they feel to the mission their jobs serve.[10]

Most people want to feel that, thanks to their efforts, the world is a little better off by nightfall than it was when the day started. You just have to figure out how what they do makes it happen that way. The first thing is for you to understand how your company improves the world—and how your job serves that mission.[11] After connecting your job to the mission, you can help your employees make the connection. For example, hospital custodians are not just sweeping floors— they are helping to save lives. The "connection" can come from the product itself, the guests, the community co-employees, even the employees' families and the dream future that the employees' jobs with your company are helping them realize.[12]

Focus: The Individual

LEARNING OBJECTIVE: Explain why individual strategies of motivation are essential.

The starting point is your individual workers—one by one. The idea that everybody works for one thing, like money, is no longer credible. Employees are glad to have the paycheck, but whether they are willing to work hard for that money or for something else or for anything at all is what you want to determine. Because everybody is different, you are going to need an individual strategy of motivation for each person—not a formal program, just a special way of dealing with each one that brings out their best efforts and offers them the greatest personal satisfaction.

Getting to know your people takes an indirect approach. People are not going to open up to the boss if you sit down with them at the coffee break and ask them questions about what they want from their jobs. They will tell you what they think you want to hear, and they will probably feel uncomfortable about being quizzed.

You may have hired them for one reason, but they probably come to work for altogether different reasons, which they may think is none of your business. They have taken the job as a vehicle for getting where or what they want, but that is a hidden agenda. For some people it is money, for some it is pride, for some it is status, for some it is something to do *until*. If you can find out what kind of satisfactions they are looking for, it will help you to motivate them.

You can learn about them best by observing them. How do they go about their work? How do they react to you, to other workers, to customers? What questions do they ask, or do they ask any? How do they move—quickly, slowly, freely, stiffly? How do they look as they speak or listen? Notice their gestures and facial expressions. What makes them light up? What makes them clam up? Pay special attention to what they tell you about themselves in casual conversation. This may be an entirely new approach for you, but people-watching is really quite interesting, and you can quickly become good at spotting clues.

Clues to what? Needs and desires, discontents and aspirations. Frustrations, drive, and achievement. Ability and performance, too, and whether performance is up to par for that job and whether this person has abilities the job does not call on. But primarily needs, desires, and responses, because these are the motivators you want to channel into high performance that will satisfy both you and them.

Observing your people has a purely practical purpose. You are not going to try to psychoanalyze them, probe for hidden motives, delve into what really makes them tick. You can't. That takes years of training you don't have, and a great deal of time you don't have, either. Furthermore, you shouldn't.

High morale helps keep employee turnover low. At this hotel, the employee turnover rate is 23 percent per year.

Andrey_popov/Shutterstock

If you are wrong in your amateur analysis, your employees will consider you unjust, and if you are right, they will feel vulnerable—you know them too well. Either way, it is going to interfere with motivation rather than improve it. Your approach, in contrast, should be practical, pragmatic, and experimental; you could even call it superficial.

You observe your people and get ideas of what you might do to motivate this or that person to perform better for you, as well as get more personal satisfaction from the work. You try out an idea, and if this person does not respond, you try something else. What they respond to is what is important and what you have to work with. The personal whys—the inner needs—are simply clues that you sometimes use to reach the what-to-do.

John Kotter, former Matsushita Professor of Leadership at Harvard Business School, has observed that only companies whose employees are "intellectually and emotionally convinced that their business creates something that adds value to the world" will survive in the new economy.[13] But in an economic climate characterized by rapid change and job anxiety, can companies legitimately expect their employees to bring their hearts and minds to the workplace? And if so, what do employers need to do to or be willing to do to make this happen?[14] We should realize that fear and coercion don't work and that compensation alone is not the answer.

But giving employees information about how compensation decisions are made, and giving them the freedom to question it, is important. There's no quicker way to demonstrate your commitment to loyalty and trust than to let employees know what you know. For example, say your company faces a huge task of reducing costs. If you share the real situation with employees, you are more likely to receive several good ideas for cost-cutting that will add up to substantial savings.

Employees want their work to connect to a greater purpose—yes, people want to work in order to bring more good to the world. Hence, the importance of sustainability. Efforts to increase sustainable operations are likely to be embraced by employees, as they can identify with the cause. By focusing on values beyond profitability, companies can actually increase profitability.

Motivational Methods

LEARNING OBJECTIVE: List the best ways to keep employees motivated.

Imagine that you are a first-time supervisor and the youngest in the department. With 10 senior employees around you, how do you secure and maintain their trust as their boss?

Answer: Put not your trust in people, but put your people in trust. Really, if your subordinates don't trust you, then you won't be able to get their full cooperation.[15] The problem is that you just don't earn the trust of people overnight. It requires a long process on your part, along with a considerable amount of time and energy. There are seven best approaches:

1. *Empower the workers.* Give them the chance to participate in problem-solving and decision-making processes, whenever possible.

2. *Share vital information.* If they know that you trust them with sensitive information, they could easily reciprocate by performing better in their jobs.

3. *Work objectively with everyone.* Don't play favorites. People have different personalities, but just the same, they should be given equal treatment, as this promotes teamwork.

4. *Be a decisive boss.* Avoid being wishy-washy. If you don't have a ready answer, then be honest, and promise to immediately return to the concerned worker with the right answer or decision.

5. *Show appreciation to people's good deeds as soon as it becomes apparent.* Don't drag your feet. But be specific by citing what you appreciate.

6. *Maintain a two-way, personal, and eyeball-to-eyeball communication with everyone on a regular basis.* Even with technology around us, there is still no substitute to having an active dialogue with people.

7. *Be polite.* No matter the amount of pressures that you are carrying, you need to be nice.[16]

❋ LEADING ENERGY

The major challenge facing leaders and human resource professionals over the next 20 years will be leading human energy. The challenge of energy management, though, is not just an individual endeavor; it extends to organizations—big time. Many people, if not most people, seldom bring their best efforts to work; they seem to save them for evenings and weekends.

If you ask managers, "Of all the people you have met in your career, what percentage of them are fully engaged at work?" most say, "Less than 10 percent."[17] Authors James G. Clawson and Douglas S. Newburg comment, "One does not need nationwide polls or leadership seminar participants' opinions to observe this phenomenon; virtually every working establishment is full of people going through the motions." The authors add: "How many times in your career has your manager/supervisor asked you how you want to feel today?"[18] Most HR professionals and leaders *assume* that professionals will do what they have to do and not let their feelings get in the way. And that is the problem—feelings affect our performance—plain and simple! Sadly, this is true and presents a challenge to human resource professionals and supervisors.

The dilemma in motivating associates is knowing when the effect of goals becomes demotivating. For example, a company was having a motivational retreat when a senior vice-president came to give a pep talk. He said, "Our stock value is currently $95, and if it's not up to $125 by December 31, the CEO and I will not be getting our bonuses, so you need to get your rears in gear!"[19] Imagine how the associates felt. But we also need to realize that similar examples happen all the time: If a company gets a 10 percent increase in sales one year—yes, you guessed—it wants 12 percent the next. You've just worked your butt off and now they want more! So, what's the answer?

❋ DEALING WITH SECURITY NEEDS

It is relatively easy to spot people with high security needs. They look and act anxious, uncertain, and tentative. They may be among those who ask you how to do everything, or they might be too scared even to ask. Fear and anxiety are demotivators; they reduce motivation. When security needs are not satisfied, people cannot function well at all; in fact, these people are among those who leave during the first few days on the job.

Here is where Maslow's theories come in handy. If you see that someone has a need for security and you can help that person satisfy that need, you ease that person along to a higher motivational level. To satisfy these needs, you do all the things that we have been recommending in earlier chapters. You tell them what to do and how to do it; you tell them exactly what you expect. You train them. All these things provide a reassuring structure to the work that protects them from the uncertainties of working. It reduces their mistakes and builds their motivation and confidence.

You let them know where they stand at all times. You support them with coaching and feedback and encouragement. You give them positive reinforcement for things they do right, and you retrain them to help them correct their mistakes. You do not solve their problems, you do not cuddle and coddle; you help *them* to do their jobs *themselves*.

You keep on making positive comments about their work even when they are fully trained and you are satisfied with their performance. It is natural for a supervisor to stop paying attention to a worker once things are going well, but even a short absence of approving comments can trigger doubt and uncertainty again in workers who are insecure. Recognition, even if it is only a big smile and a passing, "Hey, keep up the good work!" is an affirmation that life on the job is, after all, not uncertain and threatening. Above all, you must avoid any use of fear as a motivator. This is the last thing that these people need.

Evaluate their work frequently, and give praise for things done right, especially for improvement of any sort. Use improvement to build confidence: Accentuate the achievement and the potential—"See how far you have come; see where you can go from here." Show them that you expect them to do well. Your confidence will give them confidence. And if you can build confidence, you may eventually activate self-motivation and aspiration. Satisfaction of primary needs allows these higher-level needs to emerge.

❄ DEALING WITH SOCIAL NEEDS

Everybody has social needs. You might not think of work as being a place to satisfy them, but it often is. For many people, a job fills the need to be with others, the need to be accepted, the need to belong. These are powerful needs. Often, they will fall into Herzberg's category of hygiene factors: They cause dissatisfaction when they are unsatisfied, but they do not motivate when satisfied. But for some people they can be motivators, too.

For example, consider the homemaker who gets a job because she wants to talk to people who are more than three feet tall. If you hire her as a cashier or a switchboard operator, she probably won't be very good at it because this is not what she came to work for. But if you make her a desk clerk or a server or a sales rep where she can talk to people all day long, she could easily become a higher achiever.

Whether or not social needs can be turned into motivators, it is useful when people find such needs being satisfied on the job, both in terms of their individual development and in terms of the general work climate. People whose social needs are unmet may just not work very well, or they may even provoke trouble and conflict.

What can you do to help meet people's social needs? There are two specific needs you can work on, and it takes hardly any of your time. One is the *need for acceptance*. We have talked about this before: You build a person-to-person relationship and you treat each person as a unique individual who has dignity and worth. You respect their idiosyncrasies (unless they interfere with the work): You speak softly to Peter because that is what Peter responds to. You scream and yell at Paul because that is your unique way of relating to Paul and you both know it, and Paul will think he doesn't matter to you anymore if you treat him any other way.

You deal with each person differently, but you treat each according to the same standards, whether she is good-looking or plain, whether his mother is on welfare or owns the biggest bank in town. Each one is a person who has value and worth, and you treat them all that way. "Mind and body in motion." Staff at different levels need to feel that they belong, are appreciated, and have a purpose with the company.

You also make it clear that you value each person's work and that it is important to the organization no matter how menial it is. The well-made bed, the properly washed salad greens, and the sparkling-clean restroom all please customers; the crooked bedspread, the gritty salad, and the empty tissue holder send customers away.

This attention to detail can be as important to the success of the hotel or restaurant as the expertise of the sommelier or the masterpieces of the chef. You can make people feel that they are an essential part of the entire organization, that you need them, that they belong there. A sense of belonging may be your most powerful ally in the long run—and it helps the long run to happen.

This *need to belong* is the other social need that you can do a lot to satisfy. Things you should be doing anyway help to satisfy this need, such as making people feel comfortable in their jobs by training them, coaching them, telling them where they stand, evaluating their work frequently. Open communications also encourage belonging; people feel free to come to you with suggestions or problems.

Keeping people informed about changes that affect them is a way of including them in what is going on—and if you leave anybody out, you reduce that person to a nobody. You can also include people in discussions about the work, inviting their ideas, feelings, and reactions. If you can build a spirit of teamwork, that too will foster a sense of belonging.

One's peer group also nurtures belonging. You need to be aware of social relationships among your workers and to realize that these relationships are just as important as their relationship with you, and sometimes more so. Often, peer pressure is more influential than the boss is. You need to have the group on your side—if it ever comes to taking sides—and that is best done through good relationships with each person.

These people work under you, and they look at you as their boss. They expect you to be friendly and to sit down with them if they invite you, but they do not expect you to be one of the gang. In fact, your uninvited presence for more than a moment or two may act as a constraint to their socializing.

Groups and group socialization are a normal part of the job scene. Often, groups break into cliques, with different interests and sometimes rivalries. You should not try to prevent the formation of groups and cliques. But if competition between cliques begins to disrupt the work, you will have to intervene. You cannot let employee competition interfere with the work climate.

❋ REWARDING YOUR EMPLOYEES

Incentive pay, bonuses, and various kinds of nonmonetary rewards can be very effective motivators if they activate people's needs and desires or are related to their reason for working. One of the problems, of course, is that what motivates one person leaves another indifferent, yet to treat people fairly you have to have rewards of equal value for equal performance.

These methods of triggering motivation begin with the carrot principle of dangling a reward for good performance. When people need or want the reward, they will work hard in expectation of getting it. If they do not want the carrot, it has no effect.

Once the reward is achieved, the cycle must start again: The desire must be activated by the *expectation of reward*, as Herzberg points out. No expectation, no achievement, and performance slumps back to a nonreward level unless people begin to derive satisfaction from the achievement itself. However, there is no doubt that rewards are useful motivators. In many jobs, the boring repetition of meaningless tasks precludes a sense of achievement that is fulfilling, and rewards may be the only resource you have for motivating.

The entire system of rewards, both monetary and otherwise, must be worked out with care, not only for getting the maximum motivation but also for fairness in the eyes of the employees. The performance required to achieve the reward must be spelled out carefully, and the goal must be within reach of everyone. People must know ahead of time what the rewards are and must perceive them as fair or they will cause more dissatisfaction than motivation.

How do you make rewards into effective motivators if people's needs and desires are so different? Somebody with eight children to feed might work very hard for a money reward or the chance to work more overtime. Another person might outdo himself for an extra day of paid vacation. Still another would do almost anything for a reserved space in the parking lot with her name on it in great big letters right near the door. Such rewards might be suitable prizes in an employee contest, with the winner being allowed to choose from among them.

Rewarding associates contributes to a positive work climate.
Courtesy of Wyndham Hotels Group.

You might get people involved by letting them suggest rewards (keeping the final decision to yourself or letting the team decide—within limits). Any involvement increases the likelihood of sparking real motivation. Actually, any reward can be more than a carrot.

It can be a recognition of achievement, of value, of worth to the company. It can build pride; it can generate self-esteem. It can also be a goal. Once employees earn a reward, if it gives them satisfaction, they will probably go for it again. Then you have activated motivation from within, with commitment to a person's own goal. And that, in miniature, is what successful on-the-job motivation is all about—fulfilling individual goals and company goals in the same process. The more both the employees' and the company's goals overlap, the greater motivation will be.

❋ DEVELOPING YOUR EMPLOYEES

Another way of maintaining a positive work climate is to help your people to become better at their jobs and to develop their potential. This may be one of the most critical things you do. A large percentage of the people in foodservice and hospitality enterprises are underemployed, and as managers we really do not utilize the skills and abilities of the people who work for us as fully as we could. Your goal should be to make all your people as competent as you can because it makes your job easier, it makes you look better to your superiors, and it is good for your people.

You can develop your beginning workers through training, feedback, encouragement, and support, as well as by providing the right equipment and generally facilitating their work. By the way you deal with them, you can also give them a feeling of importance to the operation, a sense of their own worth, and a feeling of achievement and growth. Concrete recognition of improvement, whether it is an award, a reward, or merely a word of praise, can add to the pride of achievement.

If you have people with high potential, you should do all these things and more. You should try to develop their skills, utilize any talents you see, challenge them by asking for their input on the work, give them responsibilities, and open doors to advancement to the extent that you can. One thing is certain: If you have trained someone to take your place, it will be a lot easier for your company to promote you. But if none of your subordinates can fill your job, the company is less likely to move you up because it needs you where you are.

Developing your people also helps morale. It gives people that sense of moving forward that keeps them from going stale, marking time, moving on a treadmill. It is also important to your acceptance as a leader to have people feel that you are helping them to help themselves.

You develop employees by involving them. Employees who are asked to influence what happens at work tend to develop a sense of ownership; this feeling of ownership breeds commitment. Employees can become effectively involved in many managerial activities, such as evaluating work methods, identifying problems, proposing suggestions, and deciding on a course of action. Employees can tell you better than anyone else how their own jobs should be done.

For instance, McGuffey's Restaurants, a dinner-house chain based in North Carolina, asks for employee input on ways to improve service and also asks employees to elect representatives to an associate board at which employees' concerns will be addressed. In some cases, when you involve employees, you are actually empowering them. Empowering employees means giving them additional responsibility and authority to do their jobs. Instead of employees feeling responsible for merely doing what they are told, they are given greater control over decisions about work. For example, in some restaurants, servers are empowered, or given the authority, to resolve guest complaints without management intervention. A server may decide, in response to a guest complaint, not to charge the customer for a menu item that was not satisfactory.

At McGuffey's Restaurants, the company gives employees their own business cards, which they can use to invite potential customers in for free food or beverages. The company even lets employees run the restaurants two days a year, during which they can change the menu or make other changes.

Following are some guidelines for *empowering* your employees:

- Give employees your trust and respect.
- Determine exactly what you want employees to be empowered to do.
- Train employees in those new areas. Be clear as to what you want them to do.

- Create an environment in which exceptions to rules, particularly when they involve customer satisfaction, are permissible.
- Allow employees to make mistakes without being criticized or punished. Instead, view these times as opportunities to educate your employees.
- Reward empowered employees who take risks, make good decisions, and take ownership.

Finally, you should also continue to develop yourself. Chances are that you won't have much time for reading and studying, but you should keep pace with what is going on in other parts of your company and in the industry as a whole—read trade publications and attend trade association meetings. You can also watch yourself as you practice your profession, evaluating your own progress and learning from your mistakes. Make a habit of thinking back on the decisions you have made. What would have happened if you had done something differently? Can you do it better next time?

Focus: The Job—Providing an Attractive Job Environment

LEARNING OBJECTIVE: Describe the most effective ways of creating an attractive job environment.

The employee's job environment includes not only the physical environment and working conditions but the other employees, the hours, rates of pay, benefits, and company policies and administration. You may recognize these as *hygiene* or *maintenance factors.*

As Herzberg pointed out, such job factors do not motivate. But any of them can cause dissatisfaction and demotivation, which can interfere with productivity and increase turnover. So it behooves the leader to remove as many dissatisfiers as possible. To the extent that you have control, you can provide good physical working conditions: satisfactory equipment in good working order; adequate heating, cooling, and lighting; comfortable employee lounges; plenty of parking; and so on.

You can see that working hours and schedules meet employees' needs as closely as possible. If you have anything to say about it, you can see that wages and benefits are as good as those of your competitors or better, so that your people will not be lured away by a better deal than you can give.

There is not much that you can do about company policy and administration if it is rigid and high control, except to work within its limits, stick up for your people, and do things your own way within your sphere of authority. We will assume that the management philosophy is not based on fear and punishment or you would not be there yourself.

✳ PUTTING THE RIGHT PERSON IN THE RIGHT JOB

If you get to know your associates, you are in a good position to figure out what jobs are right for what people. People with high security needs may do very well in routine jobs: Once they have mastered the routine they will have the satisfaction and security of doing it well. Putting them in a server's job would be a disaster. Putting people-oriented associates in routine, behind-the-scenes jobs might be a disaster, too.

Many cooks enjoy preparing good things for people to eat. Even when they must follow other people's standardized recipes, there is the satisfaction of being able to tell exactly when a steak is medium rare, of making a perfect omelet, of arranging a beautiful buffet platter. Bartenders often enjoy putting on a show of their pouring prowess. These people are in the right jobs.

Pride in one's work can be a powerful motivator. Some people get a great sense of achievement from tearing into a room left in chaos by guests and putting it in order again, leaving it clean and inviting for the next guest. They, too, are in the right jobs. The professional dishwasher we have mentioned several times obviously took great pride in his work and wore his occupation as a badge of honor. He belonged in his job, and it belonged to him, and in a curious way it probably satisfied all levels of needs for him.

❋ MAKING THE JOB INTERESTING AND CHALLENGING

People do their best work when something about the work involves their interest and stimulates their desire to do it well. People who like what they are doing work hard at it of their own accord. People who don't like their jobs drag their heels, watch the clock, do as little as they can get by with, and are called lazy by the boss.

Different things about work appeal to different people. Some are stimulated by working with guests: They get a kick out of making them welcome, serving them well, pleasing them, amusing them, turning an irate guest into a fan by helping to solve a problem. Some people are miserable dealing with guests and enjoy a nice, routine job with no people hassles, where they can put their accuracy and skill to work straightening out messy records and putting things in order. Some people like jobs where there is always some new problem to solve; others hate problems and like to exercise their special skills and turn out products they are proud of.

What these people all have in common is that something about the content of their job both stimulates and satisfies them. Stated in theoretical terms, it satisfies their higher needs, those related to self-esteem and self-fulfillment. Specifically, people work hard at jobs that give them opportunity for achievement, for responsibility, for growth and advancement, for doing work they enjoy doing for its own sake.

There are two ideas here that you as a leader can use in motivating your people. One is to put people in jobs that are right for them, as just discussed. The other is to enrich people's jobs to include more of the motivating elements. Of course, there are limits to what you can do, but the more you can move in this direction, the more likely you are to create a positive work climate and motivate associates.

Workers who are bored are underemployed: The job does not make use of their talents, their education, and their abilities. They are only there until they find a more interesting and challenging job. Not only will you have to train their replacements sooner or later, but you are not making

Making the job interesting and challenging is a key part of a supervisor's job.

use of abilities right now that could contribute a great deal to your department and to the entire organization. Furthermore, as we said in Chapter 1, supervisors have an obligation to develop their people.

job loading
Adding more work to a job without increasing interest, challenge, or reward.

You cannot move associates into better jobs unless jobs are available, but you can look for ways to enrich their current jobs by building some motivators into them. This does not mean asking them to take on additional, but similar, tasks—that is called *job loading*. *Job enrichment* means shifting the way things are done so as to provide more responsibility for one's own work and more opportunity for achievement, for recognition, for learning, and for growth.

You might start by giving people more responsibility for their own work. Relax your control; stop watching every move they make. Let them try out their own methods of achieving results as long as they do not run counter to the standards and procedures that are an essential part of the job. In other words, decrease controls and increase accountability. This must all be discussed between you, and there must be a clear understanding, as in any delegation agreement.

From there you can experiment with other forms of job enrichment. You can delegate some of your own tasks. You can rearrange the work in the jobs you are enriching to add more authority and responsibility for the workers. You can give new and challenging assignments. You can assign special tasks that require imagination and develop skills.

If, for example, you find that you have creative people in routine kitchen jobs, let them try planning new plate layouts or garnishes. If someone who majored in English is working as a payroll clerk, let her try her hand at writing menu fliers or notices for the employee bulletin board or stories for the company magazine. Look for people's hidden talents and secret ambitions and use them, and keep in mind recommending them when more suitable jobs are available.

Another idea that is being tried out in a number of industries is replacing the assembly-line method of dividing the work into minute, repetitious parts by giving a worker or group of workers responsibility for an entire unit of work or complete product, including quality control. Is there a way of avoiding assembly-line sandwich-making that would give each worker or a group of workers complete responsibility for one kind of sandwich, letting them work out the most efficient method? Could you give a cleaning team responsibility for making up an entire corridor of rooms, dividing the tasks as they see fit?

There are many jobs in which the work is going to be dull no matter what you do. But even in these a shift in responsibility and point of view can work near miracles. A concerted program of job enrichment for cleaning and janitorial services carried out at Texas Instruments is an example of what can be done with routine low-skill tasks. These services were revamped to give everyone a role in the planning and control of their work, although the work itself remained the same. Extensive training embodying Theory Y principles was given to supervisors and working foremen, while worker training included orientation in company goals and philosophy and their part in the overall operation. A team-oriented, goal-oriented, problem-solving approach encouraged worker participation in reorganizing, simplifying, and expediting the work.

Increased responsibility, participation, and pride of achievement generated high commitment as well as better ways of doing the work. In the first year's trial the cleanliness level improved from 65 percent to 85 percent, the number of people required dropped from 120 to 71, and the quarterly turnover rate dropped from 100 percent to 9.8 percent. The annual savings to the company was a six-figure total.

The average educational level of these workers was fourth or fifth grade, proving that Theory Y management is applicable up and down the scale. A major program such as this takes a long time to develop and implement and is out of the reach of the first-time supervisor working alone. But it shows what can be done when dedicated leadership and enlightened company policies activate employee motivation.

Any job enrichment effort is likely to produce a drop in productivity at first as workers get used to changes and new responsibilities. It takes a coaching approach to begin with and a lot of support from the boss. It is also essential to initiate changes slowly and to plan them with care. Too much responsibility and freedom too soon may be more than some associates can handle, either out of inexperience or because of the insecurities involved. Again, it is a situation in which your own sensitivity to your workers is a key ingredient.

Focus: The Leader

LEARNING OBJECTIVE: Discuss why the leader is the key to a positive work environment.

Ultimately, it is the leader who holds the keys to a positive work climate. It is not only the steps she or he takes, the things she/he does to spark motivation; it is also the way that leaders themselves approach their own tasks and responsibilities—their own performance of their own jobs.

If they themselves are highly motivated and enthusiastic about their work, their people are likely to be motivated, too. If they have high expectations of themselves and their people, and if they believe in themselves and their people, the people will generally come through for them. It is motivation by contagion, by expectation, by example.

In some operations the manager conveys a sense of excitement, a feeling that *anything is possible, so let's go for it!* You find it sometimes in the individual entrepreneur or the manager of a new unit in a larger company. If the manager is up, the people are up, too, and it is an exciting place to work. It is not unusual for people who have worked for such managers to end up as entrepreneurs themselves, putting their own excitement to work in an enterprise of their own.

Tony's Restaurant in St. Louis is a case in point. Owner Vincent Bommarito's enthusiasm, high standards, and involvement with employee development and performance, coupled with an anything-is-possible approach, have spawned at least 20 restaurants owned and operated by former employees. Of course, there are the added incentives of ownership in such cases, but it really begins with the excitement and enthusiasm of the original restaurant experience.

At the opposite extreme, leaders who are not happy in their jobs, who are not themselves motivated, will have unmotivated associates who are faithful reflections of themselves—management by example again. You cannot motivate others successfully if you are not motivated yourself. And if you are not, you need a change of attitude or a change of job.

If you give 75 percent of your effort to your job, your people will give 25 to 50 percent. If you put forth a 100 percent effort, your people will give you 110 percent. If you expect the best of people, they will give you their best. If you expect poor performance, poor performance is what you will get. If you tell people they can do a certain thing and they believe in you, they can do it—and they will. If you tell them it is beyond their ability, they won't even try.

This contagious kind of motivation can run back and forth between supervisor and workers; they can motivate you if you will let them. If you have good relationships with your people, they can spark your interest with new ideas about the work. They can help you solve problems. Their enthusiasm for the work will sustain your own motivation in the face of setbacks and disappointments. When a "we" attitude prevails, it builds belonging, involvement, and commitment.

❀ SETTING A GOOD EXAMPLE

role model
A person who serves as an example for the behavior of others.

Whether you are aware of it or not, you set an example for your workers; they are going to copy what you do. The psychologist's term for this is **role model**. If you expect the best work from your people, you've got to give your best work to your job. Again, if you give 100 percent of your time and effort and enthusiasm, chances are that your workers will give you 110 percent. But if they see you giving about 75 percent and hear you groaning about your problems, they will give you only 25 to 50 percent of their effort. So if you want a fair day's work from your people, give a fair day's work to them: **management by example** it is sometimes called.

management by example
Managing people at work by setting a good example—by giving 100 percent of your time, effort, and enthusiasm to your own job.

Giving your best means keeping your best side out all the time. Everybody has a good side and a bad side, and most of us are vulnerable to a certain few things that can turn that bad side out and cause us to lose our cool. This is disastrous when you are a role model, particularly if you are supervising people who deal with customers.

If you lose your temper with a group of workers and shout at them, they are going to carry the echo of your voice and the feelings it arouses in them right into the hotel lobby or the dining room or the hospital floor. They are going to be impatient and hostile and heedless of the guests' needs. And there goes the training you have given them in guest relations.

Your good side is as influential as your bad side. If you want your people to treat guests courteously and serve them well, treat your associates courteously and well. If bad moods are contagious, so are good moods. Enthusiasm is contagious. If you would like your associates to enjoy their work, be enthusiastic yourself. Is that a big order? Sometimes. But if you can do it, it works. Set your sights high; expect the best of your associates.

On the one hand, if you expect their best, they will usually give you their best if you approach the subject positively. If you show them you believe in them and have confidence in their ability to do the job, if you cheer them on, so to speak, they will attach the same value to their performance that you do. They will take pride in their work and in their own achievements. On the other hand, if you suddenly tell them to improve their work, without warning and in a critical way, implying that they are slackers and don't measure up, they are likely to resent the criticism and resist the demand.

✳ ESTABLISHING A CLIMATE OF HONESTY

A positive work climate requires a climate of honesty. We have talked about honesty as one of the things that workers expect of a leader. It means that you are honest with them when you talk to them about their performance, potential, and achievements and mistakes. It means that you keep your promises and give credit where credit is due.

It means that you do not cheat, lie, or steal from the company: You do not take food home from the kitchen or booze from the bar, you do not take money from guests in return for a better room or a better table. You are a role model and you do not do these things, not only because they are unethical, but also because you want your associates to be honest; they are going to imitate you. This is another example of leadership by example.

You do not say one thing and do another. Nothing confuses an associate more than a supervisor who gives good advice but sets a bad example. You are consistent and fair. You do not manipulate; you are open and aboveboard; you can be trusted. A climate of honesty encourages the growth of loyalty. If you are loyal to the company that employs you and are honest and fair and open with your associates, they will develop loyalty to both you and the company.

If you put down the company, you destroy your entire work climate because your workers will begin to believe that the company is a lousy place. If you feel like running down the company now and then, keep it to yourself. If you feel like that all the time, get out. You cannot do a good job as supervisor with those feelings bottled up inside.

CASE STUDY: Kitchen Nightmares

Congratulations, you have just been promoted to sous chef in the kitchen of a popular restaurant! You are reflecting on your good fortune and assessing the situation, including who you will be supervising, when you realize that there is an older chef who is underperforming. According to the human resource files he has been warned about his performance repeatedly, but a recent evaluation still shows a serious need of improvement. He does not respond well to criticism and has not shown any effort to improve his skills.

In addition, two of the younger employees are often late for work—sometimes showing up as late as 45 minutes past the start of their shifts. Finally, there's a lot of quarreling in the kitchen that seems to stem from generational gaps between young and older employees. You end your shift each night discouraged and uneasy. You know that you need to step in, but you're unsure of how to get started.

Case Study Questions
1. How will you deal with the older chef who is underperforming?
2. How would you handle the younger employees' tardiness?
3. How would you get the employees to work better as a team together?

KEY POINTS

1. Employees want their leader to be qualified to supervise, be experienced, take charge, treat people fairly and equally, communicate, and treat people as human beings.
2. Motivation is the "why" of human behavior.
3. There are various theories of motivation: Use fear (McGregor's Theory X); combine fear with incentives (carrot-and-stick motivation); give money (economic person theory); give them consideration (human relations or social person theory, Maslow's hierarchy of needs); satisfy employee work needs, such as a need for growth or achievement (McGregor's Theory Y); and give positive reinforcement when a worker does something right (behavior modification).
4. Factors that limit your use of motivational techniques include the boring nature of many jobs, company management policies, extent of your authority and resources, the employees themselves, and constant time pressures.
5. A positive work climate is one in which employees can and will work productively.
6. Morale is a group spirit surrounding getting a job done.
7. Motivational methods include vision, goals that individuals have contributed toward, expectations, empowerment, sharing information, frequent one-on-one communication, resources, appreciation and recognition, fun, advancement opportunities.
8. In order to build a positive work climate, you need to focus on the individual, the job, and yourself (the leader) by getting to know your people, dealing with security and social needs, rewarding and developing your people, providing an attractive job environment, providing a safe and secure work environment, making the job interesting and challenging, setting a good example, and establishing a climate of honesty.

KEY TERMS

behavior modification	morale
dissatisfiers	motivation
economic person theory	motivators
hierarchy of needs	negative reinforcement
hygiene factors	positive reinforcement
job loading	role model
maintenance factors	self-actualization
management by example	social person theory

REVIEW QUESTIONS

Answer each question in complete sentences. Read each question carefully and make sure that you answer all parts of the question. Organize your answer using more than one paragraph when appropriate.

1. Name five expectations that employees often have of their leader.
2. Briefly discuss five motivational theories that make the most sense to you.
3. Which motivational theorist thinks that most people will "become all they are capable of becoming"?
4. What limits the leader from using motivational theories to their fullest?
5. Compare the terms *positive work climate* and *morale*. In what ways are their meanings similar, and in what ways are they different?
6. To what do the terms *demotivator* and *dissatisfier* refer? How do demotivators and dissatisfiers affect productivity? How can the leader avoid them?

7. What is meant by *develop your people*?
8. Describe leadership by example.

ACTIVITIES AND APPLICATIONS

1. Discussion Questions

- When you go to work, what are some of your expectations and needs? What is positive about the work climate? What is negative?
- Why can't motivation theory be reduced to a set of rules that a supervisor can apply to maintain or increase productivity?
- What motivates you when you are working? Does your supervisor make any effort to determine what motivates your actions and use this knowledge to increase your productivity or try a new task?
- Do you think that one motivational theory is especially better than others in motivating workers to perform well? Defend your answer.
- Several factors are mentioned in this chapter to help build a positive work climate. Which three are most important to you? Can you think of any other factors? Refer to Figure 8.4 for more factors.

2. "Dear Boss" Letter

Write a letter to a future leader about how you want to be treated. For example, you could ask to be listened to, to be thanked, to be challenged, and so on. List at least ten things describing how you want your future boss to work with you.

ENDNOTES

1. Robert Trainor, executive chef with Hilton, http://hotelexecutive.com/business_review/249/function.mysql-connect. Retrieved December 4, 2014.
2. Charles Wardell, *Building Front Line Morale in Motivating People for Improved Performance* (Boston: Harvard Business School Press, 2005), pp. 22–23.
3. www.restaurant.org/Manage-My-Restaurant/Workforce-Management/Retaining-Employees/Keep-Restaurant-Employees-Motivated. Retrieved December 4, 2014.
4. Adam Tobler, *Making Work Meaningful in Motivating People for Improved Performance* (Boston: Harvard Business School Press, 2005), pp. 37–38; Michael MacCoby, *Why Work?: Motivating the New Workforce*, 2nd ed. (Alexandria, VA: Miles River Press, 1995).
5. Ibid.
6. Abraham Maslow, *Motivation and Personality* (New York: Harper & Row, 1954).
7. Frederick Herzberg, B. Mausner, and B. Snyderman, *The Motivation to Work* (New York: Wiley, 1958).
8. Joseph W. Weiss, *Organizational Behavior and Change: Managing Diversity, Cross Cultural Dynamics, and Ethics* (Minneapolis, MN: West, 1996). As cited in Charles R. Greer and Richard W. Plunket, *Supervision: Diversity and Teams in the Workplace* (Upper Saddle River, NJ: Prentice Hall, 2003), p. 234.
9. Betty Schoenbaum, personal communication, June 1, 2010.
10. Martha I. Finney, *The Truth about Getting the Best from People* (Upper Saddle River, NJ: Pearson Education Financial Times Press, 2008), p. 10.
11. Ibid.
12. Ibid., pp. 11–12.
13. Loren Gray, *Enlisting Hearts and Minds in Motivating People for Improved Performance* (Boston: Harvard Business School Press, 2005), pp. 27–28.
14. Ibid.
15. Reylito Elbo, "In The Workplace," *Business World Manila* (August 11, 2006), p. 1.
16. Ibid.
17. James G. Clawson and Douglas S. Newburg, The Motivator's Dilemma, in *The Future of Human Resources Management*. Edited by Mike Losey, Sue Meisinger, and Dave Ulrich. (Hoboken, NJ: Wiley, 2005), p. 15.
18. Ibid., p. 18.
19. Ibid., p. 16.

Chapter 9

Supervising Teams, Team Building, and Coaching

One vital factor is necessary in order to be successful in the hospitality industry: having an effective team. But, what is an effective team, and how do we turn groups into teams and make them winning teams? Many hospitality corporations realize that their main competitive advantage is their employees.

One hospitality product is much the same as another until we add personal service. We have all likely experienced a hospitality service that was less than what was expected and, hopefully, many more of the opposite.

Why is it that in one place the employees are standing around talking amongst themselves and not attending to their guests' needs? Yet, in another, there is a group synergy, with employees helping and encouraging each other to excel? Figure 9.1 illustrates the synergy created by a team. In this chapter, we examine teams and teamwork and how to establish winning teams, a vital part to achieving success in the hospitality industry. Successful concepts such as total quality management (TQM) and empowerment are presented with industry examples to reinforce the learning.

After completing this chapter, you should be able to:

- Explain the difference between groups and teams.
- Discuss the ways in which a supervisor builds a cohesive team.
- Explain how supervisors can build successful teams.
- Describe the steps in installing a TQM process.
- List effective ways for supervisors to empower employees.
- Identify major team challenges that supervisors face.
- Name the recommendations outlined in the eight-step coaching model.

What Is a Team?

LEARNING OBJECTIVE: Explain the difference between groups and teams.

group
A number of people working together, or considered together because of common characteristics.

team
A group of individuals who share a common goal and the responsibility for achieving it.

Teams are very different from groups. A **group** is defined as a number of people working together, or considered together because of similarities. If working together, they interact to achieve a certain objective. The group usually shares information but remains neutral.

A **team** is a group of people organized to work together interdependently and cooperatively to meet the needs of guests by accomplishing purposes and goals.[1] Teams are task-oriented work groups; they can evolve or be appointed, either formally or informally (which is discussed further in the following section).

The team attempts to achieve a positive collaboration among its members. A successful team will work well with each other, achieve set goals, and each member will have a feeling of self-worth. The successful team will also be adaptive, flexible, and able to deal with conflicts as they arise.

formally appointed team
A team that has a formally appointed leader who may have more influence and decision-making authority than other team members.

delegation
The act of giving a portion of one's responsibility and authority to a subordinate.

informally appointed team
A team that evolves on its own.

interdependency
Reliance on others to accomplish a task such as work responsibilities.

team morale
Confidence and enthusiasm that come from a team working in harmony.

teamwork
The cooperative actions that a team performs.

team players
Individuals who participate in the collective effort to get a job done efficiently.

A **formally appointed team** has an appointed team leader. The team leader possesses the power to influence others and may have more decision-making authority than others. The power to influence others is not the only difference between team members and leaders.

A head server is a good example of a formally appointed team member within a restaurant. Power may be *delegated* to this server from management. **Delegation** is when one gives a portion of his or her responsibility and authority to a subordinate. For example, the leader may delegate the head server to do nightly checkouts or voids throughout the evening.

An **informally appointed team** will evolve on its own. It has a rotation of leadership. The group leader does not have formal power over the group. The informally appointed team has some advantages over the formally appointed one. For instance, one person probably does not possess every quality needed to be the perfect leader.

With the rotation of leadership, everyone has a chance to show the qualities that they possess. Formally appointed team leaders may also lose popularity among the group because of their connection with management. With an informally appointed team, this is not likely to happen due to the fact that when their turn comes, everyone is linked with management.

People join teams for many different reasons. One main reason for joining a team in the hospitality industry is to accomplish tasks as efficiently and swiftly as possible. It would be a lot harder to survive a night as a server if you tried to do everything on your own. In actuality, it would be virtually impossible to expedite, deliver, and serve food while clearing, resetting tables, and waiting on people!

Being part of the team assures you that you have others to fall back on if the going gets tough. People may also simply join a team to feel like they are a part of a whole. They may want to feel that they contribute something to the overall success of the team. This may help to develop, enhance, and/or confirm some underlying identity needs.

A team that will be highly successful consists of members who care for and trust each other. They know how to listen to each other as well as express their own ideas. This will form **interdependency** within the team. The interdependence leads to a team collaboration. They find that working together is more effective than working apart.

Efficiency will increase, as well as **team morale**. Team morale is another factor in having a successful team. A team with high morale has harmony among its members. They work well together, know how to communicate openly, and trust each other. In order to have high team morale within the team you must have teamwork, as well as team players.

Teamwork is the actual action that a team performs. It is defined as the cooperative effort by a group of persons acting together as a team. In order to have teamwork in the hospitality industry, you must have **team players**. Team players are individuals who participate in a collective effort and cooperate to get the job done efficiently.

This may range from clearing tables for coworkers on a busy night to taking orders for them because they seem overburdened. One common form of teamwork in the restaurant industry is the rule of having "full hands" going in and out of the kitchen whenever possible.

It is interesting to note that with self-managed teams, the dynamics change if a member leaves or transfers to another "store" (as in restaurant) or hotel. Is this true only for self-managed teams? The new member takes time to adjust to the dynamics and culture of the group. Because we frequently work in groups in the hospitality industry, the ability to work with a team is a major requirement for selection of the associate. Being a team player is more important than being an independent-minded superstar. Ask any team coach.

Working Together

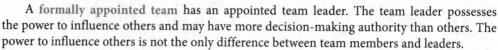

LEARNING OBJECTIVE: Discuss the ways in which a supervisor leads a cohesive team.

team norms
Implicit as well as explicit rules of behavior that result from team interaction.

Now that we have learned the differences between a team and a group, let's consider how team norms affect work behavior. In the hospitality industry, teams as well as **team norms** are constantly evolving. Team norms are defined as implicit, in addition to explicit, rules of behavior. Norms occur inevitably within every type of group—or, should we say, *team*—interaction. They are how team members communicate and conduct themselves in the workplace. Norms work best when the team is allowed to create them. Teams will resent it if preexisting norms are imposed on them or are appointed to them.

This makes it sound like norms should be stopped because they are inherently negative. However, norms can be led in a positive direction. Positive team norms are behaviors that are agreed upon and accepted within the group. They range from communication to performance.

The team should have a positive norm for open communication, as well as wanting to strive for peak performance. For example, a team might agree that if a team member is running late, the other team members will cover for him or her. This can help service overall by ensuring that one person's delayed bus won't delay service for customers. However, a supervisor must keep an eye out for the employee who decides to come in late frequently. Negative norms can develop by abusing team norms.

One way a leader may increase positive team norms in the hospitality industry is by giving rewards for high sales. This could be a weekly or monthly contest where all the servers get a reward when sales reach a goal. The rewards could range from a dinner on the house to a gift certificate. This creates a positive norm among the team members and allows them to have fun, while all of them are striving for the same goal.

Negative team norms are behaviors that are against the interest of and are not accepted by the overall group. An example of a negative team norm is an employee who feels that he does not need a preshift meeting; therefore, he always comes to work late.

This employee should not just be made an example of in this book, he should also be made an example of at work. As a supervisor, it is your duty to evaluate anything or anyone that negatively affects your team. You will never be able to stop negative norms from arising, but you can assess them so that the team may move forward.

project teams
Teams that are brought together for the completion of a project.

In hospitality companies, there are work-area teams such as a dining room team in a hotel restaurant. There may also be **project teams** where a member of the dining room team joins a project team for a period of usually about two months to work on a special project. The project could be creating a new menu or making suggestions for reconception of the theme of the outlet.

Working together becomes all the more important when we consider that the entire staff needs to have a professional service attitude with each guest the moment they walk through the door until they leave.[2]

It's called cooperation—sharing the load to ensure that your guests leave smiling. And if you've never worked in the hospitality industry, you don't know just how important cooperation is on the job.[3]

✻ COHESIVE TEAMS

cohesive team
A team that communicates well and has well-defined norms, respect, unity, and trust among its members.

Why are some teams more efficient than others? Think of it as putting pieces of a puzzle together. Each member of the team is interconnected and represents a piece of this puzzle. In order for the puzzle to be put together correctly, you must have cohesion. Building a **cohesive team** is a major factor in the success of any hospitality company. A cohesive team communicates well with each other and has well-defined norms, respect, unity, and trust among its members.

As in all teams, the members of cohesive teams have strengths and weaknesses; hopefully, what one member lacks another will make up for. This cohesion will result in a team that works well and fits together well. When members of the team fit well together, there is more of a chance that the team will reach its peak performance. If a team lacks cohesion, performance will be hindered because the group will not have any sense of unity.

To build a cohesive team, goals and objectives need to be set. Through close interaction with one another, the team will learn each other's strengths and weaknesses and how each member works. Interaction and communication among the members of the team will eventually lead to group norms, respect, unity, and trust.

In the restaurant industry, everyone should have the same objective and goals in mind ("Let's make this shift as smooth and efficient as possible, and have fun while we are doing it"). It is also easier to give negative feedback when there are agreed-upon goals.

The first step in developing a cohesive team is careful selection of team members. When selecting team members, supervisors should take care to pair employees with peers they get along well with.[4]

The best team-building experiences in the world are the ones that allow passionate, dedicated, and talented people to get the chance to give their best toward a common goal. If you help your employees find their purpose at work, not only will your bottom-line profits, productivity and morale increase, but your absenteeism may also drop, saving you more money.[5]

If you let your employees take the lead TQM style, you'll be amazed at how far they'll take you on the strength of their vision, especially if you focus on "What's in it for me?" (WIIFM):[6]

- Let them know you want to intentionally pass on some of the power. Get their buy-in.
- Brainstorm with the entire group on what shared leadership will look like in your team.
- Find out what their personal hidden beliefs are about leadership and management.
- Work with your team members to discover what additional training they need to exercise their new leadership responsibilities well.
- Learn to consider your team members as an advisory board.

Author and philosopher Johann Wolfgang von Goethe wrote, "Treat people as if they were what they ought to be, and you will help them to become what they're capable of being." So if we treat our team members as if they are already leaders, they will rise to the occasion.[7]

A consultant to hospitality industry clients begins with the purpose or main aim of the company or restaurant. The president might say the main purpose is X but, when the different levels of the organization are asked, the answer is frequently Y, especially from the lower levels of the organization. That is, management and employees are not on the same page. Other factors that are considered are the degree of overlapping values and goals. Figure 9.1 shows the degree to which an employee's personal values and goals overlap with those of the company. The more overlap there is, the more motivated the employees will likely be.

✳ LEADING A KITCHEN TEAM

Chef Gary Colpits at Manatee Technical Institute says that kitchen leadership and teamwork begin with respecting everyone and making sure that everyone is and feels a part of the team. Everyone has to have a say and can help make decisions. Chefs need to let their team members know that team members' opinions and feelings matter. As leaders, we must be sensitive to their needs and concerns.

Before a shift, the kitchen team comes together for a "battle plan" briefing so that everyone is on the same page. These briefings underline the importance of communications; imagine what

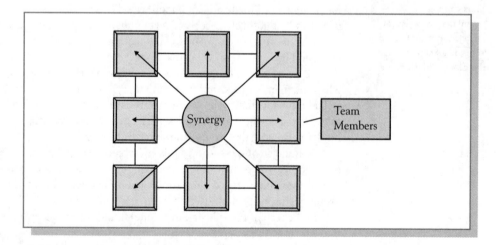

FIGURE 9.1: Team members can create a synergy (when the output is greater than the sum of the individual input). This is accomplished by group members encouraging each other to accomplish goals.

chaos and misunderstanding there would be without communication. Every station discusses its action plan for what needs to be accomplished, from *mise en place* to making their own prep list—because it might be necessary to start with the dessert if it needs to gel.

Each station needs to *prep to par* using standardized recipes. Then the chef/kitchen leader must demonstrate **participative leadership**, leading by example, working with the team—not only to show team members that the leader can do it but also to encourage and coach them in a "we can have fun but it is work" approach.

participative leadership
A system that includes workers in making decisions that concern them.

❋ LEADING A RESTAURANT SHIFT

Every shift is unique and presents different obstacles to overcome. Nevertheless, leaders must be ready to lead their staff through a successful shift. There are many things to do before and after the doors open and close! This is where checklists come in handy. The following is an example of an opening checklist for a restaurant.[8]

- [ ] Arrive in facility, take care of alarms.
- [ ] Check outside for trash, litter, etc.
- [ ] Check A/C or heat, ice machines, walk-in coolers.
- [ ] Survey the interior for general cleanliness.
- [ ] Check all registers/computers.
- [ ] Proceed to office, check notes in log.
- [ ] Complete safe audit, sign in daily log, and confirm banks from previous shift.
- [ ] Confirm deposit amount and prepare change order.
- [ ] Review previous day's daily sales report, record sales, set the budget, and make forecasts (expected number of sales and guests based on the previous year) for the day.
- [ ] Review private party function sheets.
- [ ] Review staff schedule and determine preshift meeting topics (discussed in the next section).
- [ ] Recheck manager's log for items that need attention.
- [ ] Meet with chef to review preparation lists and specials.
- [ ] Prepare seating chart for lunch.
- [ ] Enter lunch specials into computer.
- [ ] Issue bar bank to bartender.
- [ ] Check bus stands, bar, and restrooms for cleanliness and proper setup.
- [ ] Complete line check with chef.
- [ ] Conduct a preshift meeting with the front-of-the-house staff.
- [ ] Do final walk-through of dining area.
- [ ] Set music volume, lighting level, and thermostat.
- [ ] Unlock the front door.

GO TIME!

❋ CONDUCTING PRESHIFT MEETINGS

preshift meetings
Interactive meetings that offer the leader managing a shift the opportunity to address problem issues and lay out the strategy for a successful shift.

Preshift meetings offer the leader managing a shift the time and opportunity to motivate employees. Preshift meetings should be interactive, allowing for questions to be answered, but straightforward and to the point. A typical preshift meeting should last 10–15 minutes. It should cover the following:

- Any issues should be addressed (avoiding names).
- Service practices should be emphasized.

- New products ought to be covered.
- Promotional items and specials should be highlighted.
- Provide the shift's forecast (number of expected guests).
- Note items to upsell for the shift (such as a certain wine).
- Employees should know the status of any incentive programs in place (such as the employee who sells the most food for the week gets a dinner on the house).

The leader of the preshift meeting should note these guidelines:

- Outline the topics to be covered beforehand.
- Make sure that all staff members (to be in attendance at the meeting) are ready and in place 15 minutes prior to the shift.
- Hold the meeting somewhere that will decrease the number of distractions.
- Have someone designated to answer the phones.
- Show enthusiasm to help motivate the staff!

Throughout the shift, managers must be present and available for their staff. Managers should never hide in the office to complete paperwork, which can be done at another time. They need to be involved with their staff and in the guest experience. There is nothing worse for a server than being in the weeds with unhappy guests and not a manager in sight.

Leaders are present when needed and are in touch with what is going on in the entire restaurant, from the front-of-the-house to the back-of-the-house. Priorities when leading a shift include the following: safety and sanitation; driving revenue and repeat business, or service and selling; delivering on the brand promise; conflict resolution and prevention, among both guests and team; and connecting with, not merely interacting with, every customer.[9]

✳ USING CLOSING CHECKLISTS

As the evening shift begins to come to a close, the manager refers to a closing checklist that may include the following:

- ☐ Close stations/cut appropriate staff.
- ☐ Collect server checkouts, check sidework and closing duties.
- ☐ Check with chef to ensure orders for next day's deliveries are set up.
- ☐ Close kitchen and bar.
- ☐ Lock the door.
- ☐ Collect remaining server and bar checkouts.
- ☐ Check bar for cleanliness and restock.
- ☐ Check out dining room cleanliness.
  - Tables and chairs
  - Trash
  - Restrooms
  - Bus station
- ☐ Check out kitchen with chef to see that the following items were completed:
  - Walk-ins cleaned, stocked, and organized.
  - Floors cleaned.
  - Equipment turned off and cleaned properly.
  - Dish area cleaned, stocked, and organized.
  - Sales abstract recorded properly.

Getting team members' input is crucial to obtaining their full commitment to achieving goals.

Tyler Olson/Shutterstock

- ☐ Run all register/computer reports and Z (clear) machines.
- ☐ Batch all credit card processing machines.
- ☐ Complete manager's daily sales report.
- ☐ Complete deposit and drop in safe.
- ☐ Make up banks for the next day.
- ☐ Leave all necessary notes in manager's daily logbook.
- ☐ Do final walk-through; check to see that everything is locked.
- ☐ Recheck to secure all exterior doors.
- ☐ Set unit alarm.

GO HOME![10]

One thing great managers/leaders share is continuous and *purposed motion*. They are continually moving and helping; spreading energy, confidence, and direction; and coaching. They move between kitchen, storeroom, back door, expo line, dining room, greeter area, and front door, assessing and directing flow, focus, food, and fun.[11] One restaurant manager attached a pedometer to his belt and logged nearly seven miles during a single shift—that is a lot of walking! Interaction with your staff and guests is the key to leading a successful and enjoyable shift.

❋ THREE WAYS TO INFLUENCE AN INFORMAL TEAM

There are some ways that you can influence an informal team. One question that you might ask yourself is, have you been giving the team enough and appropriate feedback? What type of feedback are you giving them? You should not give only negative feedback (or only positive feedback).

The amount of feedback given to employees generally should meet somewhere in the middle. Unfortunately, it is more common in the restaurant industry to hear when something is done wrong than when something is done right. Positive reinforcement is often neglected, but employees need it just as much as criticism. It takes only two words from you to change an employee's whole perspective: "Great job!" Therefore, employees need to be told when their actions are unacceptable as well as when they've done a job well.

feedback
Giving information about the performance of an individual or group back to them during or after performing a task or job.

In addition to feedback, you should be able to *identify the key players within the group*. Although, as already noted, there are no formally appointed leaders in the informal team, there are always some members who have more of control over the team.

The leader should identify the "unappointed" leaders and assess whether they are positive or negative impacts. If they are negatively impacting the group, then appropriate steps should be taken to address the issue. (The manager talks with the individual to find out why they are doing negative things, and to formulate a plan to get them to change their ways.)

Finally, another way to influence the informal team is **communication**. *Open* communication not only builds a trusting, open relationship with the staff but also helps to confirm that you are addressing the right issues. If you are consistent at openly communicating with staff, they are more likely to come to you with problems that are occurring within the establishment. If you had a manager who never or rarely spoke to you, the chances of you going to him or her with an issue are slim to none.

communication
Productive expression of ideas; the key to building trust and resolving conflicts.

Building Teams

LEARNING OBJECTIVE: Explain how supervisors can build successful teams.

One of the biggest challenges a leader will face is building a successful team. Before actually implementing the plan of building a team, managers should consider what they want out of the team they are about to build. What needs to be the focus of the team? What is the major goal that you want the team to accomplish? In the restaurant/hotel industry, the goal might be increased sales or simply more customer appreciation and/or feedback. After you have a clear answer to these questions, you may then start on building the team.

The first step to take is clarifying these goals to preexisting members (if there are any). Next, you should be very selective about whom you hire, and always conduct a reference check! It happens often in this industry, but you should never hire employees simply to fill a position.

You should always hire based on the idea that the applicant will provide something for the team (skills, personality, good attitude, etc.). There is rarely a shortage of applicants in this industry; more often, managers make rash decisions on hires due to a lack of time.

Keep in mind that one team member's problem affects the whole group. Essentially, if a team member has a problem that is not addressed, it will create a downward spiral. One team member's problem will end up being the entire staff's problem. Although you are working on a team, you should consider each person as an individual, and even work with him or her on issues that do not concern the group. Once team members see that you are concerned with them individually, in addition to the team as a whole, you start to build a sense of trust and confidence.

Build a positive work environment. If you are delegating tasks to team members, or simply asking the team to kick performance up a notch, provide incentives. Incentives range from actual rewards to extra positive feedback to the chance promotion. You want the team to know that you are actively looking for rising stars. Once the team knows that you have appreciation for those who work hard and that you recognize those who are slackers, chances are you will find more team members rising in performance.

❋ TURNING GROUPS INTO TEAMS

Many leaders mistakenly assume that they have a team when, in actuality, what they have is a group. A group is two or more interacting and interdependent individuals who come together to achieve particular objectives. Groups may be formal or informal. **Formal groups** are work groups established by the company.

Formal groups include committees, group meetings, work teams, and task forces. Formal groups are either permanent or temporary. For instance, the executive committee of a resort hotel

formal groups
Work groups established by the company.

is permanent and meets regularly to run the resort. A temporary committee is established to work on a particular project such as a staff appreciation banquet. After the banquet, the temporary committee, having achieved its goals, is disbanded.

As a leader, you will want your group to become an effective team. To accomplish this, you will need to understand how groups can become true teams, and why groups sometimes fail to become teams. Groups become teams via basic group activities, the passage of time, and team development activities.[12] According to a theory advanced by Bruce Tuckman, team-building goes through four stages:[13]

1. *Forming.* Group members attempt to lay the ground rules for what types of behavior are acceptable.
2. *Storming.* Hostilities and conflicts arise, and people jockey for positions of power and status.
3. *Norming.* Group members agree on their shared goals, and norms and closer relations develop.
4. *Performing.* The group channels its energies into performing its tasks.

(Tuckman added a fifth stage, *adjourning*, or breaking up the team.)

By contrast, **informal groups** are more social by nature. These groups form naturally in the workplace due to friendships and common interests. Examples are people sharing lunchtime together or forming a club. Remember, a group is based on independence; a team is based on interdependence. In order to have a team you must have trust, communication, and collaboration. There are many ways that a supervisor can attempt to change a group into a team. Here we discuss a few of the most critical.

First, as a supervisor you want to get the team's input toward establishing team goals. Working with the group to define goals involves the individual as a part of the whole, and a group with common goals is more likely to work as a team to achieve them.

Second, allow some team decision making. When a decision needs to be made about something concerning the team members, consult them and hear them out. This does not necessarily mean that the outcome decision has to be that of the team, but give them a say. They will appreciate having a voice in the workplace, but understand that the final decision comes from management.

If you have a cohesive team, then team members should all want to participate in the decision-making process. If you have a group, conformity tends to appear, and not everyone has an interest in the decision-making process or the outcome. Some ways to involve the team in decision making are to have a regular meeting at which changes in policy are discussed. Make it clear that the supervisors take employees' suggestions seriously. Figure 9.2 shows the steps involved in turning groups into teams.

Third, stress *communication*. In a team, members know each other's motives and what makes each other tick. In the typical group the members do not really know each other and might even distrust other members of the group because communication is not key. In a team, communication is the key that builds trust and resolves conflict.

Supervisors must stress the importance of communication. When team members feel comfortable enough to communicate their point of view to each other, they, in turn, are more apt to give other members of the team support and trust. If it seems that there are problems, address them. Make sure the team feels comfortable enough with each other and with management to point out problems. Letting employees know about changes, even those that don't affect them directly, can make them feel like they are working as a part of a whole.

Finally, you must have collaboration among team members. In a group the members might all have individual goals, but a successful team strives together to reach the same goal. Team members must be committed to reaching the goal. If members are not striving for the same goal then you have a group, and you may want to do some reassessing of the group members to establish a team.

informal groups
Groups that form naturally in the workplace due to friendships and common interests.

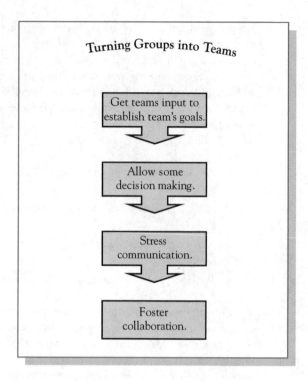

FIGURE 9.2: The steps involved in turning groups into teams.

As we stated in the Working Together section, keep in mind that even if the team is striving for the same goal, team members have strengths and weaknesses; hopefully, what one member lacks another will make up for. If the team collaborates and works through each other's strengths and weaknesses, they will have a sense of unity. When members of the team have unity, there is more of a chance that the team will reach its peak performance.

Fostering teamwork is creating a work culture that values collaboration. In a teamwork environment, people understand and believe that thinking, planning, decisions, and actions are better when done cooperatively. People recognize, and even assimilate, the belief that "none of us are as good as all of us."[14]

✳ CREATING SUCCESSFUL TEAMS

LEARNING OBJECTIVE: Discuss the building of teams, turning groups into teams, creating successful teams, and the characteristics of successful teams.

Creating successful teams depends on creating the climate for success. We know that teams must have a passion for the company's vision, mission, and goals, but supervisors need to give clear guidelines as to exactly *what* is to be done *by whom, when, where,* and *what resources* are required.

The supervisor also ensures that the resources are available when needed. The word TEAM can be used as an acronym for Together Everyone Achieves More. Team members should be selected for their attitudes and skills and trained by a "coach," not a boss. Training for group decision making and interpersonal communications as well as cross-training makes for success.

Select people who like teamwork—not everyone does—and reinforce behaviors that make for good teamwork by having formal recognition awards for those who "walk the talk." Some companies make a DVD and give copies along with framed photos taken with senior management to team members.

Other companies profile members in newsletters. Give teams an opportunity to show their work to senior management. Team selection, especially team leader selection, is important. It's best not to select the most senior member, who may be a member of the executive committee or guidance team (to use the Ritz-Carlton term) because other team members will simply agree with whatever the senior manager says. It's better to select another team member.

A good example of creating a successful team was at a major resort hotel where the servers at the Beach Club reported to two different departments. Guest comments alerted management to a *challenge* (which sounds better than *problem*) of poor timeliness in the delivery of food and beverage orders, yet the service received an outstanding score. The pool and beach attendants reported to one department and the bartenders and cooks to another. When both groups were united to form one team, there was some initial resentment, but as the teamwork improved, so did the tips. Figure 9.3 illustrates the elements of a successful team.

Team effectiveness is defined by three criteria:[15]

1. The *productive output* of the team meets or exceeds the standards of quantity and quality.
2. Team members realize *satisfaction* of their personal needs.
3. Team members remain *committed* to working together.

If you think of the interaction of participants on television shows such as *Survivor* or *The Apprentice*, you can appreciate the nuances of teams.

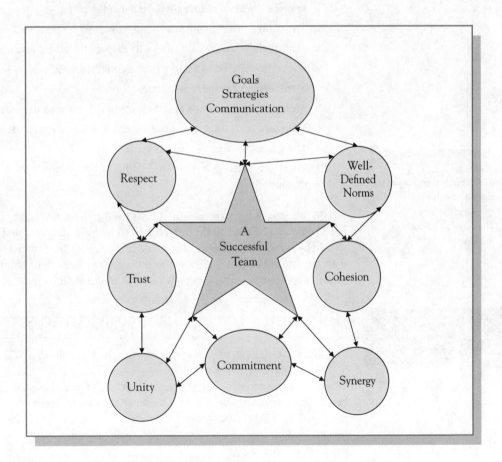

FIGURE 9.3: Elements of a successful team.

When supervising a team, I establish positive norms of behavior by having a discussion on what we want for our team norms. By encouraging all team members to participate and contribute I am more likely to get their buy-in. I also stress the importance of mutual respect and cooperation. The same technique applies to setting team goals. I stimulate discussion on the goals and am amazed to find that many of the goals set by the team exceed those set by management—so we all get bonuses!

We come together as a team for each shift and go over topics of importance—past, present, and future. I spend time on each shift ensuring that everything is going according to plan. I make a checklist with the times that the various tasks should be done by. At the end of the shift we have a quick wrap-up to discuss what went well and what could be improved.

❋ CHARACTERISTICS OF SUCCESSFUL TEAMS

Having described the creation of successful teams, we can now take a look at how a successful team looks and behaves. There are 10 main characteristics of successful teams:

1. The team understands and is committed to the vision, mission, and goals of the company and the department.
2. The team is mature—not necessarily in age—in realizing that members sometimes need to place the team before their personal interests.
3. The team works to continually improve how it operates.
4. Team members treat each other with respect: they listen and feel free to express their thoughts.
5. Differences are handled in a professional manner.
6. Members have respect for their supervisor.
7. Members are consulted and their input is requested in decision making.
8. Members encourage and assist other team members to succeed.
9. The team meets or exceeds its goals.
10. There is a synergy, where the output of the team is greater than the input of each team member.

synergy
The actions of two or more people to achieve an outcome that each is individually incapable of achieving.

In order to become successful, teams need to have the skills required for the job. They also need to be empowered to do the job and to be held accountable for their performance. Teams should be rewarded for meeting or exceeding goals. In the fast-paced hospitality industry, people with insufficient skills are quickly discovered. They need to be trained or replaced for the benefit of the other team members; otherwise, team morale will suffer.

Installing Total Quality Management

LEARNING OBJECTIVE: Describe the steps in installing a TQM process.

Given an increasingly competitive market and fluctuations in guest service levels in many hospitality organizations, it is no wonder that so many companies have adopted a **total quality management (TQM)** continuous improvement process. TQM is a concept that works well in the hospitality industry because its goal is to ensure continuous quality improvement of services and products for guests.

With TQM, the word *guest* is preferred over *customer*, the inference being that if we treat customers like guests, we will exceed their expectations. Successful and progressive companies realize that quality and service go hand in hand. A good meal poorly served results in guest

TQM being introduced to a team.

Hero Images/Getty Images

total quality management (TQM)
A process of total organizational involvement in improving all aspects of the quality of a product or service.

dissatisfaction and a consequent loss of revenue. TQM works best when top management, middle management, supervisors, and hourly employees all believe in the philosophy and concept of TQM. It is a never-ending journey of continuous improvement, not a destination.

TQM is applied in all areas of the business at every level. It works like this: A detailed introduction of the TQM concept and philosophy is given by a senior member of management to underline the importance of the TQM process. The best example of a TQM philosophy is the Ritz-Carlton Hotel Company, which has built a reputation for exceptional guest service. Horst Schulze, former president and CEO, nurtured the tradition of excellence established by the celebrated hotelier, Cesar Ritz, beginning with the motto, "We are ladies and gentlemen serving ladies and gentlemen."[16]

"We practice teamwork and lateral service to create a positive environment." The mission, "To provide the finest personal service and facilities, instill well-being, and even fulfill the unexpressed wishes of our guests," expresses the need for uncompromising service. It is no wonder that the Ritz-Carlton was the first hospitality company to win the coveted Malcolm Baldrige National Quality Award. The main reason for the Ritz-Carlton Company winning this award, not once but twice, is due to TQM (1992 and 1999).[17] Ritz-Carlton associates are empowered to "move heaven and earth" to fulfill a guest's request.

All associates are schooled in the Company's Gold Standards, which include a credo, motto, three steps of service, and 20 Ritz-Carlton basics. Each employee is expected to understand and adhere to these standards, which describe the process for solving guest problems.

Figure 9.4 shows the steps for a successful TQM continuous improvement process. Top and line management is responsible for the process. Once they commit to ownership of the process, the team participants will be energized to focus their energy on the process. Notice how step one calls for leadership (existing, or in need of leadership training). It is critical to have good leader-managers in place to maximize the effectiveness of TQM.

- *Step one* in the process: Have excellent leaders as supervisors and managers. The more successful companies develop leader-managers who can inspire the TQM teams to exceed guest expectations.
- *Step two:* Build and train teams of volunteer associates within each department and later cross departmentally in problem solving.
- *Step three:* Have the teams decide on and write down the appropriate levels of guest service and relative weighting for "their guests" because front-line associates best know the service expectations of "their" guests.

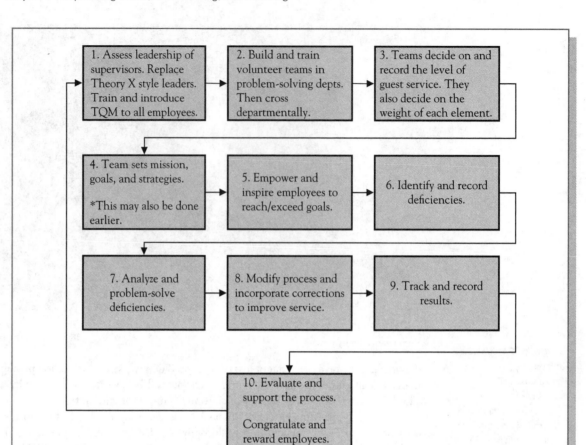

FIGURE 9.4:   Ten steps to a total quality management continuous improvement process.

Of course, management has input, but the whole point of TQM is that management has to give up some of its power and allow associates to share in the decision-making process for determining the criteria and performance levels for guest service.

For example, how many times should the telephone ring before it is answered? Or, how long should guests have to wait for a hostess to seat them at a restaurant? The answers are, answer the phone within three rings if possible and, in the absence of the hostess, appoint someone to acknowledge the guests and let them know that the hostess will be there momentarily. The list of performance criteria for each department will vary according to the type of hospitality business and the guest expectations. In a restaurant situation, guests can be asked in a survey how much they liked their food. A Likert scale of 1 to 10 or 1 to 7 (using descriptors such as "Very much" on one end and "Not at all" on the other) can be used to score each criterion. Food quality is further broken down into taste, smell, appearance, and temperature.

Other restaurant meal quality criteria include service equals courtesy, friendly, efficient, prompt, professional, and so on. The total dining experience includes ease of access to the restaurant, parking, curbside appeal, cleanliness, condition of bathrooms, decor, noise levels, lighting, ambiance, music, and so on.

Once a score has been determined for each element of the guest experience, a base has been arrived at for future comparison. The team of associates comes up with ideas and how-tos for improving the guest experience for each of the elements that is below the level of quality expected by guests.

- *Step four:* Set mission, goals, and strategies based on guest expectations. Write the company, property/unit, or department mission and goals, and create strategies to meet or exceed those goals.
- *Step five:* Empower and inspire associates to reach goals.

- *Step six:* Identify deficiencies, which are areas where service falls below expectations.
- *Step seven:* Analyze and resolve identified deficiencies.
- *Step eight:* Modify processes to incorporate corrections and to improve service to expected levels.
- *Step nine:* Track results—improvements in service, guest satisfaction, employee satisfaction, cost reduction, and profit.
- *Step ten:* Evaluate and support the process. If the goals are not being met, begin again with step one. If the goals are met, congratulate team members for their success and reward them.

Installing TQM is exciting because once everyone becomes involved, the teams find creative ways of solving guest-related problems and improving service. Other benefits include increased guest and employee satisfaction, cost reductions, and, yes, increased profit. TQM is a top-down, bottom-up process that needs active commitment and involvement of all employees, from top managers to hourly paid employees. With TQM, if you are not serving the guest, you had better be serving someone who is serving the guest. To the guest, services are experiential; they are felt, lived through, and sensed.

Empowering Employees

empowerment
To give employees additional responsibility and authority to do their jobs.

LEARNING OBJECTIVE: List effective ways for supervisors to empower employees.

More hospitality companies are empowering teams and employees to deliver outstanding guest service. **Empowerment** means ensuring that employees have the skills, knowledge, and authority to make decisions that would otherwise be made by management. The goal of empowerment is to have enthusiastic, committed employees who do an outstanding job because they believe in it and enjoy doing it. Empowerment encourages employees to be creative and to take risks, both of which can give a company a competitive edge.

There are two types of empowerment: *structured* and *flexible*. Structured empowerment allows employees to make decisions within specified limits, an example being *comping* (giving at no charge) an entrée, not the whole meal. Flexible empowerment gives employees more scope in making decisions to give outstanding guest service.

As described in the TQM section, employees are empowered to problem-solve and do whatever it takes to delight the guest (so long as it's legal). With Ritz-Carlton, associates are empowered by *owning* the guest's request. Associates can spend up to $2,000 without consulting management to solve a guest's problem. An example was when the laundry was pressing a bridesmaid's dress and accidentally burned a hole in it. The concierge took the guest to the nearby Versace store and bought a new dress on her own credit card.

Empowerment enables companies to get quick decisions to satisfy guests. An associate no longer has to find a manager to approve a request; the associate is *empowered* to handle the situation. Empowerment also means fewer levels of management are required. For example, a hotel had several floor housekeepers whose job it was to inspect all rooms. Finally, managers wised up and asked themselves the question, why aren't we doing it right to start with?

Now, certified housekeepers no longer need their rooms checked; the first person in the room after it is cleaned is the guest. This has saved thousands of dollars in salaries and benefits. Empowered employees can schedule, solve TQM problems, budget, do performance evaluations, and participate in employment selection. Today, the supervisor's role is to formulate a vision, show trust, provide resources, coach, train, and offer encouragement and help when needed.

The steps in establishing an empowerment program are similar to those for TQM—a meeting of all employees to announce (with the use of specific guest survey data) the need to increase guest satisfaction. This is followed by an introduction and explanation of empowerment. A training session goes over problem resolution, decision making, and guest service.

The program is monitored and recorded so guest and employee satisfaction—both of which hopefully increase—can be celebrated. The number of times a manager is called to deal with a request is also recorded, along with any costs involved with empowering employees to give away or *comp* a service. Hospitality companies find that the cost of reducing or comping a few charges is more than made up by the increased business they receive as a result of any "guarantee" program.

Another story that illustrates how empowerment can encourage an associate to go the extra mile is: Picture a fabulous resort hotel on a cool day in February. Two guests arrive and decide that they want to have their lunch out on the terrace, rather than in the restaurant.

A table was duly set for them, and, because it was cold, the server went to the laundry and had them put two blankets in the dryer to warm them up. When the guests were presented with the blankets, they were really impressed. It so happened that the guests were travel writers for a major newspaper, and they wrote up the story as an example of exemplary service.

Empowered employees tend to feel more in control of and have a greater commitment to their work and are also more productive than nonempowered employees. So it's no wonder that many hospitality companies such as Marriott and T.G.I. Friday's gain their associates' feedback and ideas on a regular basis. They empower their employees and they, in turn, score highly in guest and employee satisfaction surveys. It's a win–win situation.

Guest feedback is an important part of TQM. Some hospitality businesses have outside companies conduct guest surveys, asking such questions as: Did you have a sense of well-being? Did you feel cared for as an individual? Did you feel wanted as a guest? These questions are measuring the emotional attachment that the guest has with the company and brand.

Overcoming Team Challenges

LEARNING OBJECTIVE: Identify major team challenges that supervisors face.

Every team must overcome some challenges to be successful. Regardless of how much supervisors strive to overcome them, some of these challenges must work out on their own. For instance, you can implement ways that the team may gain personal development, cohesion, positive norms, and so on, but the supervisor cannot simply make them happen. The people in the team must want to gain these qualities and must want a positive workplace.

One major team challenge that management must help to overcome is negativity—including "us versus them." No matter how selective supervisors are at hiring employees, they will always come across an unexpected negative hire. You must remember that when you are interviewing someone, they always have their best face on. The first impression is not necessarily what is behind the real person. If someone is applying for a position in the company, they are probably not going to come in being negative (if they do, then you should not hire them in the first place).

Therefore, it is important (once again) for the supervisor to be an active part of the team. If one person is bringing everyone down, either the active supervisor will recognize it or a team member will be comfortable enough with the supervisor to bring it to his or her attention. This issue should be immediately addressed; otherwise it may create a domino effect.

Another major challenge that management must overcome is learning how to delegate responsibilities. Supervisors must learn how to let go of certain responsibilities, and which responsibilities are to be let go of. Even if you think that it will be much quicker and easier to just do it yourself, this is another way to gain the respect of the team.

Some examples of things that may be delegated to the team in a restaurant are reservations, server cash-outs, nightly station checks, and time for evening "cuts" (when you cut the server staff down from a full staff to a smaller staff at the end of the night). When a job is delegated, make sure you explain what to do clearly and precisely. Also make sure that the person you are delegating the job to possesses a full understanding of what is to be done and that he or she is confident in doing it.

High turnover is one of the major obstacles in the hospitality industry. You cannot have a cohesive, successful team if the team members are always changing. Although there is no clear-cut way of how to overcome high turnover, there are some strategies to reduce it. One way is to be in tune with your staff. If it seems that members of the team are distraught, take the time to talk with them. Ask what the problem is and try to reach a solution. Maybe the only problem is that they are having scheduling conflicts, and it can be resolved simply by giving them more or fewer hours.

Finally, supervisors must overcome the challenge of gaining the respect of the team. This is a tricky one because you must learn how to be their friend to gain trust as well as their leader to gain respect. A supervisor who is too friendly will get walked all over. A supervisor who is not friendly enough will not gain trust. Where do you draw the line?

The answer to this is not so simple. You should always be professional when talking with the staff. Never use inappropriate language; you never really know whom it will offend. Also, limit activities that you attend together outside of the company. Take part in organized activities, but do not make a regular appearance at the local hangout after shifts.

Coaching

LEARNING OBJECTIVE: Name the recommendations outlined in the eight-step coaching model.

In today's workplace, the *leader as a commander* approach is no longer acceptable. This type of approach follows the rule that employees will do as told what to do, how to do it, and when to do it. Often employees find this type of leadership style to be unmotivating and even somewhat demeaning. Today, employees want their voices to be heard and to be more integrated as part of the operation, not simply a warm body.

Coaching is a process involving observation of employee performance and conversation focusing on job performance between the manager and the employee. Coaching can take place informally at the employee's workstation or formally by having coaching sessions in an office. It is different from **counseling**, a process used to help employees who are performing poorly because of personal problems such as substance abuse.

The first step in coaching is to observe employees doing their jobs. Be sure you are completely familiar with pertinent performance standards and job duties. Coaching focuses on enhancing skills of the employee, productivity of the employee, and elevating employee motivation. When employees feel that they are part of a team working together to achieve a goal, they are more likely to excel in performance.

As taught today in management and education circles, the notion of a *self-fulfilling prophecy* was conceptualized by Robert Merton, a professor of sociology at Columbia University. This is the notion that once an expectation is set, even if it is not accurate, we tend to act in ways that are consistent with that expectation.[18]

If managers try to control their workers, they will manage them in a restraining way that will condition employees to do nothing unless they are directly under supervision. If they act as coaches and lead their employees in the direction to assume responsibility, they manage their employees in such a way that the self-fulfilling prophecy will prepare them to take on those responsibilities.

Therefore, if high expectations for employees are set, they will be more likely to strive toward achieving those expectations. If low expectations are set, then the employee will not be motivated to go beyond the low expectations.

Based on an in-depth study, an eight-step coaching model has been developed:

1. *Be supportive.* Be flexible, assist when needed, show understanding, listen, provide positive feedback, encourage, and be open to new ideas.
2. *Define the problem and expectations.* Give the employee the chance to explain the situation. The coach should make sure the problem is understood and then clarify expectations.
3. *Establish impact.* Make sure that the negative impact of the problem is understood.

coaching
Individual, corrective, on-the-job training that is focused on improving performance.

counseling
Counseling occurs when a counselor meets with a client in a private and confidential setting to explore a difficulty the client is having, distress he or she may be experiencing, or perhaps his or her dissatisfaction or loss of a sense of direction or purpose.

4. *Initiate a plan.* The coach and employee should collaborate to develop a plan to correct the problem.

5. *Get a commitment.* Make sure the employee knows what is expected of him or her and is willing to commit to the plan that was collaboratively developed.

6. *Confront excuses/resistance.* Do not accept excuses or resistance. Make sure that the focus is on what can be done to be successful instead of what might not work.

7. *Clarify consequences.* Be clear about what will happen if the plan is not completed.

8. *Don't give up.* Coaching is hard; it takes time. There is never one right answer or solution. Working together, the coach and the employee can resolve problems successfully.[19]

What every good leader needs is continuous feedback about his or her performance as well as their employees. Anyone who has sat in the manager's chair knows how little feedback you get during the course of a day. In fact, being promoted often means the end of virtually all feedback.[20] This does not have to be so. If the leader takes the coach approach, he or she will still be part of the team, involved in employee performance. In turn, the leader will still know what is going on in the operation, how employees feel about it, and what changes need to be made. The coach views the leadership process as a collaborative venture in which every voice is heard.

The overall purpose of coaching is to evaluate work performance and then to encourage optimum work performance either by reinforcing good performance or confronting and redirecting poor performance. Coaching therefore provides your employees with regular feedback and support about their job performance and helps you to understand exactly what your employees need to know. It also prevents small problems from turning into big ones that may require much more attention later. If coaching employees is so beneficial, why do supervisors often avoid it? Possible reasons are:

■ Lack of time. (In most cases, coaching requires only a few minutes.)

■ Fear of confronting an employee with a concern about his or her performance. (A mistake—the problem not faced may only get worse, not better.)

■ Assuming that the employee already knows that he or she is doing a good job, why bother saying anything? (Your employee would love to hear it anyway.)

Coaching a team can be fun and stimulate productive teamwork.
Monkey Business Images/Shutterstock

- Little experience coaching. (You can start practicing now.)
- Assuming that the employee will ask questions when appropriate and does not need feedback. (Many employees are too proud or shy to ask questions.)

There are many skills needed to be an effective coach. Most important is being present—not just physically being there on the job, really being there mentally. Avoid distractions and focus on building the performance of your team. The more you are focused on your employees' needs, the better rapport you will build with them.

This is the foundation of building a team that looks up to you as a coach, not a manager. Goals need to be set for you and your employees. Don't assume you or your employees already know what goals need to be reached; brainstorm and interact with them. Come to shared, clear agreements and follow up to be sure they are being kept. If the agreements are made interactively, employees will be more apt to feel responsible for their upkeep.

A coach does not tell their employees what needs to be done and how to do it. A coach asks and teaches. Resist the temptation to "give them a fish"; instead, teach them *how to fish* for their own answers. If you feel compelled to offer advice, it can be packaged in a way that puts the person fully at choice and in charge. For example, try, "I have an idea that you might find useful."[21]

If an employee is doing the job well, do not hesitate to say so. Everyone likes to be told that they are doing a good job, so praise employees as often as you can. Work on catching your employees doing things right, and then use these steps: (1) describe the specific action you are praising, (2) explain the results or effects of the actions, and (3) state your appreciation.

In some cases, you might want to write a letter of thanks and make sure that a copy goes into the employee's personnel file. You could instead use a standard form, which is quicker to complete.

When observing employees, sometimes you will see what appears to be a problem with performance. If Ted left a pallet of canned goods outside the storeroom, ask him, "Is there a reason why these canned goods are not put away yet?" If Sally is not using the new procedure for cleaning flatware, ask her, "Are you aware of the new procedure for washing flatware?" It could be that Ted stopped putting the canned goods away when a delivery of milk (which has to be put in the refrigerator immediately) came in, or perhaps Sally was on vacation when the cleaning procedure was changed. Consider the following questions before correcting an employee:

- Does the employee know what is supposed to be done, and why?
- Are there any reasons for poor performance that the employee cannot control, such as inadequate equipment?
- How serious are the consequences of this problem?
- Has the employee been spoken to about this concern before?

By asking questions before you point a finger at someone, you help maintain the self-respect of the employee. After asking questions, if it is obvious to you that a correction is needed, be careful not to correct one employee in front of another. No one likes being corrected in front of his or her peers. Arrange to talk privately with the employee to define the performance problem, agree on why it is happening, make your standards clear, and work with the employee to set goals for improvement.

Let's say that Kim's performance as a cocktail server has deteriorated noticeably in the last few weeks and there have been guest complaints of poor service, ill temper, and rudeness. If you tolerate her poor performance, it will reduce her respect for you, for the job, and for herself. So you arrange to talk to her privately and try to get her to do most of the talking. Get Kim to tell you what is wrong and let her know how you perceive the problem as well.

The goal is to resolve the problem, and you encourage her to make her own suggestions for doing this. This leads to her commitment to improve. Make sure that she understands the performance you expect, and get her to set her own improvement goals: measurable performance goals such as a specific reduction in customer complaints within a certain period of time.

- If the problem is related to the job, do what you can to solve it. For instance, one fast-food employee's poor performance turned out to be caused by a large puddle of water in which she had to stand while working. The supervisor had the plumbing fixed, and that solved the performance problem. Kim's problem might have an equally simple solution.

- If the problem involves other people on the job, the solution might be more complex, but do everything you can to resolve it. In the meantime, you have a management obligation to help Kim meet performance standards.

- If the problem is personal, it might help Kim to talk about it, but you cannot solve it for her. You can only listen in order to help her overcome its interfering with her work.

- If the problem is burnout, you might be able to motivate her with some change of duties and responsibilities that would add variety and interest to her job. In any case, she will probably respond to your supportive approach, and when she has set her own goals, she will feel a commitment to achieving them.

When goals have been set that you both agree to, establish checkpoints at which you meet to discuss progress (get Kim to set the times—perhaps once a week or every two weeks). Express your confidence in her ability to meet her goals, make it clear that you are available when needed, and put her on her own.

During the improvement period, observe discreetly from the sidelines, but do not intervene. Kim is in charge of her own improvement; you are simply staying available. Compliment her when you see her handle a difficult customer; give her all the positive feedback you can; keep her aware of your support.

You wait for the checkpoints to discuss the negatives, and let Kim bring them up. Use the checkpoints as informal problem-solving sessions in which you again encourage Kim to do most of the talking and generate most of the ideas. To summarize, make sure that a counseling session includes the following:

1. Speak in private with the employee. Be relaxed and friendly.

2. Express in a calm manner your concern about the specific aspect of job performance you feel needs to be improved. Describe the concern in behavioral terms and explain the effect it has. Do so in a positive manner.

3. Ask the employee for his or her thoughts and opinions, including possible solutions. Discuss together these solutions and agree mutually on a course of action and a time frame.

4. Ask the employee to restate what has been agreed upon to check on understanding. State your confidence in the employee's ability to turn the situation around.

5. At a later time, you should follow up and make sure that the performance concern has been addressed.

In many operations, you will be asked to document (meaning you will need to write on paper) coaching sessions. Some leaders document coaching sessions in a logbook, which is much like a diary of day-to-day events in the operation. Depending on the policy, coaching sessions may be recorded on forms intended for that purpose.

Any documentation of coaching sessions should include the date and place of the coaching and a summary of the coaching session. Although this might seem time-consuming, documentation is essential if you ever need to terminate an employee or simply to do yearly performance evaluations.

Making coaching behaviors part of what you do is essential to the process. Here is a recap of some behaviors that you, as a coach, should focus on:

1. Do not think about employees as people who need to be controlled.

2. Listen, listen, listen!

3. Develop the individual strengths of each employee.

4. Endorse effort and growth (instead of pointing out mistakes).

5. Stop providing solutions. Give your employees an opportunity to figure it out.

6. Stop making all the decisions. Delegate decisions where appropriate and engage your employees.

7. Be unconditionally constructive. Take responsibility for how you are heard, even if you meant something different but came across the wrong way.

8. Create an environment where people want to work with you and feel valued, respected, and part of a team.[22]

CASE STUDY: The Supervisor Blues

Mike graduated from school a few months ago. He applied to various companies and has decided to take a position as a supervisor at a well-known local restaurant. He's excited to lead a team and feels hopeful for the future of the restaurant.

On his first day at work, Mike was sorely disappointed in what he found. For one, he felt that his employees were only looking out for their own best interests. There was little teamwork evident and hardly any communication. Employees seemed bored and unhappy. Furthermore, they were unwilling to engage with him and often gave one-word answers to his questions.

That night, Mike drove home from his shift feeling defeated and wondering if he made the wrong decision in accepting the position. Before Mike decides if he is going to leave and find a job with less management challenges, he turns to you for some suggestions.

Case Study Questions
1. What do you think Mike should do to break the ice?
2. What are some ways Mike could try to turn the group into a team?
3. What are some reasons that this group could have such negative norms?
4. What would be some advantages for Mike if he implemented a TQM plan?

KEY POINTS

1. A group is defined as a number of persons working together, or considered together because of similarities. They share information but remain neutral. A team is a special kind of group that attempts to achieve a positive collaboration among its members.

2. A formally appointed team has an appointed team leader. The team leader possesses the power to influence others; power may be delegated to this server from management. An informally appointed team will evolve on its own, it has a rotation of leadership, and the group leader does not have formal power over the group.

3. One main reason for joining in a team in the hospitality industry is to accomplish tasks as efficiently and swiftly as possible. People may also simply join a team to feel like they are a small part of a whole (this may help to develop, enhance, and/or confirm some underlying identity needs).

4. Teamwork is the actual action that a team performs. It is defined as the cooperative effort by a group of persons acting together as a team. In order to have teamwork in the hospitality industry you must have team players.

5. In the hospitality industry, teams as well as team norms are constantly evolving. Team norms are defined as implicit, in addition to explicit, rules of behavior. Positive team norms are behaviors that are agreed-upon and accepted within the group. Negative team norms are behaviors that are against the interest of and are not accepted by the overall group.

6. A cohesive team communicates well with each other, and has well-defined norms, respect, unity, and trust among its members. To build a cohesive team, goals and objectives must be set.

7. Three ways to influence an informal team are feedback, identification of the key players, and communication.

8. Before actually implementing the plan of building a team, managers should consider what they want out of the team they are about to build. They should be very selective with whom they hire and should conduct reference checks. A supervisor should also keep in mind that one team member's problem affects the whole group.

9. Groups may be formal or informal. Formal groups are work groups established by the company; informal groups are more social by nature.

10. There are four steps to take in turning a group into a team: (1) get the team's input into establishing team goals, (2) allow some team decision making, (3) stress communication, and (4) have collaboration among team members.

11. A team that will be highly successful understands and is committed to the vision, mission, and goals of the company and the department, is mature, works to continually improve how it operates, treats each other with respect, handles differences in a professional manner, has respect for the supervisor, is consulted for input in decision making, encourages and assists other team members to succeed, meets or exceeds its goals, and has synergy.

12. Total quality management's goal is to ensure continuous quality improvement of services and products for guests. TQM is applied in all areas of the business at every level. There are 10 steps to total quality management.

13. There are two types of empowerment: structured and flexible. Structured empowerment allows employees to make decisions within specified limits; flexible empowerment gives employees more scope in making decisions to give outstanding guest service.

14. Major team challenges are negativity, learning how to delegate responsibilities, high turnover, and gaining the respect of the team.

15. The coach uses energy and positivity, not fear or status, as a form of motivation to get the job done.

 ## KEY TERMS

coaching
cohesive team
communication
counseling
delegation
empowerment
feedback
formal groups
formally appointed team
group
informal groups
informally appointed team

interdependency
participative leadership
preshift meetings
project teams
synergy
team
team morale
team norms
team players
teamwork
total quality management
 (TQM)

 ## REVIEW QUESTIONS

Answer each question in complete sentences. Read the question carefully and make sure that you answer all parts of the question. Organize your answer using more than one paragraph when appropriate.

1. Explain in detail the differences between a group and a team.
2. Compare and contrast a formally appointed team and an informally appointed team. What are some advantages and disadvantages of each?
3. What are norms? How do they evolve? When do they work best?
4. How do you build a cohesive team?

5. What are some ways to influence an informal team?
6. Explain total quality management.
7. What is empowerment? What are the two types of empowerment?
8. What are some of the major challenges a team must overcome in order to be successful?
9. Explain the differences between a coach and a supervisor.

ACTIVITIES AND APPLICATIONS

1. Discussion Questions
■ Explain delegation. Other than the examples in the text, what are some appropriate duties in a restaurant that may be delegated? How about in a hotel?
■ What are some of the main reasons that people join teams? This does not necessarily have to relate to the hospitality industry. Give examples inside as well as outside the industry.
■ Give some examples of how a supervisor can contribute to heightening team morale.
■ What are the steps in building a team?
■ What does creating successful teams depend on?
■ What does TEAM stand for?

2.1. Group Activity: Total Quality Management
You and three investors just bought a new restaurant. In groups of four, review the 10 steps of TQM to implement a plan of action for your new establishment. Have one person keep a list of the plan as the other three brainstorm. Example: The first step is to have excellent leaders. As a new owner, how will you find these leaders? Discuss it with the class.

2.2. Group Activity: Team-Building Exercise
The class is assigned to teams of about six people each to participate in a scavenger hunt. Each team must create a team name and motto.

Scavenger Hunt:
Each team must collect and bring to class the following items:
■ A photo of the group
■ A meal pass for the cafeteria
■ A copy of *USA Today*
■ A CD-ROM
■ A page of college letterhead paper
■ A $50 bill
■ A copy of this course outline/syllabus
■ An athlete who represents the school
■ A T-shirt with the school logo on it
■ A hair clip

You have 30 minutes to bring the above items to class. After the winner is declared, reflect on how well your team did. What role did each member play? Did a leader emerge? Do you now feel more of a bond with your team members?

ENDNOTES

1. http://humanresources.about.com/od/teambuilding/f/teams_def.htm. Retrieved December 13, 2014.
2. http://smallbusiness.chron.com/restaurant-customer-service-guidelines-10537.html. Retrieved December 13, 2014.
3. John Horne, President, Anna Maria Oyster Bar, personal communication, November 29, 2014.
4. http://smallbusiness.chron.com/restaurant-customer-service-guidelines-10537.html. Retrieved December 13, 2014.
5. https://hospitalitycoach.wordpress.com/2009/01/04/wiifm-getting-the-best-out-of-your-employees/. Retrieved December 13, 2014.
6. Ibid.

7. Martha I. Finney, *The Truth about Getting the Best from People* (Upper Saddle River, NJ: Pearson Education, Financial Times Press, 2008), p. 191.

8. Adapted from Chris Tripoli, "Operating Checklist," A'La Carte Consulting Group Online. www.alacarteconsultinggroup .com/article6.html. Retrieved February 2008.

9. Jim Sullivan, "The Lost Art of Guest Finesse in the Dining Room," *Nation's Restaurant News*. findarticles.com/p /articles/mi_m3190/is_10_40/ai_n16101662. Retrieved February 2008.

10. Tripoli.

11. Sullivan.

12. Thomas S. Bateman and Scott A. Snell, *Management: Leading and Collaborating in a Competitive World*, 7th ed. (New York: McGraw-Hill Irwin, 2007), p. 464.

13. B. W. Tuckman, "Developmental Sequence in Small Groups," *Psychological Bulletin*, vol. 63 (1965), pp. 384–399. As cited in Thomas S. Bateman and Scott A. Snell, *Management: Leading and Collaborating in a Competitive World*, 7th ed. (New York: McGraw-Hill Irwin, 2007), p. 464.

14. Susan Heathfield, "How to Build a Teamwork Culture. Your Guide to Human Resources," November 2005. www.human-resources.about.com. Retrieved December 15, 2014.

15. Nadler, Hackman, and Lawler, *Managing Organizational Behavior*. As cited in Thomas S. Bateman and Scott A. Snell, *Management: Leading and Collaborating in a Competitive World*, 7th ed. (New York: McGraw-Hill Irwin, 2007), p. 67. www.team-effectiveness.com. Retrieved December 15, 2014.

16. The Ritz-Carlton, "Working at The Ritz-Carlton," corporate.ritzcarlton.com/en/Careers/WorkingAt.htm. Retrieved December 15, 2014.

17. Vivian Deuschl, "The Malcolm Baldrige National Quality Award," corporate.ritzcarlton.com/en/Press/Kits/Baldrige. htm. Retrieved December 15, 2014.

18. ACCEL, "Better Management by Perception," www.accel-team.com/pygmalion/index.html. Retrieved February 2008.

19. Steven J. Stowell and Matt M. Starcevich, Center for Management and Organization Effectiveness, "The Coach: Creating Partnerships for a Competitive Edge," U.S. Office of Personnel Management (August 1996), www.opm .gov/perform/articles/030.asp#Eight. Retrieved December 15, 2014.

20. Ibid.

21. Daniel Robin & Associates, "Making Workplaces Work Better: The Eight Essential Skills of Coaching: How to Bring Out the Best in Others," www.abetterworkplace.com/075.html. Retrieved December 7, 2014.

22. Megan Tough, "Coaching—The New Word in Management," www.siliconfareast.com/coaching.htm. Retrieved December 8, 2014.

Chapter 10

Employee Training and Development

W e give many excuses for not training: We don't have the time, we don't have the money, people don't stay long enough to make training worthwhile, they don't pay attention to what you tell them anyway, they'll pick it up on the job, and so on. There is an edge of truth to all of this, but the edge distorts the truth as a whole.

When you look at the whole picture, you find that the money saved by not training is likely to be spent on the problems that lack of training causes. And those problems involve more than money; they involve guest satisfaction and the well-being of the enterprise.

In this chapter we explore the subject of training in detail and offer a system for developing a training program tailored to a particular enterprise. After completion of this chapter, you should be able to:

- Discuss the importance of training in the hospitality industry.
- Explain the value of supervisors training all new employees.
- List the principles of the adult learning theory.
- Describe the steps for developing a job-training program.
- Identify scenarios in which retraining is necessary.
- Explain how to improve employee retention rates.

Importance of Training

LEARNING OBJECTIVE: Discuss the importance of training in the hospitality industry.

training
Teaching people how to do their jobs; job instruction.

In a hospitality setting, **training** simply means teaching people how to do their jobs. You may instruct and guide a trainee toward learning knowledge (such as certain facts and procedures), skills necessary to do the job to the standard required (such as loading the dishwashing machine), or attitudes (such as a guest-oriented attitude). Three kinds of training are needed in food and lodging operations:

orientation
A new worker's introduction to a job.

1. **Orientation** is the initial introduction to the job and the company. It sets the tone of what it is like to work for the company and explains the facility and the nitty-gritty of days and hours and rules and policies. It takes place before beginning work or in the first few days at work.

job instruction
For every detail of a given job in a given enterprise, instruction in what to do and how to do it.

2. **Job instruction** is just that, instruction in what to do and how to do it in every detail of a given job in a given enterprise. It begins on the first day and may be spread in small doses over several days, depending on how much needs to be taught and the complexity of the job.

retraining
Additional training given to trained workers for improving performance or dealing with something new.

3. **Retraining** applies to current employees. It is necessary when workers are not measuring up to standards, when a new method or menu or piece of equipment is introduced, or when a worker asks for it. It takes place whenever it is needed.

Training is not a standalone entity; it is one of several elements that make for organizational effectiveness. Consider this: A study by Towers Perrin found that of 1,100 North American workers, 50 percent had negative feelings about their job and 33 percent had intensely negative feelings.[1]

Hopefully, none of those surveyed were hospitality employees! Yet, findings such as this heighten the need for training in general. Training can help by encouraging a shared purpose and meaning; if employees care about what they do, surely they will be motivated and do it better. It means tapping into employees' own moral intuitions, their sense of what is right and what is worthwhile for its own sake.[2] We should stress the economic benefits of proper training of new staff.

❋ NEED FOR TRAINING

In our industry as a whole, we do very little of all three kinds of training. There is always that time pressure and that desperate need for someone to do the work right now, so we put untrained people to work and we hassle along with semi-competent or incompetent workers. Yet somehow we expect—or hope—that they will know how to do the work or can pick it up on the job because we are not quite sure ourselves exactly what we want them to do.

We are nearly always shorthanded; we don't take time to train; we need a warm body on the job and that is what we hire and put to work. How do people manage people if they do not train them? We have mentioned the magic apron training method—the idea that if you put them to work, they can learn the job. The assumption is that anyone can make a bed. Anyone can carry bags and turn on the lights. Anyone can take orders and serve food. Anyone can push a vacuum cleaner, wash dishes, and bus tables. That is the prevalent wishful thinking.

Many employers assume that experience in a previous job takes the place of training—a busboy is a busboy; a salad person can make any salad. They depend on these people to know how to do the job to their standard and according to their methods.

Imagine a coffee shop manager recently found a replacement for Shirley, who quit her barista job in favor of a job at a new restaurant. The manager persuades Shirley to come in at 6:00 A.M. and train her replacement. This manager is practicing another common method of training—having the person who is leaving a job train the person who will take it over. This method is known as *trailing* if the new worker follows the old one around. Another method is to have a *big sister* or *big brother* or buddy system; an old hand shows a new worker the ropes, often in addition to working his or her own job.

None of these training methods provides any control over work methods, procedures, products, services, attitudes toward the customers, and performance standards. You do not control the quality of the training, and you do not control the results.

You may serve a 1-ounce martini on Thursday and a 3-ounce martini on Friday because your two bartenders got their experience at different bars. The blankets may pull out at the foot of the beds because no one showed the new housekeeper how much to tuck in. The cups may be stacked three deep in the dishwashing machine because no one trained the new dishwasher—because anyone can run a dishwashing machine, right?

The draft beer gets a funny taste because they did not have draft beer where the new bartender worked before and no one has told him how to take care of the lines. The fat in the fryer takes on a nauseating smell, but the new fry cook does not know it should be changed or filtered because on his last job someone else took care of that. (In fact, he was not even a fry cook, but you didn't check his references.)

Food costs may be high because the kitchen staff have not been trained in waste and portion control. Breakage may be high because servers, bartenders, bus personnel, and dishwashers have not been trained in how to handle glassware and dishes. Equipment breaks down frequently because no one has been trained in how to use it. Health department ratings are likely to be low because sanitation always suffers when training is poor, and pretty soon the local television station may send around an investigative reporting crew to expose your shortcomings to the entire community.

When good training is lacking, there is likely to be an atmosphere of tension and crisis and conflict all the time because nobody is quite sure how the various jobs are supposed to be done and who has responsibility for what. Such operations are nearly always shorthanded because someone did not show up and somebody else just quit, and people are playing catch-up all the time instead of being on top of their work. Service suffers.

buddy system
Training method in which an old hand shows a new worker the ropes.

Good potential trainers need to be identified and receive additional compensation for training.

Tim Pannell/Corbis

Guests complain or they just do not come back, and managers begin to spend money on extra advertising that they could have spent on training and avoided all these problems. Yes, it is true that "proper food safety training and more sophisticated equipment are more costly. But in the long run guests gain confidence in restaurants that are well run."[3] The benefits of training indeed outweigh the cost in the long run.

A small mistake or oversight made by a poorly trained employee can have enormous impact. This lesson was brought home a few years ago when a night clerk whose training was obviously incomplete turned off a fire alarm "because it was bothering me" and several people died in a fire.

For hotels and restaurants alike, the current emphasis in training is on guest service. This is because one guest room is much the same as another, just as one restaurant chair and table are similar. If all else is equal—meaning that in a restaurant the food quality is about the same—it is service that makes the difference.[4] Companies such as Hyatt are seeking to hire the best employees and offer the best service. They want their training to teach employees to be ambassadors of the company.

A training need is identified when current performance falls short of expected performance. A gap is identified between the expected results and the actual results. The gap indicates the need to train for the task being done. For example, a cook is expected to make 25 portions of New England clam chowder in 25 minutes but actually takes 35 minutes, resulting in a delay of 10 minutes. Proper training would have resulted in the cook having the correct mise en place, thus saving time and money. This indicates a training need to increase the speed of preparation to meet expected performance standards.

✳ COST OF NOT TRAINING

There appears to be no specific information on the cost of not training; however, we can quickly realize the enormous cost of losing business to the competition due to poor service, inferior food and beverage, a greasy kitchen, or a dirty hotel room. Just ask OSI Restaurants Partners, which settled a class-action discrimination lawsuit for $19 million against Outback Steakhouse in 2009.[5]

The ramifications of not training can be any or all of the following: legal; regulatory as in dealing with the health inspector or fire marshal or alcoholic beverage control department; or a drop in guest and employee satisfaction scores. We have all experienced poor or even bad service, restaurants and hotels that were not clean—and what did we do? We didn't go back. That's how it is with guests. They will go to places that have well-trained staff, where they feel welcome, safe, and clean and receive great service, food, and beverages.

✻ BENEFITS OF TRAINING

Perhaps you are already beginning to think that what is needed is some sort of system like performance standards that would define the last detail of every job so that each person could be trained to do the job correctly. You are absolutely right. Each new person would learn the same information and procedures as everyone else.

Everybody would learn the same ways of doing things. Job content, information, methods, and procedures would be standardized, and performance goals would be the same for everyone. The new employee would end up producing the same product or service to the same standards as everyone else doing that job.

Suppose that you had such a system in place and used it to train your people. *How would it help you on the job?*

- *It would give you more time to lead.* You would not have to spend so much time looking over people's shoulders, checking up, filling in, putting out fires, and improving solutions to unexpected problems.

- *You would have less absenteeism and less turnover* because your employees would know what to do and how to do it. They would feel comfortable in their jobs, and you would spend less of your time finding and breaking in new people.

- *It would reduce tensions between you and your associates.* You would not be correcting them constantly, and you would have more reason to praise them, which would improve morale. It would also reduce tension between you and your boss. When your people are performing smoothly, the boss is not on your back. You worry less, sleep better, and work with less tension.

- *It would be much easier to maintain consistency of product and service.* When you have set standards and have taught your staff how to meet these standards, the products and the service are standard, too. Guests can depend on the same comfort, the same service, the same excellence of food, the same pleasant experience they had the last time.

- *You would have lower costs*—less breakage, less waste, fewer accidents, less spoilage, better cost control. New employees would be productive sooner. You might be able to get the work done with fewer people because everyone would work more efficiently.

- *Trained personnel would give you happier guests and more of them.* The way that employees treat guests is the single most important factor in repeat business. One worker untrained in guest relations can make several guests per day swear that they will never set foot in your place again.

- *Training your associates can help your own career.* Your performance depends on their performance. And if you have not trained anyone in your job, you may never be promoted because you will always be needed where you are.

Good training will benefit your associates, too. Here are some things that it can do for them:

- *It can eliminate the five reasons why people do poor work*: not knowing what to do, or how to do it, or how well they are doing; not getting any help from the leader; and not getting along with the leader at all. Good training can get them through those first painful days and make them comfortable sooner.

- *Trained employees do not always have to be asking how to do things.* They have confidence; they can say to themselves, "Hey, I know my job; I can do my job." This gives them satisfaction, security, and a sense of belonging, and it can earn praise from the boss.

- *Training can reduce employee tension.* The boss is not on their backs all the time with constant negative evaluations, and they are not worried about how they are doing.

- *Training can boost employee morale and job satisfaction.* When employees know exactly what the leader expects from them, they tend to be more satisfied and relaxed with their jobs. Wouldn't you be?
- *It can also reduce accidents and injuries.* If you have been trained on how to lift heavy luggage or cases of food, you are not going to hurt your back. If you have been trained on how to handle a hot stockpot, you are not going to scald yourself.
- *Training can give people a chance to advance.* The initial training, even at the lowest levels, can reveal capabilities and open doors to further training, promotion, and better pay.

Good training will benefit the entire enterprise. Training that reduces tensions, turnover, and costs and improves product, service, and guest count is certainly going to improve the company image and the bottom line. Many corporations recognize this and have developed systematic training programs.

Another aspect of training is cross-training (training on more than one task or job). It can keep workers interested and motivated; cross-training cuts turnover. It creates loyal, multiskilled employees that chains need to open new locations. Cross-training increases productivity and pares labor costs, and it lays a foundation for careers rather than dead-end jobs.[6]

However, not everyone in the industry sees training as an investment that pays its way. Many managers of small operations consider training an exercise in futility because, they say, it takes more time than it is worth, because people do not stay, because people are not interested in being trained, because it does not work, because it should not be necessary. Return on individuals (ROI) means "what you put into it you get out of it," truly an investment in people.

The myth persists that people in entry-level service jobs should be able to do these jobs without training. When times are bad, with lack of volume and low guest counts, training is the first thing that a manager gives up, as though it were a frill.

It is hard to convince these people that training is worth the investment. It is difficult to measure and prove the difference that training makes because there are always many variables in every situation. Perhaps the best way to be convinced that training pays off is to compare individual operations where the training is good with those that do little or no training. The differences will be obvious in atmosphere, in smoothness of operation, in customer enjoyment, and in profit.

Among larger establishments, there are some that have gained a reputation for their training, that train people so well they are hired away by other firms. It is a nice reputation to have—a nice image for bringing in customers as well as attracting good workers. How do you measure an image? Usually, you do not have to.

On the downside, there have been instances where cutting down on training to cut expenses has proved to be false economy and has resulted in deterioration of service, decline in customer count, and eventually the demise of the enterprise.

Again, ROI means *return on individuals*. Training is an investment in people. In an industry whose every product and service depends almost entirely on individual people at work—people who deal directly with guests—investing in training those people is a major key to ROI of any sort.

❈ PROBLEMS WITH TRAINING

Leaders who do not train their people are not all stubborn fools or cynics; the problems are real. Perhaps the biggest problem is *urgent need*; you need this person so badly right now that you don't have time to train, you can't get along without this pair of hands. You put the person right into the job and correct mistakes as they happen and keep your fingers crossed.

A second critical problem is *training time*: your time and the employee's time. While you are training, neither of you is doing anything else, nor do you have that kind of time. Your time and the associate's cost the company money. A training program requires an immediate outlay of money, time, and effort for results that are down the road. This is especially a problem for the small operation with cash-flow problems and a day-to-day existence. Training is an investment in the future it cannot afford; its problems are here today.

A third problem is *turnover*—people leave just as you get them trained, and you have spent all that time and money and effort for nothing. Training may reduce turnover, but it does not eliminate it, given the easy-come, easy-go workers in the hospitality industry.

The *short-term associate* is a training problem in many ways. People who do not expect to stay long on a job are not highly motivated. They are not interested in the job and they are not interested in getting ahead; they just want the paycheck at the end of the week. They do not like training programs. They do not like to read training materials. They do not get anything out of lectures. Most of them have poor listening, reading, and studying skills. They do poorly with the general, the abstract, and the complex. They are impatient; they are looking for a *now* skill—something they can do this afternoon.

The *diversity of employees* can be a training problem. Some are pursuing college degrees; others are poorly educated. Many have never had a job before; others have been in the industry for years. (Some of these are floaters who move from one operation to another; they like to work openings, stay about a month, and move on.) Some are know-it-alls; others are timid and dependent. Some are bright; others are below average in intelligence and aptitudes. Some do not speak English. Overall, they are not a promising kitchen or dining crowd. How can you train such different people for the same jobs and expect the same performance standards?

We also have problems with the *kinds of jobs* we train for. One type is the dull, routine job that takes no high degree of intelligence or skill: vacuuming carpets, mopping floors, prepping vegetables, and running a dishwashing machine. The problem here is the very simplicity of these jobs: We tend to overlook the training. Yet these jobs are very important to the operation, and it is essential that they be done correctly.

Most housekeeping jobs, for example, involve sanitation. Yet because sanitation can be technical and at times boring, and much of it is not visible to the untrained eye, it is easy to skip over it lightly. Techniques might not be properly taught or their importance emphasized—the sanitizer in the bucket of water, the indicator on the temperature gauge in the right place, the dishwashing machine loaded so that the spray reaches every dish and utensil.

Also overlooked are techniques of doing routine tasks quickly, efficiently, and safely. Food preparation such as knife skills for cutting or proper table setting. The optimum stroke of the vacuum cleaner, the order of tasks in cleaning a room, how to handle your body when making a bed or scrubbing a tub so that you don't strain your back—these little things can make a critical difference to efficiency, absenteeism, and employee well-being.

At the other extreme is the *complexity* of jobs containing up to 200 or 300 different tasks, plus the subtle skills of customer relations. Such jobs—server, bartender, and desk clerk—are so familiar to people who supervise them that they do not stop to think how much there is for a new person to learn. Training time for these multiple tasks can be a real problem. Therefore, you skimp on the training, you rush it, or you hire experienced people and skip the training. You forgo the control, the consistency of product and service, and the high-grade performance of people you have trained to your own standards.

The final typical training problem is *not knowing exactly what you want your people to do and how.* If you do not know this, how can you train them?

What you need is a system of training that defines what your people are to do and how, trains everyone to the same standards, adapts to individual needs and skills, and lends itself to one-on-one training. Although not many people are going to take the trouble to develop a full-blown system, you can still see how its principles and techniques apply in training, and you can go as far as you find practical in applying them.

Before we look at the three different types of training, let us first consider who will do the training and how employees learn best.

Who Will Do the Training?

LEARNING OBJECTIVE: Explain the value of supervisors training all new employees.

We have mentioned various ways of assigning the training responsibility: the magic apron, having an employee who is leaving train her replacement, the buddy or big brother/sister system. They do not work because such training is haphazard and incomplete, but most of all because the wrong person is doing the training.

With the magic apron, people train themselves. They are the wrong persons to do the training. They make a mistake and are yelled at, and what they train themselves to do is what will keep them from being yelled at. They will also train themselves to do things the easiest way and, in general, to do what is in their own best interest. Often, these things are opposed to the interests of the house.

Shirley is leaving or has already left. She will do only enough to calm the coffee shop manager and get her paycheck. She will tell her replacement only what she knows, which might be very little, and only what she can cram into the shortest possible time, and she will not care whether the new person learns anything. She will also teach shortcuts and ways of getting away with breaking rules.

Big brother, big sister, and buddy will also teach only what they know, and only as much of that as they happen to think of, and they too will not care how well the new person understands. Unless they are paid extra for training, they may resent the assignment. They may also resent the new person as a competitor. In addition, they will pass on to the new worker all their own bad habits and all their accumulated gripes, and they will condition the trainee to their view of the job, the boss, the customers, and the pay.

The logical person to train your people is you, the supervisor. It is your responsibility, whether you delegate it or do it yourself. Training is one of those obligations to your people that goes with your job—giving them the tools and knowledge to do theirs. However, you have a thousand other responsibilities and your day is interrupted every few seconds.

On the one hand, if you can possibly make the time, you owe it to yourself to do the training. It is the beginning of leadership. A good teacher forms a lasting impression in the learner's mind, a special regard that will color the relationship from that point on. It gives you a chance to get to know your new people, to establish that one-to-one relationship necessary to being an effective leader.

On the other hand, someone on your staff with the right potential might be able to train new people even better than you can, considering all the demands on your time and attention. If you have established a good training program, you can delegate the training to someone like this.

Such people must be trained. They must know how to do everything to teach. They must learn the skills needed to train others: how to treat people as individuals, how to put themselves in the learner's place, how to gear the lesson to the learner, how to increase motivation, how to lead—all the good things you have been learning yourself.

It is essential for these people to receive appropriate compensation: extra pay for extra work, a promotion, whatever fits the situation. They must also want to do the training. You remain responsible for the training, and if it is not done well, it will come back to haunt you.

By the way you train, you are teaching more than rules, procedures, skills, and job standards. You are teaching basic attitudes toward work, personal standards of performance, the importance of the person, getting along with other people (both guests and colleagues), your own work values, and many subtle but lasting lessons in human relations and values.

People on their first jobs are particularly susceptible to this type of learning. Their first job will probably affect their attitude toward work performance, work relationships, and work values for the rest of their lives. It is important to be aware of this. You do not have to save their souls, but you do owe it to them to set high standards and a good example and to teach a work ethic of being on time, meeting standards, and giving their best efforts to the job. You owe it to yourself, too, and to the organization. The economic benefits of proper training are priceless. Happy staff equals a better product and performance, which equals healthy revenue.

How Employees Learn Best

LEARNING OBJECTIVE: List the principles outlined in the adult learning theory.

learning
The acquisition of knowledge or skill.

Training is a form of communication, and as in all communication, the sender (trainer) controls only the first half of the interaction. The second half, the receiving of the message—the learning—depends on the trainee. **Learning** is the acquisition of knowledge, skills, or attitudes. How do

PROFILE　Maira Sommerhauzer Horta

Courtesy of Maira
Sommerhauzer Horta

Throughout my experience in the hospitality industry, I have always played on what I call my *Brazilian jeitinho,* which embraces some qualities I have gained from growing up in Brazil. *Jeitinho* is the Brazilian optimistic attitude of being especially creative in a variety of ways suitable to work out certain situations, such as making extra efforts to satisfy guests and employees when resolving certain issues. Indeed, most things can be accomplished somehow, in order to please guests and employees, as long as you keep your goals in mind.

First of all, let me tell you a little about myself. Upon my graduation from high school I promised myself I would continue my education. In fact, I decided to study hospitality management in San Diego, California, because San Diego is a beautiful city, with a high volume of tourism. Shortly after I moved to San Diego and started going to school, I wanted to experience the hospitality industry.

I started at the front desk of a major hotel; I did not see working my way through college as a hardship, but rather as an opportunity, requiring that I manage my time and priorities and exert every effort to be productive. Although I started at a front-desk position, my curiosity and initiative took me all over many departments of the hotel industry, such as reservations, sales and marketing, and the food and beverage division. Moreover, my greatest accomplishment was to supervise one of the food and beverage departments of another major hotel in San Diego. My new target customers were East Coast companies that sent groups of employees to our hotel in San Diego, mostly for work and conferences. We provided these preferred guests with a friendly, welcoming "happy hour," featuring a complimentary buffet and a full bar, from Monday through Friday.

We usually had baseball, basketball, or football games playing on a big-screen TV during happy hour, which made an even more relaxed and fun atmosphere in which to "hang out." I found it very much fun to develop a relationship with some of these guests, especially because they always came back to our hotel. Some of them stayed for weeks, sometimes months; and some of them came back for a week every month of the year. I believe that most of these guests always came back because they got to meet a lot of people in our happy hour, and they felt very comfortable.

Next, as a supervisor, I found it effective to have a vision, or to set goals you want to accomplish during a given period of time. Then, communicate these goals to the team. Goals should be achievable given sufficient effort, and it must be possible to tell when they have been achieved. Recognizing and prizing employees who are accomplishing goals and devoting themselves to their work is crucial for employee motivation and satisfaction, leading to a higher quality of work.

Furthermore, I found it wonderful to build strong relationships with employees; you can understand them and get sincere feedback from them on many aspects of the work; and they can understand you, including what you expect from them. Supervisors make a great impact on an employee's performance; it is important to be aware of this fact. Whenever you have to correct a certain behavior or employee attitude, always mention the good things about that employee, building his or her self-esteem before criticizing the person's work.

In conclusion, I believe that effective supervising is, essentially, using common sense when making decisions in every situation. Also, treating people fairly and with respect; recognizing and prizing their good work; and knowing how to criticize someone's conduct, if needed, without hurting their feelings, is very important. Finally, supervisors should listen to employees and guests and have a good sense of humor and a little *jeitinho* to please guests and employees while complying with the set goals for the company.

---

adult learning theory
A field of research that examines how adults learn.

teaching methods
Ways in which teachers and trainers convey information to learners.

adults learn best? Many of the following tips for helping employees learn are derived from a field of study called **adult learning theory**.[7]

1. *Employees learn best when they are actively involved in the learning process.* When employees participate in their own training, they retain more of the concepts being taught. To get employees involved, you need to choose appropriate teaching methods. Teaching methods are the ways we convey information to learners. You are no doubt familiar with the lecture method because it has dominated U.S. education for many years.

Using the lecture method, content is delivered primarily in a one-way fashion from a trainer to the participants. Little or no interaction occurs between the trainer and the

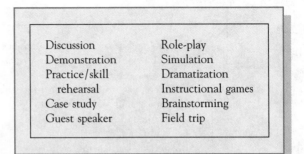

Discussion	Role-play
Demonstration	Simulation
Practice/skill	Dramatization
rehearsal	Instructional games
Case study	Brainstorming
Guest speaker	Field trip

FIGURE 10.1: Teaching methods to promote employee involvement in training.

participants. Although the lecture method might be an efficient method to get across needed information to employees, it is most inefficient at getting employees to use the information. Figure 10.1 lists teaching methods that more actively involve employees.

2. *Employees also learn best when the training is relevant and practical.* Adult learners are picky about what they will spend their time learning. They often pursue learning experiences in order to cope with life-changing events such as a job change. Learning must be especially pertinent and rewarding for them.

3. *Besides being relevant, training material needs to be well organized and presented in small, easy-to-grasp chunks.* Adults need to be able to learn new skills at a speed that permits mastery. Unlike children, adults come to the classroom with prior experience, so they need to integrate new ideas with what they already know.

   Using visual aids such as posters during training helps to focus employees' attention, reinforce main ideas, save time, and increase understanding and retention.

4. *The optimal learning environment for employees is an informal, quiet, comfortable setting.* The effort you put into selecting and maintaining an appropriate environment for training shows your employees how important you think the training is.

   You give a message that training isn't that important when employees are stuffed into your office, made to stand in a noisy part of the kitchen, or when you allow yourself to be interrupted by phone calls. Employees like to feel special, so find a private room and consider having beverages, and perhaps some food, available.

5. In addition to training in an appropriate setting, *employees learn best with a good trainer.* Figure 10.2 lists characteristics of successful trainers, some of which we have already discussed. As you go through the table, you will see that the characteristics listed also apply to successful supervisors. A good supervisor is usually a good trainer.

6. Toward the end of training, employees are generally evaluated on how well they are doing. *Employees learn best when they receive feedback on their performance and when they are rewarded* (perhaps with a certificate or pin) for doing well.

Developing a Job-Training Program

LEARNING OBJECTIVE: Describe the steps for developing a job-training program.

A good job-training program should be organized as a series of written **training plans**, each representing a learnable, teachable segment of the job. Once you have prepared such plans, you can use them for every new person you hire for the job: They are all ready to go. You can use as much or as little of each plan as you need, depending on what the new employee already knows.

1. Displays enthusiasm and has a sense of humor
2. Communicates clearly in a way that participants understand
3. Is knowledgeable
4. Is sincere, patient, and listens to participants
5. Encourages and positively reinforces all participants
6. Is organized
7. Plans the training session
8. Involves all participants
9. Presents the material in an interesting and appropriate way for participants to learn
10. Checks the outcome to see that all participants have learned and has participants sign that they have learned the training topics

FIGURE 10.2: Characteristics of a good trainer.

training plan
A detailed plan for carrying out employee training for a unit of work.

Performance standards provide a ready-made structure for a training program for a given job: Each unit of the job with its performance standards provides the framework for one training plan. In this section, we describe how to develop a training program using performance standards. Although you might not complete the system in every detail, you can apply the principles and content to any training program.

❋ ESTABLISHING PLAN CONTENT

Even if you do not have performance standards, you still have to go through pretty much the same procedures to develop a good training program. You must analyze the job as a whole, identifying all the units that make up that job classification and then the tasks that make up

Training objectives are based on performance standards.
dotstock/Shutterstock

each unit. Then you must decide how you want each unit and task done and to what standard. You then develop a procedures manual or some other way of showing how the tasks are to be carried out.

Figure 10.3 traces the progress of one training plan from its beginning to its implementation on the job. Let us follow it through, using a bartender's job as an example. The job of bartender contains a dozen or so units, such as setting up the bar, mixing and serving drinks, recording

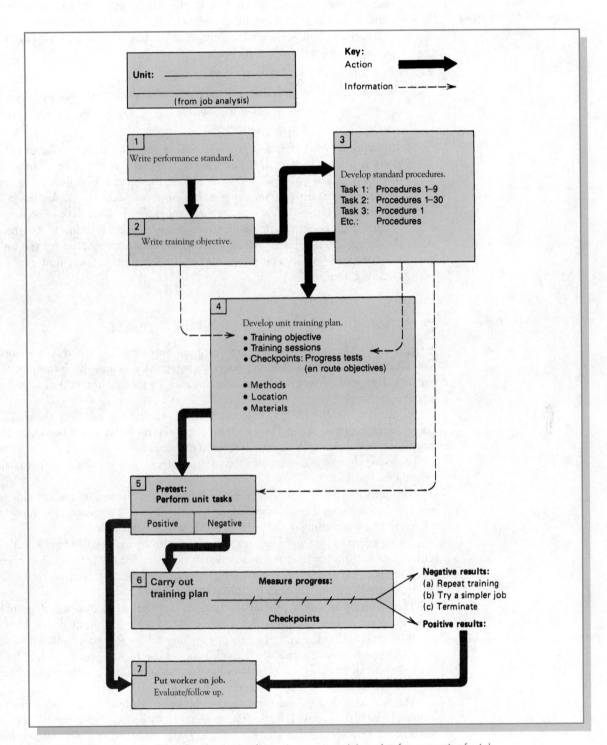

FIGURE 10.3:   Flowchart for developing and carrying out a training plan for one unit of a job.

drink sales, operating the cash register, and so on. You will write a training plan for each unit. Your first training plan will be for *setting up the bar*.

1. Write your performance standard: "The bartender will set up the bar correctly according to standard house procedures in half an hour or less."

2. Write a **training objective** derived from your performance standard: "After three hours of instruction and practice, the trainee will be able to set up the bar correctly according to standard house procedures in 45 minutes." This training objective expresses what you expect the person to do after training, the training goal. It differs from the on-the-job performance objective in three ways:

   - A time limit is set for the training.
   - The verb expresses trainee achievement rather than on-the-job performance.
   - The performance standard is lower for this learning level (45 minutes) than for the day-in, day-out performance level.

3. Incorporate your standard procedures that you may already have developed. If not, here is what you do. You list all the tasks of the unit in the order in which they are performed, and you spell out each task in the form of a procedures sheet or visual presentation stating or showing exactly how you want that task carried out in your operation. Figure 10.4 is a procedures sheet for the first task of setting up the bar. Figure 10.5 goes with it to illustrate some of the procedures. All the procedural materials taken together define the content of the instruction for the unit, and they become both guides and standards for the training. You are now ready to plan the training itself.

> **training objective**
> A trainer's goal: a statement, in performance standard terms, of the behavior that shows when training is complete.

✳ DEVELOPING A UNIT TRAINING PLAN

A training plan (Figure 10.3, step 4) sets forth not only what you will train someone to do, but how, when, where, what supplies and materials you need, and how much training you will do at one time. Figure 10.5 is an example of a training plan for the first unit of the bartender job. Let us go through that plan item by item.

Notice first that the training objective is stated, so that you can keep the goal in mind and shape the training to reach it. The unit is taught in several *training sessions*. The primary reason for this is to avoid giving the trainee too much to learn at once. Another reason is to avoid tying up the person doing the training for too long a time. In this particular case it is also to avoid tying up the bar itself.

The tasks are taught in the order in which they are performed on the job. One training session may include several tasks, some taking as little as 5 or 10 minutes. Length of time for each session will vary according to the trainee's previous experience.

An experienced bartender, for example, will learn your par-stock-empty-bottle-requisition routine far more quickly than someone who has never tended bar before. (Most operations look for experienced bartenders because training from scratch takes too long, but even an experienced bartender must be trained in *your* bar and *your* procedures.)

The training plan should provide *checkpoints* along the way, as shown in step 6 of Figure 10.3. These allow you to measure a worker's progress toward achieving the objective. They may follow groups of tasks within the unit of work, or they may follow the whole unit with a series of less demanding performance standards (a more lenient time limit, a greater margin for error). You can write special intermediate objectives for the checkpoints or set several successive levels of performance for the entire unit of work.

The *method of training* must include two elements: (1) showing and telling the trainee what to do and how to do it, and (2) having the trainee actually do it and do it right. These elements are combined in a widely used formula known as *job instruction training*, which we examine in detail shortly. There are various ways to show and tell: demonstration, movies, company websites and intranets, CDs, and DVDs.

Standard Procedures

Job: Bartender.   Unit 1:   Setting up the bar

Task no. 1:   Replenishing liquor supplies, standard house procedures

1. Count the number of full or partly full bottles of each brand and compare with the Par Stock Sheet posted at the bar. This will give you the numbers to be replaced.

2. On requisition sheet, enter name, unit size, and number needed for each brand.

3. Count empty bottles (box under bar) brand by brand and compare numbers with requisition. If they do not agree, report differences to supervisor. Supervisor will OK discrepancies or tell you what to do.

4. Sign and date completed requisition form on line 1 and have supervisor sign.

5. Lock the bar gate. Take requisition and empties to storeroom (use dolly or cart). Storeroom will count empties, issue fulls, and sign requisition.

6. Count full bottles to make sure you have received the numbers storeroom has shown on requisition. Sign and date on bottom line. Storeroom keeps requisition.

7. Take full bottles to bar. Wipe them and arrange all bottles as shown on well and backbar diagram below.

8. Set up two reserve bottles with pourers for each bottle in well.

9. Check all pourers and replace corks as necessary.

Liqueurs   Call brands   Wines
BACKBAR

WELL

Sweet/sour

Orange juice

Bloody
Mary mix

Dry vermouth   Tequila Rum Bourbon Gin Vodka

Piña colada mix

Sweet vermouth

Brandy

Scotch

FIGURE 10.4: Portion of a procedure for setting up the bar: procedures for one task (page from a procedures manual for the job of bartender).

TRAINING PLAN

Job classification: Bartender

Unit 1:   Setting up the bar

Learning objective: After 3 hours of instruction and practice, trainee will be able to set up the bar correctly according to standard house procedures in 45 minutes.

Training sessions: 1. Replenish liquor (Task 1, training time 30 minutes).
2. Replenish other supplies (Tasks 2–6, training time 30 minutes).
3. Set up draft beer, check soda system (Tasks 7–8, training time 20 minutes).
4. Prepare garnishes (Tasks 9–13, training time 30 minutes).
5. Set up register (Tasks 14–15, training time 20 minutes).
6. Set up ice bins, glasses, sinks, mixing equipment, bar top, coffee (Tasks 16–21, training time 25 minutes).

Method: Demonstrate and do (JIT), one on one. DVD on beer setup.

Location: Bar, 1 hour before opening (bar must be as left after closing)

Materials:

Liquor and all other supplies as of closing	Cocktail napkins
	Ice
Liquor empties from previous day	Bar knife and cutting board
Par Stock Sheet	Bar spoon, jiggers
Requisition forms	Mixing glasses
Well and backbar diagram	Guest checks
Pourers, replacement collars	Credit card slips
Ashtrays	Cash register
Matches	Opening bank
Picks	Dolly or cart
Snacks	Coffee machine and coffee
	Procedure sheets, Tasks 1–21

Checkpoints: After each task

FIGURE 10.5: Training plan for one unit (page from a training manual).

Training plans need checkpoints to measure the trainees' progress.

Courtesy of Monkey Business Images/Shutterstock.

The closer the training method and setup are to the on-the-job situation, the better the training. You can teach table-setting by actually setting up tables for service, and teach bed-making while you are actually making a room ready for the next guest. But there are many things you cannot teach while doing them on the job. In such cases you simulate on-the-job conditions as closely as you can: You use the real equipment and real supplies and you set up the equipment and the task as realistically as possible.

One-on-one training generally works best. In a classroom, a person does not learn as well. Everybody absorbs the material at different rates or has different problems with it. The slow learners are lost and the fast learners are bored. The classroom also causes anxiety and inhibits everyone except the know-it-all.

However, group presentations have certain advantages. They are useful for giving general information and background that may be overlooked by the individual trainer. Because group presentations can be more closely controlled, it is a good way to convey company policy so that it is always stated accurately and everyone gets the same message.

A number of chains use audiovisual presentations such as DVDs that are developed at headquarters and are sent out for use by individual stores or several stores in an area, or offered via the company website. Training in customer relations, for example, can be given in this fashion. It is effective and ensures not only a consistent message but also consistent training quality.

The *location* of the training should be a quiet place free of interruptions. Ideally, training is done in the actual job setting during off hours. Some corporations have special training facilities completely equipped to simulate the actual job environment.

Your *training materials* should include the same equipment and supplies that will be used on the job, and they should all be on hand and ready before the training starts. You must prepare your entire session in advance if training is to be effective.

Developing a written training plan helps you to think out all the aspects of the training and to orient everything to the new employee and the details of the job. Each completed plan gives you a checklist for readiness and a blueprint for action.

Like many good things prepared for a well-run operation, training plans take a long time to develop, and the manager or supervisor will have to do one plan at a time one piece at a time, probably over a long period. It may be helpful to schedule development along the lines of an improvement setting of overall goals and interim goals. Then each completed piece of the plan will provide a feeling of achievement and the momentum to continue. Of course, if you already have job analyses, performance objectives, and procedures manuals, your work is half done at the outset.

✳ MOVING FROM PLAN TO ACTION

pretest
Testing an experienced worker's job performance before training.

Now you are ready to train the associate. But before you train new employees, you must find out how much training they really need. If they have knowledge and skills from previous experience, it wastes both their time and yours to teach them things they already know. For this reason, the training of associates who have some experience begins with a **pretest** (Figure 10.3, step 5): You have them actually do the unit of work. If the unit consists of operating the dishwashing machine, you have them operate the dishwashing machine; if it is serving wine, you have them serve wine; if it is setting up the bar, you have them set up the bar.

You observe the new associate's performance and confine your training to what the person does not know, what does not measure up to your standard, what varies from your special ways of doing things, and what the person must unlearn in the way of habits and procedures from other jobs. Experienced associates should end up meeting the same standards as the people whom you train from scratch.

Not all units and tasks are suitable for pretesting. Some are too complex, and some are different every day. In this case, you can ask experienced new associates to describe how they would carry out the tasks in question and then adjust the training accordingly.

Now suppose that you are training Gloria and David, who have never had a job before. You carry out your first unit training plan (Figure 10.5), teaching each of them every task in the unit's action plan. You test them at every checkpoint to make sure that they are following you and are putting it all together and meeting your time requirements. Finally, when you have taught all the tasks in the unit, you evaluate Gloria and David by having them perform the entire unit in sequence.

Does Gloria meet the performance standard of your learning objective? If so, you move on to the second unit of the job and the second objective. Suppose that David fails to meet the standard for the first unit. You retrain him in those procedures that he is not doing correctly. If he did not meet the time requirement, you have him practice some more.

If he just can't get it all together, you might try him on the second unit anyway. If he can't do that either, he may not be able to handle the job, and you may have to place him in a less demanding job. It is also possible that your training was at fault, and you have to take a hard look at that.

formative evaluation
An ongoing form of evaluation that uses observation, interviews, and surveys to monitor training.

If you do not have a simpler job, or if he cannot learn that simpler job either, you may have to let him go. The training has not been wasted if it has identified an untrainable employee in time to save you from paying unemployment compensation. Ideally, you will put a new employee on the job (Figure 10.3, step 7) after training for all units of the job has been completed. But you may need Gloria and David so badly that you will have them work a unit of the job as soon as they have been trained for that unit. In complex jobs, it may even be easier for them to work certain units of the job for awhile before going on to learn the entire job.

Once the formal training process is completed, there is still one very important step: evaluation. Making an evaluation is the crucial process of determining whether training objectives were met. It can occur both during and after training. **Formative evaluation** uses observation, interviews, and surveys to monitor training while it is going on. **Summative evaluation** measures the results of the training after the program is completed, looking at it in five ways:

summative evaluation
A form of evaluation that measures the results of training after a program has been completed.

1. *Reaction:* Did the employees like the training?
2. *Knowledge:* Did the employees learn the information taught?
3. *Behavior:* Are the employees using the new skills or behavior on the job?

4. *Attitudes:* Do the employees demonstrate any new attitudes?

5. *Productivity:* Did the training increase productivity, and was it cost-effective?

Various techniques can be used to answer these questions. Participants can report on what they liked and did not like about the training by filling out evaluation forms. Tests are frequently given at the end of training sessions to determine whether the employees know the information and/or skills covered.

Results can also be measured through observation of employee behavior and monitoring of critical indicators such as the number of guest complaints, level of repeat business, and so on. After collecting information from various sources, the person doing the evaluation needs to compare the results to the learning objectives to determine whether the training indeed succeeded in bringing about the desired changes.

✳ JOB INSTRUCTION TRAINING

Successful training observes the flow of the learning process. During World War II, when war plants had to train millions of workers quickly, a training method was developed that took maximum advantage of the learning flow. It was so successful that it has been used in various forms ever since in all kinds of training programs in all types of industries. This is **job instruction training (JIT)**, sometimes also called on-the-job training.

The method consists of four steps:

1. Prepare the associate for training.

2. Demonstrate what the associate is to do (show-and-tell).

3. Have the associate do the task, as shown, repeating until the performance is satisfactory.

4. Follow through: Put the associate on the job, checking and correcting as needed.

These four steps are applied to one task at a time. Figure 10.6 shows the steps and the relationship to learning flow.

job instruction training (JIT)
A four-step method of training people in what to do and how to do a given job in a given operation.

	☐1 **Preparation**	☐2 **Show-and-tell**	☐3 **Performance, practice**	☐4 **Follow through**
Trainer:	Prepares learner for instruction	Shows and tells what to do, how to do it	Gives feedback: corrects, evaluates, praises	Checks, gives reinforcement
Learner:	Sees need, accepts training (phase 1)	Acquires knowledge: learns what, how, why (phase 2)	Learns to do: performs, receives feedback, practices, succeeds (phases 3, 4)	Performs on job: knows how, can do. Reward achievement, reinforce correct performance (phase 5)

Key: Instructor impact – – – ▸   Learning flow ——▸

FIGURE 10.6: Learning flow in job instruction training.

Step 1, preparing the associate (let's call him Bob), consists of several things you do to let him know what is coming, make him feel at ease, and motivate him to learn. One thing is telling Bob where his job fits into the overall operation and why it is important to the operation. Another is giving Bob a reason to learn ("It will benefit you," "It will help you to do your job." "You will be rewarded in such and such a way") A third is telling him what to expect in the training and expressing confidence that he can do the job.

Step 2, demonstrating the task, is show-and-tell: "This is what I want you to do and this is how I want you to do it." You explain what you are doing and how you are doing it and why you are doing it the way you are. You use simple language and stress the key points. You tell Bob exactly what he needs to know but no more (unless he asks). If you tell him too much, you will confuse him.

You do not go into the theory of the dishwashing machine: the temperature it has to reach, the bacteria that have to be killed, and why bacteria are such an issue. You tell him, "The dish must get very hot. The needle must be at this number." The core of the action stands out clearly—no theory, all application. You take care to demonstrate well because what you do is going to set the standard of performance, along with teaching the how-to. You cannot give a second-class demonstration and expect the worker to do a first-class job.

Step 3, having Bob do the task as you showed him, is really the heart of the training. The first time is a tryout. If he can do it correctly right off, he is stimulated. If he cannot, you correct the errors and omissions in a positive way and have him do it again, showing him again if necessary. As he does it, have him tell you the key points and why they are done the way they are; this will reinforce the learning.

Job instruction training prepares the associate for training and demonstrates what the associate is supposed to do—show-and-tell.
wavebreakmedia/Shutterstock

Have him do the task several times, correcting himself if he can, telling you why he made the correction, letting him experience the stimulation of his increasing understanding. Encourage his questions, taking them seriously no matter how simpleminded they sound to you. Praise him for his progress and encourage him when he falters or fails. Have him repeat the task until you are satisfied that he can do it exactly as you did, to the standard you have set for him. Let him see your satisfaction and approval.

Step 4, following through, means putting Bob on his own in the actual job. You do this not for individual tasks but for units or groups of units or when the worker has learned the entire job. You stay in the background and watch him at work. You touch base frequently, correct his performance as necessary, and let him know how he is doing.

Retraining

LEARNING OBJECTIVE: Identify scenarios in which retraining is necessary.

Training people for jobs does not always take care of their training needs. Further training is necessary in several instances.

One such situation arises when changes are made that affect the job. You might make some changes to the menu. You might put in a different type of cash register. Your boss might decide to use paper and plastic on the hospital trays instead of china and glasses. Your food and beverage director might decide to install an automatic dispensing system for all the bars.

When such changes affect the work of your people, it is your responsibility to tell them about the changes and see that they are trained to deal with them. If the changes are large, you might develop new performance standards, procedural sheets, and training plans and run your people through additional training sessions.

If the changes are small, such as a new kind of coffeemaker or a new linen supplier with different routines and delivery times, they still affect people's work, but in the usual daily rush it is easy to overlook letting people know or to assume they will find out and know what to do. Even posting a notice or a set of instructions is not enough; a person-to-person message is in order, with show-and-tell as called for.

It is as important to keep your people's knowledge and skills up to date as it is to train them right in the first place. They cannot do the job well if the job has changed, and it makes them feel bad to know that no one has thought to tell them of the change and show them what to do.

A second kind of training need arises when an employee's performance drops below par, when he or she is simply not meeting minimal performance standards. It could be caused by various things: difficulties involving the job itself or other people, personal problems outside the job, or simply job burnout—disenchantment with doing the same old thing day after day and lack of motivation to do it well anymore.

Suppose that Sally's performance as a cocktail waitress has deteriorated noticeably in the last few weeks and there have been guest complaints of poor service and rudeness. A manager might sharply reprimand her and order her to shape up, but threats and coercion are not going to do the trick. Nor will it help to ignore the problem. A person whose previous work has been up to standard is usually well aware of what is happening. If you tolerate Sally's poor performance, it will reduce her respect for you, for the job, and for herself.

This situation calls for a positive one-on-one approach, generally referred to as *coaching* (see Chapter 9). Coaching is a two-part process involving observation of employee performance and conversation focusing on job performance between the manager and the employee. *A few other situations also call for retraining. For example, you may notice that a worker has never really mastered a particular technique* (such as cleaning in the corners of rooms) *or has gone back to an old bad habit* (such as picking up clean glasses with the fingers inside the rims). In such cases, you simply retrain the person in the techniques involved.

In other instances, people themselves may ask for further training. If you have good relationships with your people, they will feel free to do this and, of course, you should comply. It is testimony to your good leadership that they feel free to ask and that they want to improve their performance.

Overcoming Obstacles to Learning

LEARNING OBJECTIVE: Identify effective ways of reducing learning obstacles.

When you think of the many barriers to communication that were discovered in Chapter 4, it should not be surprising that training should have its share of obstacles. Some of them are learning problems, and some have to do with teaching, the trainer, or the training program. (You can see the two halves of the communication process again: sending and receiving the message.)

One problem for the learner may be fear. Some people are afraid of training, especially if they did not finish school and never really learned how to learn. This kind of anxiety clouds the mind and makes learning difficult. Some people have fear as their basic motivation. Contrary to prevalent belief in this industry, fear is usually a barrier to learning, not a motivator. It interferes with concentration and inhibits performance. People who are afraid of the leader or the instructor will not ask questions. They will say that they understand when they do not.

You can reduce fear and anxiety with a positive approach. Begin by putting the new employee at ease, conveying your confidence that he or she can learn the job without any trouble. Everyone will learn faster and better if you can reduce their anxieties and increase their confidence. Work with the trainee informally, as one human being to another, and try to establish a relationship of trust. Praise progress and achievement.

Some people have little natural motivation to learn, such as ambition, need for money, desire to excel, desire to please, and self-satisfaction. If they don't see anything in the training for them, they will learn slowly, they will not get things straight, and they will forget quickly.

There are three key ways of increasing motivation:

1. *Emphasize whatever is of value to the learner:* how it will help in the job, increase tips, or make things easier. As you teach each procedure, point out why you do it as you do, why it is important, and how it will help.

2. *Make the program form a series of small successes for the learner.* Each success increases confidence and stimulates the desire for more success.

3. *Perhaps the most important motivator is to build in incentives and rewards for achievement as successive steps are mastered.* These can range from praise, a progress chart, and public recognition (a different-colored apron, an achievement pin) to a bonus for completing the training.

Some people are not as quick to respond or don't process information quickly. They might have trouble with the pace and level of the instruction: too fast, too much at once, too abstract, too many big words. They may be capable of learning if the teaching is adjusted to their learning ability. *One-to-one instruction, patience,* and *sensitivity* are the keys here. Often, things learned slowly are better retained.

Some people are lazy and indifferent. And if they are lazy about learning the job, they will probably be lazy about doing it. Others will resist training because they think they know it all. They expect to be bored and they pay little attention to the instruction. These are potential problem types, and they will be either a challenge or a headache.

Sometimes we do not deal with people as they are. We assume that they know something they don't know, or we assume that they don't know something they do know. Either way, we lose their attention and their desire or ability to learn what we want them to learn. To overcome this obstacle, we need to *approach training from the learner's point of view.* Instead of teaching tasks, teach associates. Put yourself in their place, find out what they know, teach what they do not know, and interest them in learning it.

Keep it simple, concrete, practical, and real. Use words they can understand: familiar words, key words they can hang an idea on. Involve all their senses: seeing, hearing, feeling, experiencing. It is said that people remember 10 percent of what they read, 20 percent of what they hear, 30 percent of what they see, 50 percent of what they hear and see, and they remember more of what they do than what they are told. Teach by show, tell, and do—hands-on.

Sometimes the training program is the problem. If it is abstract, academic, impersonal, or unrealistic, it will not get across. If you have not carefully defined what you want the trainee to learn and you have no way of measuring when learning has taken place, the trainee may never learn the job well.

If the training sessions are poorly organized, or if the training materials are inadequate or inappropriate, or if the setting is wrong (noisy, subject to constant interruptions, lacking in equipment, or other on-the-job realism), the sessions will be ineffective. If the program does not provide incentives to succeed, the program itself will not succeed.

Sometimes the instructor causes the learning problems. Trainers need to know the job well enough to teach it. They need to be good communicators, able to use words other people will understand, and sensitive enough to see when they are not getting through.

They need to be able to look at the task from the learner's point of view, a very difficult thing when you know it so well that it is second nature to you. They need patience. They need leadership qualities: If people do not respond to the trainer as a leader, they do not learn willingly from that person.

Above all, trainers must not have a negative attitude toward those they are training. Never look down on either the person or the job, and take care to avoid Theory X assumptions (people are lazy, dislike work, and must be coerced, controlled, and threatened). Assume the best of everyone.

When a mistake is made, correct the action rather than the person, and correct by helping, not by criticizing. A useful technique is to compliment before correcting. Say, for example, "You are holding the bottle exactly right and you have poured exactly the right amount of wine. What you need to do to avoid spilling is to raise the mouth and turn the bottle slightly before you draw it away from the glass" (instead of, "Look what you did, you dribbled wine all over the table; don't *do* that, I *told* you to raise the mouth!"). Emphasize what is right, not what is wrong.

Be patient. Hang on to your temper. Praise progress and achievement. Think success. Cheer your people on as they learn their jobs, and stick with them until they have reached your goals.

Turnover and Retention

LEARNING OBJECTIVE: Explain how to improve employee retention rates.

Human resource directors estimate the cost of **employee turnover** at about $4,000 for an hourly paid employee, $8,000 for a middle management position, and $12,000 for a management position. Given that many hospitality operations have a labor turnover of more than 100 percent, we can quickly calculate the cost per year. For example, a 30-employee restaurant would likely be 30 times $4,000, or $120,000. That amount would be higher if management were also considered. So, it's no wonder that hospitality HR professionals and managers are keen to reduce labor turnover, and the way they do this is to focus on retention.

Retention is the term given to keeping employees from "jumping ship" to work for a competitor or another industry, or from being let go due to a variety of reasons. In a tight labor market it is even more critical to begin a retention program with recruitment. By taking more care in recruiting the right candidate, instead of the first candidate, hospitality companies stand a better chance of improving their employee retention. Noted psychologist and author William James of Harvard University stated:

> The greatest need of every human being is to feel needed or appreciated. Recognizing and meeting that need within your workplace's retention programs will go very far in satisfying retention goals. By establishing a proper "fit" for each employee as he/she is hired, the odds of that person being successful in the new job will be significantly enhanced.[8]

Obviously, in the formulation of a retention plan there are strong links to other elements of human resources. Serious attention is paid to the basics of clear job goals in the job description. By having clear job performance goals detailing the most important things an employee must

employee turnover
The rate of employee separations in a company—usually expressed as a percentage.

retention
The extent to which employees are retained by a company—thus reducing turnover.

know or be able to do, there is a target to aim for. Additionally, each goal should be quantifiable, with measurable results and an estimated timeline for accomplishment. By doing this, we can be more sure of individual employee success.[9]

❋ FACTORS IN RETAINING EMPLOYEES

In a Society for Human Resource Management survey on retention, those who conduct exit interviews indicated that the most widely cited reason for leaving an organization is to advance to a better job. The Families and Work Institute asked, "What makes a better job?" in a survey of 3,400 nationally representative employees titled "The National Study of the Changing Workplace." This survey asked what the employees considered to be "very important" in deciding to take their current job. The following were cited as the top three reasons, based on the greatest number of respondents indicating "very important":

1. Open communications (65% of respondents indicating "very important")
2. Work–life balance (60%)
3. Meaningful work (59%)[10]

In another study by the Hay Group, over half a million employees in 300 companies were asked about important retention factors. Interestingly, pay was number 10 out of 10. Here are the top 10:[11]

1. Career growth, learning and development
2. Exciting and challenging work
3. Meaningful work (making a difference and a contribution)
4. Great people to work with
5. Being part of a team
6. Having a good boss
7. Recognition of work well done
8. Autonomy and a sense of control over work
9. Flexible work hours
10. Fair pay and benefits

❋ STRATEGIES FOR RETAINING EMPLOYEES

Here are some strategies that work toward improving employee retention.[12]

Hold 50/50 Meetings
Hyler Bracey, author of *Managing from the Heart,* suggests that "key caveats for leadership with heart are that employees want to be heard and understood, and that they want to be told the truth with compassion."[13] Bracey says 50/50 meetings are ones in which management talks for half the time about goals, vision, and mission. Employees talk the other half of the time by raising their own questions and issues. This is a good way to curb negativity and low morale and therefore improve retention.

Practice Management by Wandering Around (MBWA)
Managers need to follow Tom Peters and Bob Waterman's "excellence" strategy of management by walking around—getting out among employees to discuss important day-to-day issues.[14]

Work Side by Side with Employees
"Walk a mile in my shoes" teaches compassion for the issues faced daily by employees.

Conduct Exit Interviews
Go beyond vague reasons such as "more money" or "better opportunity" to find out the real dissatisfiers ("You're not paying me enough to put up with . . .").

Use Other Methods to Listen

Listen to employees who are consistent with the organization's culture. Suggestion systems, employee task force meetings, and employee committees may serve as excellent strategies for more effective listening. HR professionals and management must be prepared to act on employee issues; otherwise, the purpose is defeated.

Another key element in employee retention is that employees want to be recognized for a job well done. Recognition should be a part of an organization's culture and making rewards count is a must. Rewards should be immediate, appropriate, and personal, and it is better to ask employees for their input on the most desirable form of recognition.[15] Retaining the best employees is further enhanced with two additional simple no- or low-cost things: respect and rewards.[16]

A unique approach to retention is practiced by the Attrition and Retention Consortium (ARC), a group of about 20 companies that formed to share and benchmark HR metrics. By comparing their turnover to industry or average benchmarks, member organizations can determine if they have a legitimate problem. By sharing best practices relating to a comparison of employees across tenure groups—measuring quality of life, identifying key drivers of turnover, and determining ways to hold on to valued employees—everyone benefits.[17]

Hospitality human resource directors and managers realize that turnover is often higher in certain areas such as dishwashing, serving, or housekeeping. A quick inquiry will likely identify strategies that will improve retention rates in these important areas. Without good dishwashers, servers, and housekeepers, a company will not reach its potential.

CASE STUDY: A Quick-Fix Training Program

Tom is assistant manager of a restaurant with 40 people on the payroll. He reports to Alex, the manager. Tom has full charge of the restaurant on the 7-to-3 shift, calculates the weekly payroll, takes care of all the ordering and receiving, and carries out special assignments for Alex. He couldn't be any busier, but this morning, Alex handed him his biggest headache yet.

"Tom," said Alex, "things are going downhill here, and we've got to do something. Sales are off, profits are down, our employee turnover is high and getting higher, and customer complaints are going up. They complain about the food, the service, the drinks, the prices, everything. I really don't think any of our people are doing the best they could, and maybe some more training would help.

"Look into it for me, would you, and see if you can figure out how you and I between us can find time to train our people to do a better job. I want to start tomorrow—I've got two new waiters and a grill cook coming in at 10:00 A.M." And he handed Tom a copy of a book called *Supervision in the Hospitality Industry* and told him to read Chapter 10.

Case Study Questions

1. What can Tom come up with between now and tomorrow morning? Is Alex expecting the impossible of Tom?
2. What kind of training can he provide for the three new people starting work at 10:00 A.M.? What might he do that he was not doing before?
3. What should Tom and Alex consider in deciding which category of current employees should be trained first?
4. How can Alex and Tom sell the entire process to current employees and get their cooperation?
5. How can either Alex or Tom find time to carry out the training, and when should it be carried out in relation to employee time?
6. Should Tom recommend bringing in outside help? Why or why not?
7. How can the two of them determine what training is needed?
8. How long do you think it will be before they can expect perceptible results?
9. What kind of long-range, permanent training plans and policies should Tom recommend?

KEY POINTS

1. Training means teaching people how to do their jobs. You may instruct and guide a trainee toward learning knowledge, skills, or attitudes.

2. Three kinds of training are needed in hospitality operations: job instruction, retraining, and orientation.

3. Training has the following benefits: more time to manage, less absenteeism and less turnover, less tension, higher consistency of product and service, lower costs, happier customers and more of them, and enhancement of your career. By making sure that your employees know what to do, tension is reduced, morale and job satisfaction are boosted, the number of accidents and injuries are reduced, and your workers have a better chance of advancing.

4. The problems involved in training are real: urgent need for trained workers, lack of time, lack of money, short-term workers, diversity of workers, kinds of jobs and skills, complexity of some jobs, and not knowing exactly what you want your people to do and how.

5. The logical person to train your people is you, the supervisor. It is your responsibility, whether you delegate it or do it yourself. Training is one of those obligations to your people that goes with your job—giving them the tools and knowledge to do theirs.

6. Employees learn best when they are actively involved in the learning process, when the training is relevant and practical, when the training materials are organized and presented in small chunks, when the setting is informal and quiet, when the trainer is good, and when employees receive feedback on their performance and reward for achievement.

7. Figure 10.3 portrays the steps involved in developing and carrying out a training plan.

8. Once the formal training process is completed, there is still one very important step: evaluation. Formative evaluation uses observation, interviews, and surveys to monitor training while it is going on. Summative evaluation measures the results of the training after the program is completed, looking at it in five ways: reaction, knowledge, behavior, attitudes, and productivity.

9. The procedure for job instruction training is illustrated in Figure 10.6.

10. Classroom training may be used at times for job instruction or retraining. Teaching in a classroom requires certain skills: Be aware of and use appropriate body language and speech, convey respect and appreciation, use informal and familiar language, correct in a positive and friendly manner, handle problem behaviors effectively, avoid time-wasters, facilitate employee participation and discussions, and use visual aids properly.

11. Retraining is needed when changes are made that affect the job, when an employee's performance drops below par, or when a worker simply has never really mastered a particular technique.

12. Orientation is the prejob phase of training that introduces each new employee to the job and workplace. Your goals for orientation are to communicate necessary information, such as where to park and when to pick up a paycheck, and also to create a positive response to the company and job.

13. Some keys to training include the following: Use a positive approach to reduce fear and anxiety; look at ways to increase employee motivation, such as building in incentives and rewards as steps are mastered; adjust the teaching to the employee's learning ability; don't assume anything; approach training from the learner's point of view; keep it simple and practical; and make sure that the trainer is doing a good job.

KEY TERMS

adult learning theory	pretest
buddy system	retention
employee turnover	retraining
formative evaluation	summative evaluation
job instruction	teaching methods
job instruction training (JIT)	training
learning	training objective
orientation	training plan

REVIEW QUESTIONS

Answer each question in complete sentences. Read each question carefully and make sure that you answer all parts of the question. Organize your answer using more than one paragraph when appropriate.

1. Define the three kinds of training needed in hospitality operations.
2. State why leaders find it hard to train and the benefits that training can bring.
3. Why is the leader the logical person to do training?
4. How do employees learn best?
5. Discuss the steps involved in developing and implementing a training plan.
6. Give an example of formative and summative evaluation.
7. If you had to teach an employee how to clean a hotel room, how could you use job instruction training to do so?
8. List 10 tips for training in a classroom (group) situation.
9. When is retraining necessary?
10. Why is orientation important? What happens during orientation?

ACTIVITIES AND APPLICATIONS

1. Discussion Questions

- In your opinion, what is the most serious consequence of not training? What is the most persuasive reason for not training? How can you weigh one against the other?
- Why can't people simply be trained by working alongside another employee until they learn the job? In what kinds of jobs would this work best? In what situations would it be impossible or undesirable?
- Have you ever started a new job and received little or no orientation? Describe what kinds of orientation programs you've received from past or present employers. What were some of their good features? Was orientation helpful?
- A server for a well-known Italian restaurant chain has finished her training. As the supervisor, how might you evaluate her training?

2. Brainstorming: My Favorite Teacher

As a class, brainstorm the personal qualities and characteristics of teachers you've had in the past who you thought were excellent at teaching. How does your list compare to the list in Figure 10.2?

3. Group Activity: Cost–Benefit Analysis

Cost–benefit analysis is a way to evaluate training by comparing its costs to its benefits to see which are greater. In groups of four students, make a list of possible costs (such as the trainer's and trainees' salaries) involved in training kitchen staff about preventing accidents.

Next, make a list of the possible benefits to be derived from this training. Which costs and benefits did you list that can be quantified into dollars and cents? Under what circumstances might an accident-prevention program have more benefits than costs?

4. Group Activity: Develop a Training Plan

The general manager at Nighty-Night Hotels suspects a problem with sexual harassment among some staff members. He needs a training plan to educate staff on this topic. Form groups of four and discuss how you would go about using Figure 10.3 (steps 1–4) to develop such a training plan.

ENDNOTES

1. Towers Perrin and Gang and Gang Research, "Working Today: Exploring Employees' Emotional Connections to their Jobs" (2002). As cited in Nikos Mourkugiannis, "Training for Purpose," *Training Journal* (March 2007): 33.
2. Nikos Mourkugiannis, "Training for Purpose," *Training Journal* (March 2007): 33.

3. Pamela Parseghian, "For Raw Recruits, Training Is the Best Defense Against Food Poisoning," *Nation's Restaurant News*, vol. 34, no. 33.

4. Bob Haber, human resource director, Grand Hyatt Tampa Bay, personal communication, October 5, 2007.

5. Andy Vuong, "Outback Will Pay $19 Million to Settle Sex-Bias Lawsuit," *Denver Post*, December 30, 2009. www.denverpost.com/headlines/ci_14090540. Accessed June 2011.

6. Adapted from "Inservice Training: How Do Employees Learn Best?" *Hospital Food and Nutrition Focus*, vol. 6, no. 4 (1989): 1–4. Reprinted with permission from Aspen Publishers.

7. Lisa Bertagnoli, "Ten-Minute Managers Guide to Cross Training Staff," *Restaurants and Institutions*, vol. 114, no. 18 (August 2004): 26.

8. Danny W. Avery, Society for Human Resource Management, White Paper, December 2002.

9. Ibid.

10. Catherine D. Fycok, "Managing for Retention," SHRM White Paper, September 2002. www.themcintyregroup.com/index.cfm/ForClients/ClientResources/StaffingStrategies/Managing_for_Retention. Accessed June 2011.

11. Beverly Kaye and Sharon Jordon-Evans, "Retention Tag: You're It," *Training and Development*, vol. 54, no. 4 (2000): 29–34.

12. Ibid.

13. Fycok.

14. Thomas J. Peters and Robert H. Waterman Jr., *In Search of Excellence* (New York: HarperCollins, 2004), p. 289.

15. Jack J. Philips and Adele O. Connell, eds., "Excerpts from Managing Employee Retention," Society for Human Resource Management, 2003.

16. Nancy R. Lockwood, "The Three Secrets of Retention: Respect, Rewards and Recognition," *SHRM Research Translations* (December 2004).

17. Pamela Babcock, "HR Measurements Library—Competitive Practices: Collect, Use Employee Data to Drive Retention Practices," Society for Human Resource Management, May 1, 2007.

Chapter 11

Conflict Management, Resolution, and Prevention

S usan wants Friday night off. She requested it off, but she is on the schedule for Friday night anyway. Management had to put her on the shift because she is one of the restaurant's top servers and it is going to be a very busy night. Susan is not aware that management thinks of her as one of the top servers. Instead, she perceives management as trying to punish her because she had Jane cover her lunch shift last week. Here, a possible conflict with management may arise due to a misperception and a lack of communication. In this chapter, we define conflict and discuss how it arises and how to manage it in the workplace.

After completion of this chapter, you should be able to:

- Define conflict.
- Explain the key principles of conflict management.
- Recall effective methods of conflict resolution.
- Define workplace violence, listing the warning signs and appropriate preventative measures.
- Identify key principles of conflict prevention.

What Is Conflict?

LEARNING OBJECTIVE: Define conflict.

anger
Feeling of hostility, wrath, indignation, or great displeasure.

Conflict. What do you think or feel when you hear that word? If you are like most people, you probably experience some sort of discomfort.[1] Why does conflict happen? To put it simply, because we are human and today we have many more choices in our lives and careers, which means more opportunity for conflict. We all have differing opinions. As a supervisor, the conflict at hand is not the real issue. How the conflict is dealt with is.

There are various ways to define conflict, but first, let's define anger. *Anger*, according to *Webster's Dictionary*, is defined as a feeling of great displeasure, hostility, indignation, exasperation, wrath, trouble, or affliction. **Conflict**, according to *Webster's Dictionary*, is defined as discord, a state of disharmony, open or prolonged fighting, strife, or friction.[2] We are likely to find anger where there is conflict but not always the reverse. One can certainly be angry without having any form of conflict with someone.

conflict
Discord, a state of disharmony, open or prolonged fighting, strife, or friction.

In this chapter, we define conflict as a disagreement resulting from individuals or groups that differ in opinions, attitudes, beliefs, needs, values, or perceptions. Conflict arises when two or more individuals, or groups, have opposing positions on the same subject.

desires
The things that we want.

✻ THE MAIN INGREDIENTS OF CONFLICT

needs
The things that we feel are vital to our well-being.

Why do people get into conflicts in the first place? Most conflicts are fueled because one's interests or values are challenged, or because their needs are not met.[3] The main "ingredients" of conflict include desires, needs, perceptions, power, values, and feelings. **Desires** can also be thought of as "wants." These are things that we would like to have or have happen but do not *need* to have or have them happen. **Needs** are those things that we feel are vital to our well-being. Conflicts are bound to arise when needs are ignored—or we ignore the needs of others.

perceptions
How people interpret things—situations, events, people.

Perceptions are how people interpret things (situations, events, people, etc.). We all see things differently inside and outside of a situation. It is how people interpret, or perceive, a situation that determines whether a conflict will arise. Conflict may also arise when someone is

Most conflicts are fuelled because one's interests or values are challenged.

Courtesy of bikeriderlondon/Shutterstock.

power
The capacity to influence the behavior of others.

feelings
Emotions or tendencies to respond in emotional ways.

values
Deeply held beliefs that are not open to negotiation.

rejecting, or seeking to gain, **power**. The way that managers utilize their power may have an effect on the number and type of conflicts that arise. Lack of leadership, as well as overuse of power, are both potential sources of conflict.

Feelings and emotions are a main cause of conflict. Many people are unable to separate themselves from their feelings and emotions, causing things to become "cloudy," so to speak. Conflicts can also occur when people ignore the feelings of others or if the feelings of two or more parties differ over an issue.

Values are deeply held beliefs. When values are at the center of a conflict they are usually not up for negotiation, or any type of conflict management strategies. Here it might be best to just agree to disagree. These ingredients of conflict should not be viewed as negative. However, they can turn into elements that cause conflict if, and when, they are misunderstood. Figure 11.1 shows the main ingredients of conflict.

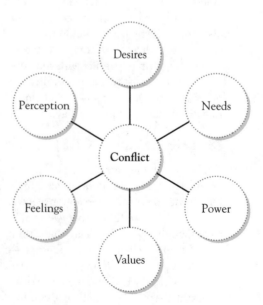

FIGURE 11.1: The main ingredients of conflict.

✺ COMMON CAUSES OF CONFLICT IN THE WORKPLACE

There are many causes of conflict in the workplace. It would be virtually impossible to cover all of them in one chapter. Here are some of the most common causes:

- Lack of communication
- Feelings of being undervalued
- Undefined/not clearly defined roles
- Poor use of managerial criticism
- Preferential treatment
- Poor management/leadership
- Impractical expectations
- Overworked employees
- Stress
- Personality differences
- Internal conflict

Conflict can sometimes appear to be with another individual when, at its center, it is not. Sometimes the person really has a conflict with him- or herself.

Workplace conflicts can be triggered by a variety of issues, such as the following factors suggested by Vicki Hess, principal of Catalyst Consulting:[4]

- *Different work methods.* Two employees have the same goal but approach the task two different ways.
- *Different goals.* Employees have goals that conflict with the goals for people in other areas of the company, such as front-of-the-house versus back-of-the-house.

Not all conflict is bad—some is positive. But all conflict has a past, present, and a future, and resolving conflicts effectively requires that all three are dealt with.

michaeljung/Shutterstock

- *Personalities.* People sometimes annoy each other just because of the way they look or act, or because of biases.

- *Stress.* On a good day, people can let issues roll off their back, but as stress increases, people often snap.

- *Different viewpoints or perspectives.* Someone might be so closely involved with an issue that he might have a different perspective from someone who sees the same issue more broadly. Hess also suggests that employees' viewpoints vary according to gender, age, upbringing, and cultural differences.

Okay, now go back and reread the common causes of conflict in the workplace. Try to find one answer to resolve them all. Did you come up with *communication* as your answer? It is the resolution to the majority of these conflicts. Poor communication is the number one topic raised by employees in questionnaires conducted in the workplace. Everyone wants to be valued. Showing a genuine interest in your employees fosters a positive workplace with open communication.

Conflict Management

LEARNING OBJECTIVE: Explain the key principles of conflict management.

We might wrongly assume that all conflict is bad for individuals and the organization. This is simply not so—some conflict is not only natural but also productive, experts say. Learning how to manage it, however, does not come naturally.[5] Every relationship and every conflict has a past, present, and future, and resolving conflicts effectively requires that we deal with all three.[6]

conflict management
The application of strategies to settle opposing ideas/goals.

Conflict management is the application of strategies to settle opposing ideas, goals, and/or objectives in a positive manner. Managers are often put in the middle of conflicts. They must know how to manage themselves, as well as the situation, positively and delicately. Managers must be able to separate their own emotions and feelings from the situation at hand. They need to be able to act, not react! There are many ways to manage conflict. For the purpose of this chapter, we use a five-step approach to conflict management, which is illustrated in Figure 11.2.

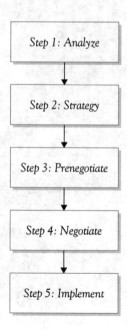

FIGURE 11.2: Five-step conflict management process.

✳ ANALYZE THE CONFLICT

The first step is to *analyze* what is at the center of the conflict. To do this the supervisors need to ask themselves questions, as well as ask those involved in the conflict. Here are a few questions to ask:

- Who is involved?
- How did the conflict arise?
- Can a positive spin be put on the situation?
- Are there any secondary issues?
- Have positions been taken?
- Is negotiation plausible?
- Is there a way to serve all interests at hand?
- Are there external constraints/influences?
- Is there a previous history of the conflict?

After the main source is identified and the source of the conflict is understood, it is helpful to brainstorm and write thoughts and ideas of resolution on paper.

✳ DETERMINE THE STRATEGY TO RESOLVE THE CONFLICT

The second step to managing conflict is to determine the type of *strategy* that will be used to resolve the conflict. Some examples of commonly used resolution strategies are collaboration, compromise, competition, accommodation, and avoidance.

Collaboration results most often when concerns for others are of high importance. This type of strategy results in a win/win outcome. Both parties cooperate with each other and try to understand the other parties' concerns, while also expressing their own. The parties both put forth a mutual effort and come to a solution that is completely satisfactory for all.

Compromise results from high concern for one's own interest or one's own group interest accompanied by moderate to high interest for the other parties involved. Both parties try to resolve the conflict by finding a resolution that partially satisfies both of them, but completely satisfies neither. This type of strategy either produces a win/win or lose/lose outcome, depending on if the solution chosen is the most effective. This varies, depending on the situation at hand.

Competition results when there is a high concern for one's own interest or one's own group. The outcome could vary from win/lose to lose/win, depending on who prevails. This strategy is not ideal, as it may cause increasing conflict. The losing party might try to even the score.

Accommodation is the result of low concern for your own interests or the interest of your group, which produces a lose/win outcome. The opposing party is allowed to satisfy their interest, while one's own interests are neglected.

Avoidance is exactly what it sounds like. The conflict is avoided by both parties and neither party takes action to resolve it. This produces a lose/lose outcome. In the hospitality industry, this strategy is generally useless because employees work in close quarters. This makes it virtually impossible to avoid each other.

✳ BEGIN PRENEGOTIATIONS

The third step to managing conflict is to start *prenegotiations*. This is a key part of the conflict management process. Being effective at negotiating is a fundamental skill for supervisors. During this step, there are several substeps. Initially, both of the parties involved in the conflict should be given the opportunity to come forth and offer a negotiation. If neither party is willing to come forth, then an outsider, in this case the leader, must step in.

Next, the situation should be *reassessed*. The key parties involved in the conflict must be willing to cooperate with each other in the resolution process. The issues should be laid out on the table. From here, what is negotiable, as well as what is not negotiable, must be determined. The parties involved should agree on what information is significantly related to the conflict, as well

collaboration
Strategy in which concern for others is of high importance and both parties try to cooperate with each other to come to a solution that is completely satisfying for both parties.

compromise
Concern for both one's own and the other parties' ideas or position, finding ways of agreeing (give and take) on positions.

competition
When there is high concern for one's own interest—two different individuals/groups become rivals.

accommodation
Strategy in which concern for your own interests is of low importance, resulting in an outcome whereby the opposing party is allowed to satisfy its interests while one's own interests are neglected.

avoidance
Strategy in which conflict is avoided by both sides, resulting in a lose/lose outcome where problems continue to remain unresolved.

as how communication and decision making will take place. All of this should be completed before moving on to the fourth step.

❋ NEGOTIATE A SOLUTION TO THE CONFLICT

The fourth step to managing conflict is to begin the *negotiation* phase. All parties must be able to express their concerns and interests; they must also be willing to listen to each other. As a manager, you will be considered the neutral third party. This means that you should not judge or favor either of the parties' ideas or suggested options. You are there to facilitate a healthy discussion and keep the parties focused on the cause of conflict and how it is to be resolved (not to assign blame to a particular party).

The parties involved in conflict should make a list of options that might help resolve the conflict, as well as satisfy their interests. After the lists of possible solutions are completed, the options should be discussed and evaluated. Which option would best resolve the conflict and satisfy the most interests should be determined together. A commitment ought to be made to carry out the agreements, and both parties must feel assured that the other will carry out their part.

❋ IMPLEMENT THE NEGOTIATED SOLUTION

The final step is for the parties to *implement* the negotiations made. As a supervisor, you need to support the resolution and continue to communicate. It is also beneficial to continue monitoring the situation in order to be certain that the agreement is in fact being carried out.

Conflict management is a very important human resources topic, because if it is handled in a professional manor it can be a win/win for both parties to improve their working relationship. I'm sure you realize that some conflict is positive, but the conflict we mostly hear about is conflict that challenges both HR professionals and hospitality leaders alike. I work with our team leaders and suggest that they immediately sit down to discuss the situation and hopefully clear the air. If it is a team issue, then the whole team needs to sit down and address the situation. If it is an

Courtesy of Shirley Ruckl

I was with the Longboat Key Club & Resort (LBKCR) as director of human resources for three years, where I directed the human resource functions for the club and resort for 550 associates. Originally from Chicago, I attended Columbia College–Chicago for Business Management/minor in Advertising Art. I relocated to Orlando in 1995 and started my career in HR as a recruiter at the Walt Disney World Dolphin Resort and then joined Universal Studios Florida as a recruiting manager, where I helped open Citywalk—the entertainment complex for Universal. Prior to coming to the LBKCR, I worked for Wyndham International for five years. I was hired as a recruiting manager, but after six months I was promoted to assistant director of HR, and then director of human resources for a 1,100-room convention resort.

Conflict management is a very important topic because if it is handled in a professional manner, it can be a win/win for both parties to improve their working relationship. I'm sure you realize that some conflict is positive, but the conflict we mostly hear about is conflict that challenges both HR professionals and hospitality leaders alike. The objective of conflict management is to resolve workplace challenges by addressing the behavior, and by being solution-focused on changing the behavior.

I work with our team leaders and coach them to promptly sit down with the associate to discuss the situation in a quiet, neutral setting. A key to successfully coaching through conflict is remembering to always leave the associate's pride intact—regardless of the situation. Depending on the severity of the issue, the team leader and/or human resources director can talk with the individual(s) concerned. The goal is to communicate the challenge(s) effectively and candidly, and agree on a solution to correct the behavior moving forward.

When conflicts are dealt with immediately and professionally, there is a very good chance of the situation working out well for both the company and the individuals concerned. Remember to keep accurate documentation, and only discuss on a need-to-know basis.

individual who is not carrying his or her weight, then escort the person to a neutral, quiet place where, depending on the severity of the issue, the team leader and/or human resource director can talk with the individual concerned—at first alone, and then with supervisors or others with whom the person has a conflict. The goal is to state the differences, agree on what the differences are, and then find ways to resolve them.

When conflicts are dealt with immediately and professionally, there is a very good chance of the situation working out well for both the company and the individuals concerned. Remember to keep records and only discuss on a need-to-know basis.

Conflict Resolution

LEARNING OBJECTIVE: Recall effective methods of conflict resolution.

Handling conflict in the workplace can be a challenging task. As a manager, you should always first keep the best interest of your company in mind, as detailed in Herb Kindler's book *Conflict Management: Resolving Disagreements in the Workplace*, and Robert Friedman's article: "Knock Out On-the-Job Conflicts, Complaints with Six Simple Steps," published in *Nation's Restaurant News*. They discuss the following guiding principles for handling conflict.

❄ PRINCIPLES FOR HANDLING CONFLICT[7]

First of the guiding principles is to *preserve dignity and respect*. This means preserving the dignity and respect of all parties involved in the conflict, including yourself. The focus should stay on resolving the conflict, not on the individual characteristics of the parties involved. As a manager, you should never talk down to an employee, especially during a conflict; this could result in them feeling like they are being attacked. If you make everyone feel respected, this will lower defenses and help the process of resolution.

Second, listen with empathy and be fully present and identify the issues.[8] As you listen, try to determine what issues created the conflict. In some cases, the real issues may be beneath the surface. The flashpoint of a festering disagreement might ignite and result in serious consequences. For example, some employees in a restaurant might be hoarding cutlery; when it is discovered that there is a shortage of spoons and another employee finds out where they are being hidden, a fight breaks out.

Don't daydream while an employee is trying to voice an opinion. Listen carefully to everyone involved and withhold any judgments until everyone has had a chance to speak. Try to see from each differing perspective, put yourself in each of the individual's shoes. Give everyone a chance to speak with you on a one-on-one basis. Give your full attention and make direct eye contact. Most important, make sure that your employees feel heard. There is nothing worse than being left with the feeling that your opinion (or words) do not matter.

Third is to *find common ground without forcing change*. Agree on what the problem issues are. Recite for the participants what you perceive to be the issues and ask them to agree with you or correct you. Appealing as it might seem, as a manager it is important not to try to force others into changing. People don't change for others, they change for themselves. They change only when they believe that they will benefit from the change. Therefore, throwing weight around as a superior will result in getting nowhere. It is also important for your employees to trust and respect you. If they believe that you are always looking out for their best interest, they are more likely to believe in you, and look up to you as their mentor.

Fourth is to *discuss solutions*: The parties involved have some idea of how they want the situation to be solved—ask them for suggestions.

Fifth is to *honor diversity, including your own perspective*. According to *Webster's Dictionary*, **diversity** is defined as a difference, variety, or unlikeness.[9] To **diversify** is to give variety to something; to engage in varied operations; to distribute over a wide range of types or classes. During this step, it is important to honor diversity, as well as to foster diversification.

diversity
Differences, variety, characteristics that are not alike.

diversify
Provide variety; distribute assets, jobs, or other things over a wide range of types or classes.

Strategy	Results from	Results in
Collaboration	High concern for others.	Win/win
Competition	High concern for one's own interest.	Win/lose or lose/win
Compromise	High concern for one's own interests and moderate to high interest for the other parties.	Win/win or lose/lose
Accommodation	Low concern for one's own interests	Lose/win
Avoidance	Conflict is avoided by both parties.	Lose/lose

FIGURE 11.3: Commonly used conflict resolution strategies.

groupthink
Phenomenon in which groups minimize conflict, resulting in a decision that excludes contrary opinions. Group members cheerlead what seems to be the prevailing idea, excluding critical evaluation.

Sixth is to *agree on the solutions and follow up.* Discuss solutions with each participant until there is agreement on the issues. Keep detailed notes or have a recorder. Then, once agreement has been reached, document it and have the participants sign it. Then follow up to see if the agreement holds or needs further discussion. Figure 11.3 shows commonly used resolution strategies.

Okay, so let's say everyone has differing viewpoints on a certain issue. This can lead to a creative way of searching for the right resolution, or it can result in feelings of isolation. All too often, the search for a resolution during a conflict is a hasty one. When we rush, we rush others into an agreement.

We don't let them have time to understand what really matters to them, or come to an independent viewpoint from that of the group, a phenomenon known as **groupthink**. Let's say a group has to make a decision and you are the only person who holds a different viewpoint. If others are quickly (and perhaps enthusiastically) getting behind the prevailing idea, you will probably end up conforming to the group and not speaking your opinion in order to avoid conflict. What you should do, of course, is to speak out and let your voice be heard.

Having an in-house alternative dispute resolution process can save a lot of problems and money.

ProStockStudio/Shutterstock

We all know the cost of a lawsuit is very high, but in the case of employment litigation, many companies find that the cost of defending themselves against the charges of unfair employment practice is extremely high, often exceeding the amount of the employee's claim of damages. Cases for unfair employment practices may drag on for years, with increased legal expenses. So it makes sense to have an in-house dispute resolution process.[10]

❋ ALTERNATIVE DISPUTE RESOLUTION

alternative dispute resolution (ADR)
Problem-solving and grievance-resolution approaches to address disputes.

Sometimes conflicts cannot be resolved within an organization, and the dispute might escalate to the point that it seems inevitable that the organization or an individual will be sued. There is an intermediary step that can be taken before that happens. **Alternative dispute resolution (ADR)** is a term for problem-solving and grievance-resolution approaches to address employee relations and disputes outside the courtroom. The purpose of ADR is to provide employers and employees with a fair and private forum to settle workplace disputes.[11] With ADR, a process is in place to offer the following options:[12]

Open-Door Policy
Employees have the opportunity to meet with managers to discuss issues.

Third-Party Investigations
A neutral third-party, from inside or outside the organization, confidently investigates complaints and proposes resolutions.

Fact-Finding
A neutral third-party person or team from outside the organization examines the facts of the complaint and presents them in a report.

Peer Review
A panel of employees, or employees and managers, works together to resolve the employee complaints.

Mediation
Through a voluntary and confidential process, a neutral third-party facilitator trained in mediation techniques negotiates a mutually acceptable settlement. The steps in the process are gathering information, framing the issues, developing options, negotiating, and formalizing agreements. Participants in the mediation process create their own solutions, and settlements are not binding.

Arbitration
Disputes are settled by an arbitrator and may be either binding or nonbinding, according to the wishes of the participants. An arbitrator or panel of arbitrators hears both sides of an issue and then makes a determination.

As Nancy Lockwood, a human resource content specialist with the Society of Human Resources Management, suggests, the advantages of ADR are that the total cost is less than the traditional means of resolving workplace disputes, legal costs are contained, the time spent on investigations is reduced, and workplace productivity is not compromised. Figure 11.4 shows the steps in an ADR process.

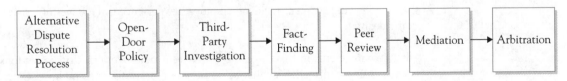

FIGURE 11.4: Alternative dispute resolution process.

Workplace Violence

LEARNING OBJECTIVE: Define workplace violence, listing the warning signs and appropriate preventative measures.

workplace violence
Any act of physical violence, harassment, intimidation, or other threatening behavior that occurs at the workplace.

Workplace violence is any act of physical violence, harassment, intimidation, or other threatening behavior happening at the workplace. The causes of workplace violence can be the result of a disciplinary measure or an employee bringing a domestic problem to work or a confrontation between two or more employees for any reason.

Workplace violence has become more problematic in recent years. The implications for HR professionals and managers are that they need to be more vigilant in creating a positive, safe, and secure workplace environment. We can see how several elements of HR come together: avoiding negligent hiring, creating a positive work environment, improving employee relations, leading, and providing an open and inclusive culture. Management must be trained to spot potentially troubled employees, and the company should have a good employee assistance program.

The possibility of workplace violence can be significantly reduced by taking a few preventative measures: increasing security by using employee name badges; reducing the number of entrances and exits to one or two; using video surveillance cameras; having metal detectors and guest and employee security checks at entrances, as hotels in Asia do; doing a complete background check on all employees; and noting and reporting any use of threats, physical actions, frustrations, or intimidation.

A company might have a problem if it decides not to employ someone because the person has a criminal record. Several state and federal laws (Title VII of the Civil rights Act) restrict employers from making employment decisions based on arrest records, since doing so might unfairly discriminate against minority groups. Obviously, however, if an applicant stole money and was arrested and found guilty, you would not employ that person in a position that has access to money.

Spotting a potentially violent person is not easy, so prevention by careful background checks prior to employment is essential. Failure to do proper background checks can open up employers to lawsuits related to negligent hiring if an incident of workplace violence takes place (see Chapter 6). Supervisors are on the front line and can share with their employees information on employee assistance programs or refer all employees to the human resource office for further assistance. Sometimes an employee with the potential for violence displays some tendencies such as being irritable or irrational, constantly complaining, or having numerous conflicts with guests or other employees. Other tendencies include substance abuse, strong interest in violence, over-aggressiveness, problems with authority, and other inappropriate behavior.

Steps to be taken before and if violence happens are:

1. Make sure all employees know how to contact their supervisor, manager, security, and police.

2. Take all threats seriously and ensure all employees have received training on how to handle a violent situation.

3. Depending on the severity of the situation, employees need to remove themselves from harm's way and contact their supervisor/manager and call security or the police.

4. Do a threat assessment to determine the severity of the situation and have any suspect removed from the premises.

5. Ensure all employees know the emergency lanes and procedures to evacuate the building.

Following a violent incident, employees experience three stages of "crisis reactions" to varying degrees:

■ *Stage one*. In this stage, the employee experiences emotional reactions characterized by shock, disbelief, denial, or numbness. Physically, the employee experiences shock or a fight-or-flight survival reaction in which the heart rate increases, perceptual senses

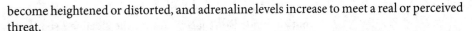

become heightened or distorted, and adrenaline levels increase to meet a real or perceived threat.

- *Stage two*. This is the "impact" stage where the employee feels a variety of intense emotion, including anger, rage, fear, terror, grief, sorrow, confusion, helplessness, guilt, depression, or withdrawal. This stage may last a few days, a few weeks, or a few months.

- *Stage three*. This is the "reconciliation stage" in which the employee tries to make sense out of the event, understand its impact, and through trial and error, reach closure of the event so it does not interfere with his or her ability to function and grow. This stage is sometimes a long-term process.

Although it is difficult to predict how an incident will affect a given individual, several factors influence the intensity of trauma. These factors include the duration of the event, the amount of terror or horror the victim experienced, the sense of personal control (or lack thereof) the employee had during the incident, and the amount of injury or loss the victim experienced (i.e., loss of property, self-esteem, physical well-being, etc.). Other variables include the person's previous victimization experiences, recent losses such as the death of a family member, and other intense stresses.[13]

Conflict Prevention

LEARNING OBJECTIVE: Identify key principles of conflict prevention.

Just think of some of the recent outcomes of lawsuits. Millions of dollars would have been saved by companies if they had proper conflict prevention in place. In 2007, a jury awarded $11.6 million in punitive damages against the New York Knicks for sexual harassment and retaliatory discharge.[14] In March 2010, a McDonald's franchise owner paid $90,000 to settle a federal discrimination lawsuit on behalf of a worker with an intellectual disability. The worker was repeatedly harassed by supervisors, managers, and coworkers because of his disability, and at one point was threatened by a coworker with a box cutter.[15]

Conflict is bound to arise in any atmosphere that requires *interdependency* between people and work. Preventing it is substantially more effective than having to undo it! What can be done to prevent such conflict from arising? Well, the conflict itself is not really the root of the problem. The root is a *lack of direct, properly handled conflict*. As already mentioned, communication is key in conflict management; it can also be thought of as the key to conflict prevention.

Be prepared to handle conflict. It is inevitable, and you should not be surprised when one comes your way. By preparing and thinking ahead of situations, you might be able to foresee a conflict before it happens. Conflicts arise in any situation that involves a decision being made that affects other people. If you think one might be brewing, talk to each of the individual parties. This might mean taking time out and putting other things that need to be done on the back burner, but if it results in the conflict being diffused, it will be well worth it in the long run and you will not have to deal with the impact the conflict would have had on the overall work environment.

As a manager, you should *pay close attention* to your employees. By paying attention, you might be able to diffuse a conflict before it actually takes place. If you do not pay close attention, you'll probably never see that a conflict exists until your employees are infuriated with each other; at this point, it's too late to diffuse it before it begins. Now you must manage it!

After becoming an active, responsive, and empathetic listener and learning to speak and act with commitment and integrity, the next challenge in resolving conflict is to work through the powerful, intense negative emotions that keep you from listening with an open heart and mind.[16]

Prevent conflict by *listening actively*. The easier you are to talk to, the more likely employees will come to you with their problems. Always take the time to see the conflict from every perspective. Never side with one person before hearing everyone's side of the story. This is the worst thing a supervisor could ever do. Remember, there are always three stories; your story, my story, and the actual story!

If it seems like a situation may lead to a conflict, you should *speak up* before the situation gets out of hand. Don't just stand on the sideline listening; diffuse the situation. Express concern before circumstances become intolerable. This may lead to the parties stepping back and reassessing the situation.

Always remember to *keep a sense of humor*. Once again, the more approachable you are, the more likely you will hear about or notice a problem before it begins. So remember, lighten up! Everyone in the organization will benefit from the implementation of these conflict prevention techniques. Conflict in the workplace has many negative effects; dealing with it early can prevent these effects from escalating and possibly creating more conflict. In many organizations, conflict is unidentified and never dealt with, and these organizations suffer. Identification and resolution result in success!

CASE STUDY: Conflict Management

At Cool, the new restaurant in town, Jim is the closing manager. Usually, his team gets along quite well. They are friendly and professional. They typically work well together and many are friends with each other outside of work.

On Thursday night, however, he observes two of the kitchen staff having a loud verbal interaction with two of the servers. Immediately, he rushes over to the group. Stepping in to stop the argument, he tells them all to report to the office after their shift. He is furious at their unprofessionalism. When the four employees arrive in his office, they are still arguing.

Jim is at a loss for words. Finally, he says, "OK, sit down and write down exactly what happened and how it can be fixed." One hour later, all four of them had an account of the incident and some suggestions for dealing with the problem.

Case Study Questions

1. Is this a good approach to conflict management? Give reasons.
2. What should Jim do next?
3. Do you think the situation can be resolved in this manner?
4. If you were Jim, how would you deal with the four employees?

KEY POINTS

1. Conflict happens, because we are human. Conflict is bound to arise in any atmosphere that requires interdependency between people and work.
2. As a supervisor, the conflict is not the real issue. How the conflict is dealt with is.
3. A conflict is a disagreement resulting from individuals or groups that differ in opinions, attitudes, beliefs, needs, values, or perceptions.
4. The main ingredients or sources of conflict include desires, needs, perceptions, power, values, and feelings.
5. Communication is the key to resolving and preventing most workplace conflicts.
6. Conflict management is application of strategies to settle opposing ideas, goals, and/or objectives in a positive manner.
7. The first step in conflict management is to analyze what is at the center of the conflict.
8. The second step to managing conflict is to determine the strategy to resolve the conflict; common strategies are collaboration, compromise, competition, accommodation, or avoidance.
9. The third step to managing conflict is prenegotiation.
10. The fourth step to managing conflict is to negotiate.
11. The fifth step to managing conflict is to implement the agreed-upon negotiations.

12. There are six guiding principles for handling conflict.
13. Supervisors are on the front line in preventing or identifying workplace violence.
14. Preventing conflict is substantially more effective than having to manage it after the fact.
15. Identification and resolution of conflict result in success.

 ## KEY TERMS

accommodation

alternative dispute resolution (adr)

anger

avoidance

collaboration

competition

compromise

conflict

conflict management

desires

diversify

diversity

feelings

groupthink

needs

perceptions

power

values

workplace violence

 ## REVIEW QUESTIONS

Answer each of the questions in complete sentences. Read each question carefully and make sure you answer all parts of the question. Organize your answer using more than one paragraph when appropriate.

1. What is conflict?
2. How is conflict caused?
3. How can employers manage conflict?
4. How can conflicts be resolved?
5. Can conflicts be prevented?

 ## ACTIVITIES AND APPLICATIONS

1. Discussion Questions
- Can workplace violence be avoided? If so, how?
- Is there good conflict? If so, what?
- As an employee, what should you do if you notice another employee acting in a weird manner?

2. Group Activity: Workplace Violence

Develop a plan and a policy for the prevention of workplace violence at a major hospitality company.

3. Group Activity: Conflict Management

Do a quick survey of your classmates to find out what their employers are doing in the area of conflict management. Report on your findings and compare them with those in the text.

 ## ENDNOTES

1. Craig Runde and Tim Flanagan, *Becoming a Conflict Competent Leader: How You and Your Organization Can Manage Conflict Effectively* (San Francisco: Jossey-Bass, 2007), p. 7.
2. *Merriam Webster Dictionary: Home and Office Edition*, (Springfield, MA: Merriam Webster, 1955), pp. 20,110.
3. Anita Naves, *Power Principles for Peaceful Living* (Bloomington, IN: Author House, 2006), p. 19.

4. Vicki Hess, "Conflict Management Contributes to Communication," Society for Human Resource Management: Workplace Diversity Library—Employment Issues. January 2007.

5. Ibid.

6. Morton Deutsch, Peter Coleman, and Eric Marcus, *The Handbook of Conflict Resolution: Theory and Practice* (San Francisco: Jossey-Bass, 2006), p. 161.

7. This section was adapted from Herb Kindler, *Conflict Management: Resolving Disagreements in the Workplace* (Boston: Thomson, 2006), pp. 3–4.

8. Robert Friedman, "Knock Out On-the-Job Conflicts, Complaints with Six Simple Steps," *Nation's Restaurant News*, vol. 40, no. 37 (September 11, 2006), p. 30.

9. *Webster's*, s.v. diversity.

10. Stephen Barth, "Why In-house Dispute Resolution Makes Sense," *Lodging Hospitality*, vol. 58, no. 7 (May 15, 2002), p. 19.

11. Nancy R. Lockwood, SPHR, "Alternative Dispute Resolution," *Society for Human Resource Management*, SHRM Research, February 2004.

12. Ibid.

13. USDA, "The USDA Handbook on Workplace Violence Preventions and Response" (December 1998), www.usda.gov/news/pubs/violence/wpv.htm. Retrieved March 28, 2011.

14. Allen Smith, "Browne Sanders Scores Big in Harassment Lawsuit Against Knicks," *Society for Human Resource Management Workplace Law Library—Employee Relations*, October 5, 2007.

15. U.S. Equal Employment Opportunity Commission, "McDonald's Franchise to Pay $90,000 to Settle EEOC Disability Discrimination Lawsuit," Press release, March 2, 2010. eeoc.gov/eeoc/newsroom/release/3-2-10.cfm. Retrieved April 28, 2011.

16. Adapted from: Kenneth Cloke and Joan Goldsmith, *Resolving Conflicts at Work* (San Francisco: Jossey-Bass, 2005), p. 74.

Chapter 12

Discipline

There is more to discipline than meets the eye. Discipline is not a black-and white issue; there are many shades of gray. It is a fluctuating product of the continual interplay between the supervisor and the people supervised within the framework of the rules and requirements of the company and the job.

As a whole, the hospitality industry is not famous for disciplinary success. Often, discipline is administered across a crowded room at the top of the lungs, and disciplinary measures make a direct contribution to the high rate of employee turnover in the industry.

In this chapter, we explore the subject of discipline from several points of view. After completion of this chapter, you should be able to:

- Define the four essential elements of discipline.
- Compare and contrast the different approaches to discipline.
- Explain the appropriate ways to administer discipline in the workplace.
- Discuss reasons for termination and best practices for termination interviews.
- Explain the purpose of employee assistance programs.
- List direct and indirect costs of workplace accidents.
- Describe the various forms of harassment in the workplace.
- Evaluate the role of supervisors in establishing and maintaining discipline.

A clear and effective disciplinary policy offers many benefits, including clear guidelines for employee behavior, good morale for other employees, and protection against lawsuits.[1]

Essentials of Discipline

LEARNING OBJECTIVE: Define the four essential elements of discipline.

If you were to walk around your work area one day and ask your employees what discipline is, it is very likely that the most frequent response would be that discipline means punishment. Does discipline really mean punishment? Let's take a closer look.

> **discipline**
> A condition or state of orderly conduct and compliance with rules, regulations, and procedures; or action to ensure orderly compliance with rules and procedures.

The word **discipline** has two somewhat different but related meanings. One refers to a *condition* or *state* of orderly conduct and compliance with rules, regulations, and procedures. If everyone follows the rules and procedures and the work moves along in an orderly fashion, we say that discipline is good in this department or operation.

But if people are not following the rules and procedures, and maybe do not even know what the rules and procedures are, and the work is not getting done, people are fighting, the place is in chaos, and nobody is listening to what the supervisor is trying to say, we say that the discipline is terrible.

The second meaning of the word *discipline* refers to *action* to ensure orderly conduct and compliance with rules and procedures. When people break rules, you discipline them—you take disciplinary action. Disciplinary action, depending on your policy, may or may not include punishment such as a written warning or suspension. If your employees are relatively self-disciplined, it is not necessary to discipline often.

In this chapter we are concerned with both kinds of discipline. We are concerned with maintaining a condition of discipline, and we are concerned with the most effective kinds of

Ensure that associates know what the rules and procedures are by telling and showing them what to do.

wavebreakmedia/Shutterstock

disciplinary action to ensure compliance to rules. *Both sides of discipline are the responsibility of the manager, and discipline, in both senses of the word, is essential to managerial success.*

The discipline process contains three steps:

1. Establishing and communicating ground rules for performance and conduct
2. Evaluating employee performance and conduct through coaching, performance appraisals, and disciplinary investigations (Coaching and performance appraisals are discussed in Chapter 7.)
3. Reinforcing employees for appropriate performance and conduct and working with employees to improve their performance and conduct when necessary

As a manager or supervisor, you are involved in each step of this process, as discussed throughout this chapter. Let's start by looking at the *four essentials of successful discipline*:

1. A complete set of rules that everyone knows and understands
2. A clear statement of the consequences of failing to observe the rules
3. Prompt, consistent, and impersonal action to enforce the rules
4. Appropriate recognition and reinforcement of employees' positive actions

The first essential—a complete set of rules—consists of all the policies, regulations, rules, requirements, standards, and procedures that you and your workers must observe in your job and theirs. These should include:

- Company policies, regulations, and directives that apply to your department and your people. Of particular importance to you are company policies and procedures relating to disciplinary action (see Figure 12.1).

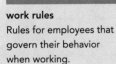

work rules
Rules for employees that govern their behavior when working.

- **Work rules** relating to hours, absences, tardiness, sick days, meals, use of facilities and equipment, uniforms and grooming, and conduct on the job. Rules might pertain to smoking, drinking, and dealing with customers, patients, or guests, for example.

- Legal requirements and restrictions, such as health code provisions, fire and safety regulations, and liquor laws.

- Job requirements, performance standards, and job procedures for each job you supervise.

- Quality and quantity standards required. These include, for example, standardized recipes, portion sizes, drink sizes, and guestroom amenities.

All this material will form a basic operations and procedures manual for your department. From it you can prepare a manual for new employees and plan their first-day orientation (Chapter 6). Then you can use it in developing your training programs (Chapter 10), incorporating all the rules, procedures, and penalties that the workers must know so that they start out well informed.

You can use your manual as a reference for verifying the proper ways of doing things and for settling any disputes that arise. Keep it in loose-leaf form so that you can update it easily when policies, regulations, and procedures change.

It is your responsibility to see that all your people know the rules and procedures that apply to them and to their jobs. These rules and procedures form a set of boundaries or limits for employee behavior, a framework within which they must live their occupational life. You might compare it to a box or a fence that encloses them while they are on the job (Figure 12.2).

Most employees really want to do a good job, and if they know what they are supposed to do and not do, most of them will willingly stay in the box and abide by the rules. Knowing the rules and the limits makes most people more comfortable in their jobs.

The second essential is to make very clear the consequences of going beyond the limits, of not following the rules and procedures. If there are penalties for breaking the rules, people must know from the outset what the penalties are. This information should be stated in matter-of-fact terms: "This is what we expect you to do; this is what happens when you don't." It should not take the form of warnings and threats. There should be no hint of threat in either your words or your tone of voice.

The penalties for breaking rules are usually written into your disciplinary policy and procedures (refer to Figure 12.1). The policy and procedures may prescribe the specific disciplinary action for each rule violation each time a given employee breaks that rule, or the penalties may be more loosely defined.

Knowing the consequences has its own security: People know where the boss stands, and they know what will happen if they go beyond the limits. Even when the penalties seem severe, and even when people do not like their supervisor personally, you often hear them say, "At least with the boss, you know where you stand."

The third essential is to enforce the rules promptly, consistently, and impersonally and to comply with the rules yourself. It is very common for supervisors to threaten punishment—"If you are late once more, I'm gonna fire you"—and never carry out the threat. After a while, other people see that the threat is never carried out and they begin to think, "Why should I be here on time?" And pretty soon, the leader has lost control and the associates are setting the rules and standards. Once you have made a threat, you have no choice but to carry it out or back down.

The principle applies not only to threats but also to rule-breaking in general. If you pay no attention when people break rules, if you walk on by and do nothing, everyone will begin to break the rules and discipline will crumble. And if you break rules yourself, people will have no respect for you because you are applying a double standard. They will think, "What is good enough for you is good enough for me, too," and you will have problems with compliance. There won't be any ground rules left.

Many people suggest the hot stove as the perfect model of administering discipline:[2]

- It gives *warning.* You can feel the hot air around it.

- Its response is *immediate.* The instant you touch it, it burns your finger.

Discipline Policy and Procedure

Policy: It is necessary to establish rules of conduct to promote efficient and congenial working conditions and employee safety. Further, it is our intention to provide equality in the administration of discipline when these rules of conduct are violated. Discipline is to be administered fairly without prejudice and only for just cause.

Procedure: In order that all disciplinary actions by supervisors are consistent, one of the following actions will be used according to the seriousness of the offense.

1. Oral warning with documentation
2. Written warning
3. Suspension
4. Termination

An employee will be subject to disciplinary action ranging from oral warning to discharge for committing or participating in any of the acts listed below. The normal level of discipline is also listed. All suspensions, terminations, or exceptions must have the approval of the Director of Human Resources.

1. False statements or misrepresentation of facts on the employment application—termination
2. Absence for one day without notifying the department manager prior to the start of the shift—written warning
3. Absence for two consecutive work days without notifying the department manager prior to the start of the shift—suspension
4. Absence for three consecutive work days without notifying the supervisor prior to the start of the shift—termination
5. Excessive absenteeism with or without medical documentation—within a calendar year—
    6 absent incidents—oral warning with documentation
    8 absent incidents—written warning
    9 absent incidents—suspension
    10 absent incidents—totaling 13 days or more—termination
6. Excessive lateness—within a calendar year—
    8 latenesses—oral warning with documentation
    12 latenesses—written warning
    16 latenesses—three-day suspension
    20 latenesses—termination

FIGURE 12.1: A discipline policy and procedure.

- It is *consistent*. It burns your finger every time you touch it.
- It is *impersonal*. It reacts to the touch, not the person who touches.

These are all sound guidelines to follow with any approach to discipline. You give *warning* by making sure that people know the rules and the consequences—what to do and how to do it and what happens if they don't. Your response is *immediate*; by tomorrow the mistake or transgression is past history and the worker has gotten away with something and three others have seen it happen and will try it today. You are *consistent*: You hold everybody to the same rules all the time; and you are *impersonal*: you are matter-of-fact, you don't get angry, you don't scold, you don't preach, you simply act as an adult. You deal with the specific incident, not with the person's bad attitude or thick skull. You eliminate your personal feelings about individual people: you do not prejudge someone you don't like, and you do not let favoritism creep in.

But there is more. Although impersonal, discipline ought to be carried out as part of a positive human relations approach to the people who work for you. In disciplining, you must focus on things that people do wrong, but your people can handle this negative feedback better if you use a lot of that positive reinforcement we talked about in the last chapter. Don't be like the manager who said, "Every time you do something wrong, I'll be there to catch you. But when you do something right—well, that's what I pay you for."

7. Falsification of time sheets, recording another employee's time or allowing others to do so—termination
8. Failure to record own time when required—oral warning with documentation
9. Leaving work area without permission—written warning
10. Leaving the facility without permission during normal working hours—written warning
11. Stopping work early or otherwise preparing to leave before authorized time, including meal periods—oral warning with documentation
12. Sleeping on the job—suspension
13. Failure to carry out job-related instructions by the supervisor where the failure is intentional—suspension
14. Threats or intimidation to managers, guests, or other employees—termination
15. Use of abusive language to managers, guests, or other employees—suspension
16. Stealing or destruction of company or guest's property—termination
17. Not performing up to performance standards—oral warning with documentation
18. Disorderly conduct during working time or on company property—suspension
19. Violations of sanitation and safety regulations—level 2, 3, or 4 depending on situation
20. Reporting to work unfit for duty—written warning
21. Possession or use of alcohol or nonprescribed drugs during working time or on company property—termination
22. Possession of explosives, firearms, or other weapons during working time or on company property—termination

Multiple or Cumulative Violations

1. Subsequent violations of a related nature should move to the next higher step in the discipline pattern (e.g., a related violation following a written warning will call for a 3-day suspension, etc.).
2. Violations of an unrelated nature will move to the next higher level after two disciplinary actions at the same level (e.g., after two written warnings for unrelated violations, the next unrelated violation would call for a suspension rather than another written warning).
3. Cumulative violations that occur more than 12 months before the violation in question will not be used to step up the discipline for an unrelated violation.
4. The above listed violations are the basic ones and are not intended to be all-inclusive and cover every situation that may arise.

FIGURE 12.1: (continued)

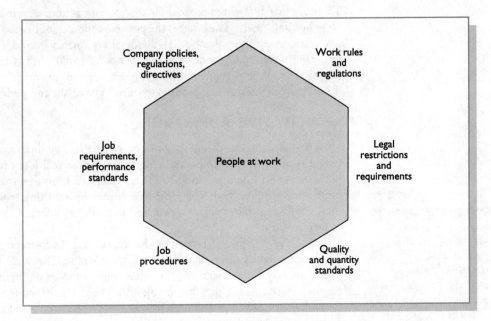

FIGURE 12.2: A framework of policies, rules, restrictions, standards, procedures, and requirements within which employees carry out their work.

You've Done A Great Job!

For: Denise Smith, Cook

From: Joe Brown, Chef

Date: 8/22/16

Thank you for giving two of our long-term guests Mr. and Mrs. Jones the extra-special treatment last night. They were most unhappy about their meal until you came out of the kitchen to help them make another selection. You did a great job of reassuring them and keeping their visit enjoyable. Thanks for going the extra mile.

FIGURE 12.3: Positive action memo.

The fourth essential is to recognize and reinforce your employees' positive actions. Discipline is not just making sure that employees follow rules, but includes recognizing those who are following rules and performing up to standards. Recognizing your employees need not be a laborious process; it can be as simple as saying, "Thanks for taking care of our guests in such a prompt and courteous manner," or filling out a positive action memo (Figure 12.3). Other ways to recognize your employees are discussed in Chapter 9.

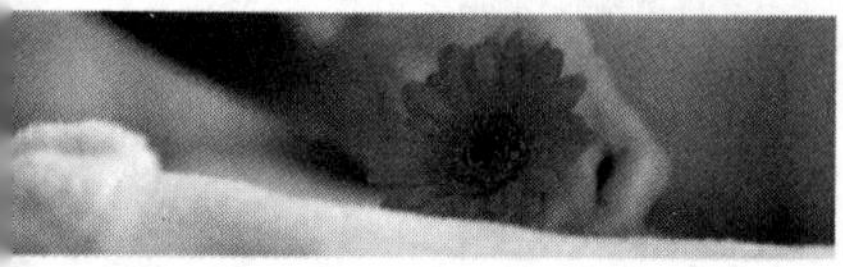

Approaches to Discipline

LEARNING OBJECTIVE: Compare and contrast the different approaches to discipline.

There are two different approaches to disciplinary action. One is the negative approach of discipline by punishment. The other is the positive approach of discipline by information and corrective training. Philosophically, they divide along Theory X and Theory Y views of people and leadership styles. Many leaders believe the word *discipline* has to do with punishment. Actually, it doesn't. Discipline pertains to improving employee performance through a process of assisting the employee (at least the first time) to learn so he or she can perform more effectively.[3]

❈ NEGATIVE APPROACH

Most people associate discipline with punishment. The theory is that if you enforce the rules by punishing people who break them, those people will learn not to break the rules and the punishment will be a warning to others that will keep them in line. It is the old theory of motivation through fear. The punishment may be anything from a public dressing-down or threat of dismissal or private reprimand to penalties tailored to fit the *violation*, culminating in termination.

Negative discipline has been used a great deal in the foodservice and lodging industries. It is commonly used by the rigid, high-control, autocratic, Theory X–style leader who believes that people are lazy and irresponsible and that you have to be on them all the time. Never mind the reason it happened: If they break rules, they've got to be punished; it's the only way to get it through their heads. It is also used by managers who are civilized and friendly but simply believe that punishment is the way to enforce rules.

negative discipline
Maintaining discipline through fear and punishment, with progressively more severe penalties for rule violations.

The fear-and-punishment approach has never worked very well. Punishing one person may deter others from breaking rules, but it does not correct the behavior of the person punished. Punishment simply does not motivate employees to shape up and do their work in an orderly and obedient manner. It may motivate them to avoid the punishment a second time—"Hey, you got me once, but you will never get me again"—but from then on, they will do just enough work to get by.

Fear and punishment are, in fact, *demotivators*. People who are punished feel embarrassed, defensive, angry, and hostile. It often arouses a desire to get back at the boss and to get the other workers on their side. They look for ways to cause trouble for the boss without getting caught, and the boss is probably going to have to punish them again and again. Punishment almost never turns a first offender into a good worker. It is, however, likely to turn that worker into an adversary.

Leaders who are rigid rule followers are usually very conscious of their right to punish and their duty to control, so they go by the book: If a rule is broken, punishment follows. Rigidity is the strongest feature of this kind of discipline; it is consistent. It does deter rule-breaking, and it maintains a certain kind of controlled order.

However, punitive leaders tend to have chronic discipline problems, which they are likely to blame on their "no-good workers." They do not recognize how their own shortcomings as leaders have contributed to the problem: They probably haven't explained the rules, communications are poor, people don't like the constant negative feedback, don't like working for them, and so on. Some Theory X leaders are really very insecure people, and their inability to control their workers' behavior makes them even more insecure. They vent their anger and frustration on their workers, reassuring themselves that the workers, not they, are to blame.

In a fear-and-punishment approach to discipline, there is a traditional four-stage formula for disciplinary action:

1. An *oral warning*, stating the violation and warning the employee that it must not happen again

2. A *written warning*, stating that the offense has been repeated and that further repetition will be punished

3. *Punishment*—usually suspension without pay for a specific period, typically one to three days

4. *Termination* if the employee continues to repeat the offense after returning to work

progressive discipline
A multistage formula for disciplinary action.

This four-stage formula is called **progressive discipline** because of the progressive severity of each stage. (The term does not in any way imply a forward-looking or humanitarian approach.) The stages are similar to those specified in most union contracts and written into most company policy manuals. The formula is not confined to hard-line Theory X leaders; it is widely used with hourly employees in all types of industries.

Over all the years that negative discipline has been used, it has never been successful at turning chronic rule-breakers into obedient and cooperative employees. There is nothing in it that will motivate change that will help anyone to become a better employee. It generally creates adversarial relationships and a sort of underground power struggle between worker and boss that is harmful to the work climate and the general morale. This is a power struggle that the supervisor must win if relationships with other workers are to be successful.

❋ POSITIVE APPROACH

If you stop thinking discipline equals punishment and start thinking discipline equals rule compliance, you can begin to see that other ways of enforcing the rules are possible. For example, what is the most frequent cause for breaking rules or going against company policies or failing to follow procedures? Up to 90 percent of the time, people do not know that they are doing something that they are not supposed to be doing. They didn't know that you should not leave the hollandaise sauce sitting all day on the back of the range. They didn't know they shouldn't let the patient in Room 302 have the sugar packets left on other patients' trays. They didn't know that champagne had to be chilled. They didn't know that they had parked in the general manager's parking space. They didn't know that guests weren't allowed in the wine cellar and you taught them that the customer is always right.

So when rules are broken, the action you take is to inform and correct. Even though you have handed out employee manuals and have told people the rules and trained them in their jobs, there are still things they don't know, or don't understand, or don't recognize in a new situation, or forgot, or they saw somebody else doing something and thought it was all right. So the positive approach to discipline is continuous education and corrective training whenever the rules and procedures are not being observed.

The philosophy behind the corrective approach is a Theory Y view of people: By and large, people are good, they will work willingly, they want to learn, they welcome responsibility, and they are capable of self-direction and self-discipline. They will do their job right if you tell them what you want them to do. The approach to discipline is educative and developmental: You inform people why the rule or procedure is important and how to carry it out correctly. The goal is to turn workers into productive employees who are self-motivated to follow the rules and procedures.

This approach to discipline is really an extension of the coaching process—observation, evaluation, and continued training as needed. It approaches rule-breaking as a problem to be solved, not as wrongdoing to be punished. It does not threaten people's self-respect, as punishment does; rather, it enlists their efforts in solving the problem.

There will still be some people who go on breaking rules, people who are irresponsible or lazy or hostile or who just don't care. So there must still be some last-resort disciplinary action if rule-breaking persists. But persistent rule-breaking doesn't happen nearly as often as it does with the punishment approach.

For chronic rule-breakers, there is a three-stage formula for disciplinary action that parallels the stages of negative discipline. However, it is not punitive; rather, it places the problem of correction squarely in the hands of the offender. The employee now has the responsibility for discipline.[4]

- *Stage 1: Oral reminder*. In a friendly way, you point out the rule violation as you see it happen. You talk to this person—let us say that it is Jim—formally about the seriousness of the offense, the reason for the rules, and the need to obey them. You listen to what Jim has to say in explanation and express confidence that he will find a way to avoid repeating the action.

- *Stage 2: Written reminder*. This follows further rule-breaking. You discuss privately, in a very serious manner, the repeated or continual violation of the rules, and you secure Jim's agreement that he will conform to company requirements in the future. Your attitude is that of counselor rather than judge or law enforcement officer; you avoid threatening him.

  Following this meeting, you write a memorandum summarizing the discussion and agreement, which both you and Jim sign. It is wise to have a third party present at this discussion to act as a witness if needed later. This memo goes into Jim's permanent file.

- *Stage 3: Termination*. Since Jim has broken not only the rules but also the agreement, there is a clear reason for the termination.

positive discipline
A punishment-free formula for disciplinary action.

The punishment-free formula for disciplinary action is known as positive discipline. Figure 12.4 compares it stage by stage with the negative discipline method. Positive discipline works. Many people who use it report that about 75 percent of the time, employees decide to come back and follow the rules. They may not maintain their turnaround indefinitely, but three months or even three weeks of productive behavior is preferable to finding someone new—and it is infinitely better than the hostile employee you are likely to end up with after an unpaid layoff.

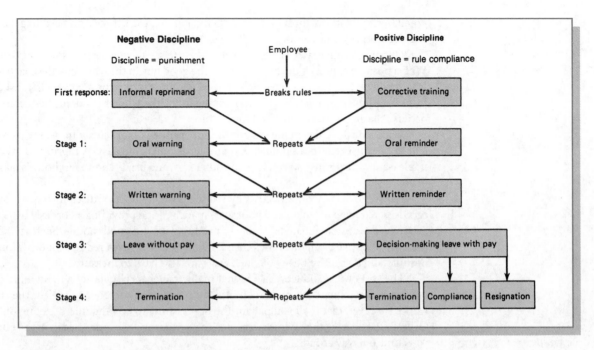

FIGURE 12.4: Negative and positive disciplinary action compared.

✳ ADVANTAGES OF THE POSITIVE APPROACH

Using a positive approach from the outset has distinct advantages over the negative system. Honest mistakes, infringements of rules, and violations of policies and procedures are educated out of people's work habits early, before they have time to become issues demanding confrontation. Many discipline problems simply do not happen. The negative consequences of punishment

It is better to avoid confrontational situations by adapting a positive proactive approach to policy and procedure violations.

michaeljung/Shutterstock

do not fester their way through the work climate. The worker feels no need to get even. The boss and the worker do not become adversaries.

Under a negative system, the worker is likely to see the manager as someone to avoid and fear. With a positive approach the boss becomes the good guy in the white hat, the coach and counselor who facilitate the employee's work. There is an opportunity for a good relationship to develop. Even if a problem reaches the point of taking leave, the worker is not likely to come back hostile because the employee has no need to save face.

With a positive discipline system the supervisor is more likely to deal with problems early and to be consistent in discipline. A reminder is quick and easy. A reprimand takes time and is unpleasant; you are busy so you look the other way, and pretty soon everybody is taking advantage of you.

Positive discipline lowers costs by reducing the number of disciplinary incidents, reducing turnover, reducing mistakes and poor workmanship, and providing an orderly work environment and a positive work climate favorable to productivity and good morale. Such savings are hard to measure. Punitive discipline raises costs by increasing turnover, reducing motivation, and causing hostility and disruptive behavior. Such costs are also hard to measure.

The cost of the paid leave is one that many managers boggle at: Why should you pay a rule-breaker to stay home and think about it, on top of paying someone to replace the rule-breaker at work? However, that is all you pay for the opportunity to end the rule-breaking once and for all. Overall, considering the savings of the positive system and the hidden costs of the punitive system, you come out way ahead.

One of the most important contributions of a positive discipline system is that the **decision-making leave with pay** does turn some people around permanently. It brings them face-to-face with themselves and puts their future in their own hands. This can become a new starting point for them. The supervisor can then play a key role by supporting all attempts to improve and by giving encouragement, positive reinforcement, and recognition for success. This is one of the few ways of transforming a hostile employee into a responsible and productive worker. It is a very rewarding kind of supervisory success.

✳ SHIFTING FROM NEGATIVE TO POSITIVE DISCIPLINE

The biggest problem of using the positive approach is in shifting from one approach to the other. As a person starting out in your first supervisory position, you might not have this problem. But supervisors who are used to administering penalties and punishments often have trouble shifting gears. To begin with, they may have difficulty accepting the idea of paying an employee to stay home and think things over. It seems like a reward for bad behavior, and it seems unfair to the people who follow the rules and are working hard for their day's pay.

The second problem is shaking loose the habit of thinking in terms of punishment and substituting the attitude of educating and helping people to avoid breaking rules. Leaders may *believe* in punishment. They may have been brought up with this type of discipline both at home and at school. It is hard to begin to teach, to help, to develop a rule-breaker when you have always reprimanded, warned, threatened, and punished. It requires an entirely new set of attitudes as well as a new tone of voice.

Administering Discipline

LEARNING OBJECTIVE: Explain the appropriate ways to administer discipline in the workplace.

In the hospitality industry, we sometimes find that management is reactive rather than planned—that is, reacting to events as they happen, dealing with problems as they come up. Certainly, enforcing the rules is one of the most reactive aspects of the supervisor's job. Even when the general outlines and the essentials are clear in your mind, each instance of enforcing the rules and procedures makes its own special demands, and positive and negative approaches seem less clear-cut

decision-making leave with pay
The final step in a positive discipline system in which the employee is given a day off with pay to decide if he or she really wants to do the job well or would prefer to resign the position.

and obvious. Consider this: A server at a restaurant was fired after making bad comments about a guest of the restaurant on her blog. Do you agree or not with the restaurant management's decision?

✸ ADAPTING DISCIPLINE TO THE SITUATION

uniform discipline system
A system of specific penalties for each violation of each company rule, to be applied uniformly throughout a company.

Many companies have a **uniform discipline system** that prescribes the specific disciplinary action for each rule violation each time a given employee breaks that rule. Figure 12.1 gives you an example. A system such as this provides a companywide set of directives that tells the manager exactly what to do. It takes the subjectivity out of disciplinary action and gives support to the supervisor, especially when drastic action is needed.

Yet even with a company system there is a good deal of room for your own method of administering it. Seldom are discipline situations black and white. You have to investigate the facts and exercise your own judgment in light of all the relevant factors. Human skills and conceptual skills are involved.

Usually, disciplinary action should be adjusted to circumstances. One of the things that you should consider is the intent of the rule breaker. Was it an accident? Was the person aware of the rule or requirement? Was it a case of misinformation or misunderstanding? Could it have been your fault?

Another consideration is extenuating circumstances such as severe personal problems or a crisis on the job. Still another consideration is the number of times that a person has done this type of thing before. Another is the seriousness of the offense. What are its consequences for the product, the customer, your department, the company? What will be the impact of your response as a deterrent to others?

You may handle different people differently for the same violation: You may take a hard-line approach with a hostile troublemaker but treat an anxiety-ridden first-time offender gently. This does not mean that you are being inconsistent; you are enforcing the rules in each case; you are not permitting either of them to go against regulations. As with everything else in this profession, you must be able to adapt your discipline to your own leadership style, to your workers and their needs and actions, and to the situation at any given time and all the time.

✸ SOME MISTAKES TO AVOID

One of the biggest mistakes that new managers make is to start off being too easy about enforcing the rules. They want people to like them, and they let people get away with small things that are against the rules, and perhaps even a few big things, just because they think people will like them for it.

For example, you see one of your people lighting up a cigarette five minutes before closing and you just look the other way. If you were promoted from an hourly position, you may still have the worker's view of the rules, and you can empathize with that person's feelings about that cigarette and that rule against smoking on the job. You may even disapprove of some of the rules you are supposed to enforce. It is very easy to let many things slip by.

This is just about the most difficult way you can start off your supervisory career. By saying nothing when a violation occurs, you are actually saying, "It's okay to do that." Right off, some people will begin to test you, to see how far they can go before you take any action, and pretty soon you will have a real problem on your hands getting people to follow rules and meet standards.

It is always easier to start out by strictly enforcing every rule and regulation than it is to try to tighten up later. People feel betrayed when you switch from leniency to enforcement, and they suddenly decide that you are bad, mean, and tough to work for. Besides, it is your obligation as a manager to enforce the rules. Even when you think rules are unfair, you do not take it upon yourself to change them by not enforcing them.

Rules can be changed, but the way to do it is to go through channels and get the changes approved. Often, supervisors look the other way because they are simply too busy to cope with discipline. You just don't have time for this today, you don't want the hassle, and you've got to get the work out, so you let things slide.

Sometimes, correcting people's behavior just doesn't seem to do any good. You go over and over a procedure with a certain worker but nothing seems to change, and you reach a point where you start taking those little white pills your doctor gave you for indigestion, and finally, you stop wrestling with the problem and do nothing.

Still another reason for doing nothing is that if you fire this person, you might get a new employee who is even worse. Another reason is that you don't think your manager will back you up, so you just don't take any action.

Sometimes, leaders are too slow to respond to an emerging problem. There are a number of reasons for this:

- The tendency to see an emerging problem (e.g., a first instance) as an accident and something not worth addressing
- The desire to have harmony
- The perception that discipline is a cause of disharmony
- The simple dread of dealing with uncomfortable issues

The reason why delay is problematic is that it sends a message that undesirable behavior will be accepted or even not noticed.[5]

All these reasons for letting people get by with rule-breaking or substandard performance add up to the same problem: It gets harder and harder to maintain discipline. And it gets harder and harder to manage your people in other ways as well because you lose their respect. In effect, *they* gradually take charge of the way they do things. The work suffers, quality of product and service suffers, customers complain, costs go up, and you are failing at your job as a manager.

Another mistake in disciplining is to act in anger. Anger will make the worker defensive and hostile, and you will seldom use good judgment in what you say and do. You won't stop to get the facts straight, you may be harsh and vindictive, or you might make a threat you can't carry out or do some other thing you will regret later. If you overreact, if you are wrong, you lose face and you lose some of your control over your people.

Threatening to take any action that you do not carry out is very common, and it is always a mistake. It is like looking the other way: It invites testing and rule-breaking. You have to stick to whatever you say you will do: You have no alternative.

Putting somebody down in front of other people is another way to ask for an uncooperative employee. It is one thing to correct someone quietly in the presence of others: "I just want to remind you that we always use the guard on the slicer." This approach informs and teaches on the spot, and although it is a form of public discipline it does not belittle, embarrass, or humiliate. But yelling, threatening, or making a fool of someone in front of others will certainly have the familiar consequences of resentment and hostility.

A different kind of mistake is to exceed your authority in taking disciplinary action. You must know exactly what your job empowers you to do as well as everything there is to know about company policy and practice. If there is a company system of procedures and penalties, you must follow it.

If you are thinking of terminating someone, make certain that you have the authority to do so and find out the termination procedures your company requires. It could be quite embarrassing if you threatened to fire someone and then found out you couldn't. And it would be a disaster all the way around if you fired somebody and then had to take that person back.

Another critical error is to try to evade the responsibility for taking disciplinary action by shifting it to your boss or the personnel department or by delegating it to someone under you. If you do this, you simply become a straw boss and your people will have no respect for you. Discipline is an obligation of supervision, and your success as a manager depends on it.

Unexpected discipline will always meet with resistance and protest. This often happens when a rule has not been enforced for some time and the manager suddenly decides to tighten up. It is important to give warning about either a new rule or a new policy of enforcing an old rule, with a clear statement of the consequences of breaking it.

Some other things to avoid are:

- Criticizing the person rather than the behavior. Keep personalities out of discipline.
- Waiting too long to take action. The longer the gap between incident and action, the more likely the action is to be interpreted as a personal attack.

- Touching someone when you are disciplining. It can be interpreted as intent to do physical harm or as sexual harassment.

- Being inconsistent. You must avoid partiality, and your actions must be fair in your workers' eyes.

❋ TAKING THE ESSENTIAL STEPS

The set of procedures that follows is one you should use when you are confronted with a serious infringement of regulations or a less serious but chronic failure to observe the rules of the operation or the requirements of a job. There are six formal steps to enforce compliance. These steps apply no matter what approaches to disciplinary action—negative or positive—you intend to follow. They apply to each stage of the disciplinary sequence (see Figure 12.4) and they amplify what should take place at each stage:

1. *Collect all the facts.* Interview any employees involved and any witnesses, especially other supervisors or managers. Write it all down. Make every effort to sort out fact from opinion, both in what others say and in your own mind. Avoid drawing conclusions until you have the full picture. Use these questions as a guide:

    a. Was the employee's action intentional? Was it an accident? Was it the result of misinformation or misunderstanding? Could it have been your fault?

    b. Was the employee aware of the rule or requirement?

    c. Were there extenuating circumstances, such as severe personal problems or a crisis on the job?

    d. How serious is the offense? What are its consequences for the product, the guest, your department, the company?

    e. What is the employee's past record? Is this the first time something like this happened, or has it happened before? How long has the employee worked here?

    f. Did you witness the violation? If not, what kind of evidence do you have? Are your sources other management personnel? Do you have enough evidence to justify action?

2. *Discuss the incident with the employee.* Do this as soon as possible after the incident; after all, justice delayed is justice denied. Plan to sit down with the employee in a quiet setting where you will not be interrupted. Also, line up a witness to sit in on the interview. Here are the steps to follow:

    a. Tell the employee that you are concerned about the incident that took place and that you would like to hear his or her side after you describe the facts as you see them.

    b. Stay calm and, without assigning blame, go over what you know. Also explain the consequences of the action. For example, if a server failed to clean up her station at the end of the shift, explain how this affects the other servers and the guests.

    c. Now ask the employee to tell you his or her side. Listen actively, encourage, stay calm, and do not get into an argument. Ask questions as needed. This step is crucial because it gives the employee *due process*. **Due process** is the opportunity to defend yourself against charges.

3. *Decide on the appropriate action, if any, to be taken.* To do this, you must consider what both you and others have done in similar cases. Your action should be consistent with that of others throughout the company or you may inadvertently set a precedent. You want to make sure that any action you take is not discriminatory. Before deciding on any action, be sure to consult your boss, and you may also be required to discuss this with a representative from the human resources department.

4. Take the appropriate actions, such as a written warning, and develop an improvement plan with the employee using these steps:

    a. Explain to the employee the action you are taking, in a serious but matter-of-fact tone of voice, avoiding any trace of vengefulness or anger. Also state clearly the consequences that will follow if the behavior recurs.

due process
Formal proceedings carried out using established rules and procedures to protect an individual's rights.

b. Ask the employee to identify some actions that he or she can take so that this does not happen again. Develop a mutual improvement plan and a date by which it is expected that the improvement will be made. Make it clear that you are willing to work with the employee but that it remains the employee's responsibility to make the changes.

c. If your policy requires it, ask the employee to sign a disciplinary report (see Figure 12.5). This is normally done so that you have written proof that the employee was informed of the contents of the report.

DISCIPLINE REPORT

Employee: _____ Date: _____ Time: _____ Place: _____
Incident as you saw it:

Employee's account:

Witness name: _____ Position: _____
Witness's account of incident: (Must be a manager)

Extenuating circumstances:

Past record:

Details of disciplinary interview:

Action taken:

Date: _____ Supervisor: _____

FIGURE 12.5: Report form for developing a written record of a disciplinary incident.

From time to time, you will have an employee who refuses to sign the report because he thinks that his signature will signify that he agrees with the contents of the report, or in other words, he is guilty as charged. When this happens, explain that the signature signifies understanding of, not agreement with, what is stated. If the employee continues to refuse to sign, you should write "Refused to Sign" on the report.

d. Close on a positive note by stating your confidence in the employee to improve and resolve the issue. Also express your genuine desire to see improvement.

5. *Make sure that you have everything written down.* Why is it important to have everything in writing? There are several reasons. First, in the event that the employee ever takes you to court or an unemployment compensation hearing, you will need written documentation to help build your case. Second, the process of writing helps you see the situation more objectively and focus on job-related issues. Third, because your documentation includes the employee's improvement plan, it helps the employee to improve.

When documenting, be specific about what the employee did and the circumstances surrounding the situation. Focus on observable, verifiable facts; be nonjudgmental. Document facts, not opinions or hearsay—they have no place in documentation. Include who, what, where, when, and how. Document accurately and thoroughly, including information obtained during your investigation and during the disciplinary meeting. Always document as quickly as possible; otherwise, you will forget many of the details. Also include objectives for future performance and the consequences for not meeting the objectives.

6. *Follow up.* You do everything you can do to help the worker meet your expectations while staying on the lookout for further troublemaking of any kind. If the behavior does not meet your stated expectations, you must take the next step as promised.

Termination

LEARNING OBJECTIVE: Discuss reasons for termination and best practices for termination interviews.

✳ SALVAGE OR TERMINATE?[6]

If you had performance standards, trained people carefully, and have evaluated and coached and corrected them on a more-or-less daily basis, then you would spot pretty quickly people who are never going to make it in their jobs. They aren't exactly rule-breakers, they just don't perform well, are absent a lot, or do some dumb thing over and over and you just can't get them straightened out.

If you hired them, your best bet is to terminate them before their probationary period runs out. But you might think that you can turn them around, so you work and work and work with them—finally, you have to admit that they are hopeless and you are stuck with them. You also inherit some of these people.

What are you going to do? Should you fire Jerry for continuing to overpour drinks even though you have showed him every single day the right amount to pour? How can you get rid of Kimberly when you can't even figure out what to pin her trouble on because it's something different every day? Can you terminate Alfred for being an alcoholic when he has been here five times as long as you have and is twice your age? Should you terminate any of them, no matter how bad their work is?

Sometimes managers will try to **dehire** people by making them want to leave the job or look for something else. In this approach, a manager gives other people all the work and leaves this person with nothing to do or in other ways hints that it would be wise to look for another job.

It is a destructive way of handling a person, and it does not work very well for the manager, either. Legally, the practice is open to all kinds of discrimination lawsuits and should be avoided. You have no control, and you have to wait for this person to take the step of leaving while you go on paying wages for little or no work done. It is both kinder and better either to terminate outright or to keep on trying to salvage this worker.

dehire
To avoid termination by making an employee want to leave, often by withdrawing work or suggesting that the person look elsewhere for a job.

Courtesy of Ryan Adams

There are many examples of good supervision that I could cite. However, there is a distinctive difference between what is taught and what is realistic. The first three words that I think of in terms of good supervision are legal, ethical, and moral. Books tend to give black-and-white examples with black-and-white results. This is not reality in the hospitality industry.

A good supervisor must be respectful of everyone on every level. You are in the middle of the line-level employees, the middle managers, and the top management. You have to be able to juggle the concerns of your direct manager with the challenges of your staff and still try to impress the top-level managers when they are around. This can be a difficult balancing act.

You have to draw a line—often the most difficult decision for a supervisor. Most supervisors have climbed their way up the ladder and are often viewed as one of the crew. You must distinguish yourself as a manager or else your staff will try to take advantage of you. Everything you do is being watched on every level. Now, more than ever, it is essential to put your foot down and establish your mark. It is often said that "being a manager is not a popularity contest"—this could not be more true. Being a supervisor can often be a lonely position.

When an employee comes to you, stop what you are doing and give the person your undivided attention. Look the employee in the eye and really hear what he or she is saying. When the person has finished talking, paraphrase what has been said so that you understand and can empathize with the feelings expressed so that he or she is validated, and then offer counseling to help resolve the concern. This is easier said than done. As a supervisor, you always have to be on top of your game.

This means knowing policy and procedures and knowing the union's contract if your business is unionized. You have to be an expert in human resources. If you do not know, you always have your manager to fall back on or the HR department. If you are not sure, ask; it is better to find out than to make an error that can cost big money. If you do not know the answer to a particular question, tell the employee that you will get back to him or her with the answer. Make sure that you follow up or the person will feel that you are incompetent and untrustworthy. If this occurs, you will have to make huge strides to improve that relationship.

Trust and empower your people. If you really want to shine and make your staff feel good about their jobs, trust them. Delegate side duties and additional responsibilities. Make sure that you give them the proper praise for their work; they will not forget that you did not praise them for their help and may be reluctant to offer in the future. Let your staff make decisions.

Give them the guidelines to make decisions and train them. Role-playing is a great way to prepare them for the chaos they will experience. This will enable you to take on more responsibility from your managers, and your staff will be confident to respond to smaller problems that would otherwise tie up your day.

In discipline situations, never have the paperwork already filled out. Make notes, but do not just present the write-up without talking to your employee first. They will feel like you did not even give them a chance to explain. You assumed they were guilty right off the bat and they will feel like they never had a chance.

Always check on your staff and make sure that you take time out of your day at some point to make small talk. Often, when an employee makes the transition to supervisor, he or she forgets how he or she wanted to be treated when an employee. It is easy to adopt negative or undesirable traits of managers of the past. Always be objective and never lose the person you are—that is what got you there in the first place. Always be on time to meetings or at least call in advance to notify someone that you cannot attend.

Make your manager's concerns your priority, your staff's needs mandatory, and the vision and values of your company your guiding principles. If you do not believe in what the company stands for, it is difficult to share that vision with others. Remember, you are not just a manager, you are a leader. This means taking your staff to a place they would not have gone otherwise. As a manager, what gets done gets measured. Not that this is right—it is a fact and, thus, a big part of your evaluation.

From the productivity point of view and your own frustration level, it would probably be far better if you simply terminated all these very poor performers. But there are other considerations. Length of service is one. The longer people have been working for a company, the harder it is to fire them. Company policy (and if there is a union, union rules) comes into play here; you may not have a choice. Seniority is one of the most sacred traditions in U.S. industry.

A person's past record is another consideration. On the one hand, you might have people who are chronically late to work, and if they always have been, they probably always will be. On the other hand, if a person has had a good performance record and there is a sudden change, whether it is coming in late or some other drop in performance, that person is probably salvageable. Another consideration is how badly you need a person's skill or experience. In a tight labor market, even somebody who does not meet standards is better than nobody.

It is very difficult to fire someone who desperately needs the job even if the person is terrible at it. You might bring yourself to do it if you had to look forward to 30 years of coping with this substandard worker. If the person in question is a senior, you cannot terminate him or her on the basis of age, but you can terminate the person for inability to do the job to the standard required, once you have given the employee ample chance to improve and have been through the steps of verbal warning, written warning, and so on.

It is also difficult to fire someone who you are pretty sure will make trouble about it. In this case, it is wise to consult with your boss or the HR department or both. (Some companies have a policy that only the HR director and the general manager or department head may terminate an employee. In any event, there should always be a warning of the termination, in case of later legal action.)

Perhaps the most difficult question to figure out is the effect on your other workers of terminating someone. On the one hand, they might have resented that person's poor work and will be glad to see you hire someone who they do not have to fill in for all the time. On the other hand, they might have been imitating this person and slowing down the whole operation—a bad example is always easier to follow than a good one. Or they might be fond of this character and will be angry if he or she is terminated. Some of them will be worried and upset about whether the same thing will happen to them. A termination is always something of an upheaval, and you might have to cope with some repercussions.

You have to consider the cost and trauma of hiring a replacement against the cost and trauma of retaining this person. You also have to consider whether your authority entitles you to terminate. In fact, you should consider this first.

If you decide to salvage, you have a few options open. You can try people in different jobs. You can look for special talents and interests and try to motivate them with some form of job enrichment. You can counsel the alcoholic to go to a clinic for rehabilitation. You can investigate the case of the sudden performance drop and try coaching this person back to the old level. Or you can grin and bear it.

Given the risk of legal complications—which can cost an organization $15,000 plus to defend whether it is right or wrong—the job of terminating someone has become more difficult and important than ever. Every year, thousands of employers are hauled into court by former employees claiming they were fired illegally.

Attorney James P. McElligott offers these suggestions to help avoid the nightmare of a wrongful termination lawsuit:[7] establish clear performance expectations; maintain clear communications; document everything; give the employee a chance to respond; do not discriminate—almost everyone belongs to a "protected group"; conduct regular employee evaluations; and deal promptly with performance problems.

❋ JUST-CAUSE TERMINATIONS

If you think that it is an appropriate time to terminate an employee, first make sure that it isn't something for which you can't fire the employee, such as discrimination. Figure 12.6 lists inappropriate reasons for firing an employee. You can, however, fire employees for *just cause*, meaning that the offense must affect the specific work the employee does or the operation as a whole in a detrimental way. Before terminating anyone, ask yourself the following questions:

1. Did the employee know the rule, and was he or she warned about the consequences of violating the rule? Are these understandings confirmed and acknowledged in writing?

2. Were management's expectations of the employee reasonable? Was the rule reasonable?

3. Did management make a reasonable effort to help the employee resolve the problem before termination, and is there written proof of this?

4. Was a final written warning given to the employee explaining that discharge would result from another conduct violation or unsatisfactory performance?

1. Because of race, color, gender, or national origin (Title VII of the Civil Rights Act of 1964)
2. In retaliation for filing discrimination charges (Title VII of the Civil Rights Act of 1964)
3. For helping other employees who have been discriminated against by a company for exercising their civil rights (Title VII of the Civil Rights Act of 1964)
4. For testifying against a company at an Equal Employment Opportunity Commission hearing (Title VII of the Civil Rights Act of 1964)
5. Because the employee is over 40 years old (Age Discrimination in Employment Act of 1967, as amended)
6. For forcing retirement or permanent layoff of an older employee (Age Discrimination in Employment Act of 1967, as amended)
7. Because she is pregnant (Pregnancy Discrimination Act of 1978)
8. In retaliation for reporting to a state or federal agency unsafe working conditions (Occupational Safety and Health Act of 1970)
9. For filing an OSHA complaint (Occupational Safety and Health Act of 1970)
10. For testifying against the company in an OSHA-related hearing or court-related action (Occupational Safety and Health Act of 1970)
11. Because an employee has a disability (Americans with Disabilities Act of 1990)
12. To avoid a pension or benefit plan, such as group health insurance (Employee Retirement Income Security Act of 1974)
13. For "whistleblowing" illegal acts by the employer or other employees (Whistleblower Protection Act of 1989)

FIGURE 12.6: Inappropriate reasons for terminating an employee.

5. In the case of misconduct, did the employee act in willful and deliberate disregard of reasonable employer expectations? Was the situation within his or her control? If the situation was out of the employee's personal control, he or she cannot be charged with misconduct.

6. Was management's investigation of the final offense done in a fair and objective manner, and did it involve someone other than the employee's direct supervisor? It is best that the employee's supervisor not function alone and fill the roles of accuser, judge, and jury. Is there substantial proof that the employee was guilty?

7. Is dismissal of the employee in line with the employee's prior work record and length of service? When an employee has many years of service that are documented as satisfactory or better, he or she is generally entitled to more time to improve before being dismissed.

8. Did the employee have an opportunity to hear the facts and respond to them in a nonthreatening environment? Was the employee able to bring someone into the disciplinary interview if requested?

9. Has this employee been treated as others in similar circumstances? Has this rule been enforced consistently in the past? If the rule has not been enforced consistently, you may have to forgo terminating the employee and instead, go back a step, such as to suspension. In the case where a rule that hasn't been enforced starts to be enforced again, you have to inform employees beforehand of the change.

10. Is the action nondiscriminatory? Has equal treatment been given to members of protected groups (minorities, women, employees over 40 years of age) and nonprotected groups?

just-cause termination
Employee termination based on the commission of an offense that affected detrimentally the specific work done or an operation as a whole.

These questions are only guidelines for determining **just-cause termination**. Even if you can answer yes to every question presented here, there is still no guarantee that you won't wind up in court. If you decide to terminate, all the basic procedural steps spelled out for disciplinary action apply to this final decision. You state the problem in writing, collect the facts, make your decision, and take the action.

✻ THE TERMINATION INTERVIEW[8]

Few people like telling an employee that he or she is terminated, fired, or dismissed. The best way to reduce your nervousness and make it less stressful for the employee is to prepare for the termination interview by using something like the termination interview checklist shown in Figure 12.7. The checklist can help you to take the following steps:

1. Select a good time and place to conduct the interview.

2. Determine who will be present at the meeting (you should have at least one person as a witness), as well as whether the employee needs to be escorted out of the building.

3. Develop your opening statement and practice it.

4. Determine how best to respond to possible employee reactions.

5. Determine the final pay, severance pay, and benefits to which the employee is entitled.

6. Develop a list of clearance procedures to be performed at the end of the interview. The timing and place for the interview are important. Although conventional wisdom says to fire an employee at the end of a workweek, that is not necessarily the best way.

   By firing the employee at the beginning of the workweek, you give him or her a chance to start immediately looking for a new job, instead of complaining and becoming upset at home. Nor is it wise to terminate employees near major events such as holidays, birthdays, or dates of their anniversary of beginning work with the company. Try to arrange a time when the employee can clean out his or her locker without other employees present.

The meeting should take place in a private room so that if the employee should become unruly, guests and coworkers will not be disturbed. A room in the HR department is ideal. Having prepared for the interview, it is time to speak with the employee. Although you might be nervous, it is probable that the employee is nervous as well. The employee probably knows what's going on and is anxious to get it over with, too. Employees who are fired are often relieved and go on to find new jobs that are much more satisfying. In some cases you might even feel that you did the employee a favor. There are six steps for a termination interview:

1. Do not beat around the bush. Avoid small talk and tell the employee that he or she is being dismissed, and why. Do so in a firm, calm manner. Avoid a discussion of all the details leading up to this decision. This is not the time for that; there have surely been plenty of previous counselings. Instead, state the category under which the discipline problem falls, such as excessive absenteeism, and mention the last step taken and how it was made clear that the employee faced termination for one more offense.

   Clearly communicate that the decision has been made and there is no possibility for negotiation. Explain that the decision is a joint decision in which others, such as the general manager and HR director, have been involved. Reinforced in this manner, your authority will not be questioned as readily.

2. At this point, listen to and accept the responses of the employee. Be prepared for any type of response, such as anger, tears, amazement, or hostility. Figure 12.8 (page 310) describes how to react to four different types of responses.

3. Now is the time to say something positive to the employee to maintain his or her self-esteem, which at this point is probably sagging. Make a statement about something that you, and others, really like about the employee.

4. Move on to a discussion of final pay, severance pay, and benefits to which the employee is entitled.

5. Finally, explain your clearance procedures and give the employee clear instructions about what to do after your discussion. Do you want the employee to go straight to his or her locker, clean it out, and leave quietly? Is there someone who will escort the employee out of the building?

TERMINATION INTERVIEW CHECKLIST

Employee Name: _____

Employee Position: _____ Department: _____

Employment Date: _____ Termination Date: _____

1. **Decide on where the interview will take place and when.**

   (Avoid firing at end of workweek or near the time of major events such as birthdays and holidays. Choose a time when the employee can clear out his or her belongings without other employees present. Select a quiet place where you will not be interrupted and the employee will not disturb others upon leaving.)

   _____

   _____

   _____

   _____

2. **Determine who will be present at the meeting and whether the employee needs an escort out of the building.**

   (There should be at least one person as a witness, and probably a representative from human resources as well. If the employee is a member of a bargaining unit, use the contract as your guide.)

   _____

   _____

   _____

3. **Develop an opening statement and practice it.**

   (Write out exactly what you are going to tell the employee. Make sure you state clearly that the employee is being terminated and include a brief summary of the reason(s). Make it clear to the employee that there are no options.)

   _____

   _____

   _____

4. **Determine how best to respond to possible employee reactions.**

   (Think about how you expect the employee to respond and how you can handle each type of response. Your goal during the interview is to let the employee respond to the termination, within limits, and then proceed with getting the necessary information across.)

   Possible Reactions: _____

   _____

FIGURE 12.7: Sample termination interview checklist.

6. End the interview by standing up and moving toward the door. Avoid making any physical gesture, such as shaking hands or putting your hand on the employee's shoulder, which could be taken the wrong way by someone who is upset and angry. Try to close on a positive, friendly note.

Keep the pace moving during the termination interview. The entire interview should be over in about 10 to 20 minutes.

Your Responses: _____

_____

Record below any threats the employee makes about legal action during the
interview.

_____

_____

_____

**5. Determine the final pay, severance pay, and benefits to which the employee is
entitled, and communicate them.**

Pay:

_____ Pay in lieu of notice   _____ Workdays
(only in the case of unsatisfactory
work performance)

_____ Accrued vacation   _____ Workdays

Payment by: _____ Lump sum _____ Payroll checks

Benefits:

Entitled to: Explained:

_____   _____ Group health insurance continuation

_____   _____ Vested rights in pension plan

OTHER: _____

6. Clearance procedures.

Make sure employee hands in the following:

_____ Locker keys

_____ Uniforms

_____ Name tag

_____ ID card

Completed by: Date:

_____   _____

FIGURE 12.7:   *(continued)*

After the interview, be sure to document important details, including the employee's reaction, any threats that may have been made, and any comments made about the fairness of the
decision. Routine terminations lacking clear resolutions can come back to haunt a business. Be
sure to keep everything confidential; inform only those employees who must know. If you tell the
employee's former coworkers about the dismissal, you are leaving yourself open to be sued
for slander.

Type of Emotional Response	What You Can Do
1. Crying	• Let the employee cry it out. • Do not apologize for your actions. • Show concern by offering a tissue, something to drink, or a moment of privacy if appropriate. • Staying calm and businesslike, think about the next step in the interview.
2. Shouting and cursing	• Keep your own emotions under control and maintain a calm and cool demeanor. • Make it perfectly clear to the employee that you will continue the conversation only when the shouting stops. Use your normal tone of voice; do not show irritation. • Tell the employee that you would like him or her to know the arrangement for termination pay and benefits.
3. Unresponsive	• Be empathic, but also continue the interview. • Do not ask the employee questions such as "How could you be shocked at this news?" Do not play counselor when the employee withdraws. • Confirm all details in a letter.
4. Employee leaves after your opening statement	• Tell the employee that you really do not want him or her to miss hearing the arrangements made for termination pay and benefits.

FIGURE 12.8: How to react to employees' emotional responses.

Employee Assistance Programs

LEARNING OBJECTIVE: Explain the purpose of employee assistance programs.

employee assistance programs (EAPs)
A counseling program available to employees to provide confidential and professional counseling and referral.

Counseling programs called **employee assistance programs (EAPs)** are an expansion of traditional occupational alcoholism programs, which began appearing years ago. Larger companies are more likely than smaller companies to have EAPs. Companies such as Marriott, KFC, and Lettuce Entertain You offer counseling and referral services to some or all of their employees.

An EAP is an employer-paid benefit program designed to assist employees with personal problems. EAPs offer outsourced counseling and referral to a range of professional services. The rationale behind the programs is that getting a valuable employee "back on track" is worth doing for them, their families, and the company. Supervisors need to be able to recognize troubled employees and to recommend them to an EAP.

The EAP handles a wide range of problems:

- Emotional
- Family
- Marital
- Mental health
- Stress
- Financial
- Substance abuse
- Legal
- Workplace
- Elder care

Signs of employees in need of help are increased tardiness, fatigue, missed goals, inappropriate behavior, medical problems, psychological problems, stress, and increased sick days. If there is a performance-related work problem, then supervisors and human resources can investigate the situation and require all concerned to submit a report. EAPs work with employee discipline and counseling to retain employees who need temporary assistance.[9]

The approach to take is called intervention rather than confrontation. Most larger companies have an EAP program that employees can use free of charge for counseling and for various types of assistance such as legal, financial, and family needs. Another reason for companies to offer EAPs is because legally alcohol abuse is classified as a sickness and should be treated as such to allow an employee a chance to get help and hopefully recover.

intervention
An orchestrated strategy that emphasizes convincing someone to discuss problems with a counselor, as opposed to confronting them directly with their shortcomings.

Employee Safety

LEARNING OBJECTIVE: List direct and indirect costs of workplace accidents.

The National Safety Council has reported that every workday a fatality happens every two hours and a debilitating accident occurs every two seconds! Besides the costs of pain and suffering and lost wages to the injured employee and the cost of lost work time to the employer, there are other direct and indirect costs to consider when an accident occurs:

- Lost time and productivity of uninjured workers who stop work to help the injured employee or simply to watch and talk about the incident (Productivity normally decreases for a number of hours, but if morale is negatively affected, it could be much longer.)
- Lost business during the time that the operation is not fully functioning
- Lost business due to damaged reputation
- Overtime costs to get the operation fully functioning again
- The costs to clean, repair, and/or replace any equipment, food, or supplies damaged in the accident
- Cost to retrain the injured employee upon return to work
- Increased premiums for worker's compensation
- In the case of a lawsuit, legal fees and possible award to the injured employee

Accidents will happen, but some accidents are more common than others. Of course, not only employees become involved in accidents; guests do also. Here are the most common causes of workplace accidents:

- Slips and trips
- Improper handling (lifting, lowering, pulling, etc.)
- Traffic accidents (being hit by a moving vehicle, objects falling from a moving vehicle, etc.)
- Electrical accidents/burns

Slips and trips account for one-third of all workplace accidents. Improper handling also accounts for one-third of all workplace accidents!

It is the employers' responsibility to minimize the risks associated with workplace accidents, as when push comes to shove most of them can be seen as being due to employer negligence. This means the employer must have workplace accident prevention strategies in place at all times.

To underline the importance of workplace safety, consider the "general duty clause" of the Occupational Safety and Health Act (OSHA) of 1970, which states that "each employer shall furnish to each of its employees a place of employment that is free from recognized hazards that are causing or likely to cause death or serious physical harm to its employees." Failure to abide by this and other laws is asking for trouble. Yet, in a workplace survey conducted by the Society for Human Resource Management, more than 50 percent of respondents expressed a level of concern that workplace violence might occur at their organizations.[10]

Harassment

LEARNING OBJECTIVE: Describe the various forms of harassment in the workplace.

Supervisors must be aware of legal issues regarding acceptable personal interactions in the workplace. **Harassment** is subjecting another person to intimidating, hostile, or offensive behavior. All forms of harassment based on national origin, race, color, religion, gender, disability, or age are illegal. Sexual harassment is illegal, as well.

The Supreme Court made clear that employers are subject to vicarious liability for unlawful harassment by supervisors. The standard of liability set forth in these decisions is premised on two principles: (1) an employer is responsible for the acts of its supervisors, and (2) employers should be encouraged to prevent harassment and employees should be encouraged to avoid or limit the harm from harassment.[11] In order to accommodate these principles, the Court held that an employer is always liable for a supervisor's harassment if it culminates in a tangible employment action. However, if it does not, the employer may be able to avoid liability or limit damages by establishing an affirmative defense that includes two necessary elements:

1. The employer exercised reasonable care to prevent and correct promptly any harassing behavior.

2. The employee unreasonably failed to take advantage of any preventive or corrective opportunities provided by the employer or to avoid harm otherwise.

Although the *Faragher* and *Ellerth* decisions addressed sexual harassment, the Court's analysis drew on standards set forth in cases involving harassment on other protected bases. Moreover, the Commission has always taken the position that the same basic standards apply to all types of prohibited harassment. Thus, the standard of liability set forth in the decisions applies to all forms of unlawful harassment.

Harassment remains a pervasive problem in U.S. workplaces. The number of harassment charges filed with the EEOC and state fair employment practices agencies has risen significantly in recent years.

✳ SEXUAL HARASSMENT

As a manager you need to be able to recognize and confront **sexual harassment**. The Equal Employment Opportunity Commission (EEOC) issued guidelines on sexual harassment in 1980, indicating that it is a form of gender discrimination under Title VII of the 1964 Civil Rights Act. The EEOC states that sexual harassment consists of "unwelcome advances, requests for sexual favors, and other verbal or physical conduct of a sexual nature when: (1) submission to such conduct is made, either explicitly or implicitly, a term or condition of an individual's employment, or (2) submission to or rejection of such conduct by an individual is used as the basis for employment decisions affecting the person."[12] This definition of sexual harassment is known as the **quid pro quo** definition. *Quid pro quo* means that something is given in exchange for something else. In this type of sexual harassment, submission to or rejection of a sexual favor is used as the basis for employment decisions regarding that employee. The employment decision might be an increase in pay, a promotion, or keeping your job. Only management or supervisors can engage in quid pro quo harassment.

There are about 76,000 EEOC cases a year, of which about 27,000 are based on race, 23,000 based on sex, and 22,500 based on retaliation. Additionally, 12,500 sexual harassment charges and 5,000 pregnancy discrimination charges are made and $274 million in monetary relief was gained by the charging parties.[13]

An example of sexual harassment occurred at a Caesars Palace property, where the EEOC asserted in a 2005 lawsuit that male supervisors would demand and/or force female workers to have sex with them under threat of being fired. Women, predominantly Spanish speakers, were

harassment
Intimidating, hostile, or offensive behavior toward someone, or the creation of an intimidating, hostile, or offensive environment for someone based on the person's national origin, race, color, religion, gender, disability, or age.

sexual harassment
Unwelcome advances, requests for sexual favors, and other verbal or physical conduct of a sexual nature when compliance with any of these acts is a condition of employment, or when comments or physical contact create an intimidating, hostile, or offensive working environment.

quid pro quo
Giving something (such as a job promotion) in exchange for giving something else (such as sexual favors).

forced to have sex and submit to other lewd acts. To make matters worse, an EEOC press release reported the following:

> *Management failed to address and correct the unlawful conduct, even though women complained about it. Further, the EEOC said, when workers complained about the unlawful conduct, they were retaliated against in the form of demotions, loss of wages, further harassment, discipline or discharge.*

In 2007, Caesars Palace paid $850,000 to settle the suit.[14]

Another type of sexual harassment is **hostile work environment sexual harassment**. In this case, comments or innuendoes of a sexual nature or physical contact are considered a violation when they interfere with an employee's work performance or create an "intimidating, hostile, or offensive working environment." In this situation, the harassment must be persistent and so severe that it affects the employee's well-being.

If severe and pervasive enough to interfere with work or learning, the following types of conduct may create a hostile environment:

hostile work environment sexual harassment
A type of sexual harassment in which comments or innuendos of a sexual nature, or physical contact, are considered a violation when they interfere with an employee's work performance or create an intimidating, hostile, or offensive working environment.

- Jokes or insults
- Flirting
- Comments about a person's body or sex life
- Sexually degrading comments
- Repeated invitations for dates
- Abusive language directed at a person because of his or her belonging to a protected class
- Sexually crude hand gestures, leering at the body, sexually suggestive winking, standing too close
- Display of posters, cartoons, etc., regarding sexually suggestive themes, race, religion, etc.
- Pornography
- Sexually suggestive "gifts"
- "Stalking" behavior
- Touching, hugging, kissing, or patting
- Intentional and repeated brushing or bumping against a person's body
- Restraining or blocking a person's movement[15]

Some behaviors that are acceptable in certain contexts are inappropriate in the workplace or classroom, particularly if an objection is expressed. Whether or not the behavior is contrary to law or university policy depends on the circumstances of each case.

A final type of sexual harassment is **third-party sexual harassment**. Third-party sexual harassment involves a customer or client and an employee. The customer or client may harass an employee, or the other way around. A male customer may harass a female bartender.

third-party sexual harassment
A type of sexual harassment that involves a customer or client and an employee.

For example, in 2009 the Equal Employment Opportunity Commission sued a Charleston, South Carolina, temporary staffing agency because female workers assigned by the agency to a construction company were subject to continued sexual harassment that included inappropriate comments, groping, and requests for sex. The workers requested reassignment to a different work site, but their requests were denied. The women ultimately resigned and brought the issue to the attention of the EEOC. Lynette A. Barnes, EEOC Charlotte District Office attorney, said, "This case is a reminder to employers that they must provide a harassment-free environment for their employees, regardless of who the perpetrators of the harassment are. Third parties can create a sexually hostile work environment for a company's employees, and that is just as illegal."[16]

As another example, consider a café, where you have coffee and log on to the Internet. What if a customer comes in, and each time, he logs on to Porn.com, and each time the waitress who serves him sees that he's logged on to this site. If the manager at this café knew about this—either by virtue of seeing it or virtue of her complaining—the manager would need to take action if the waitress felt it was harassment.

The following examples of sexual harassment include an example of quid pro quo, hostile work environment, and third-party sexual harassment. See if you can determine which is which:

1. Beth is a new employee who works as a cook's assistant in a crowded kitchen. The men in the kitchen are constantly making crude, sexually oriented comments and jokes, and leave their X-rated magazines in full view of anyone walking by. Beth feels very intimidated and ill at ease. Unfortunately, the situation doesn't improve over the first two months, and Beth feels too stressed to continue working there.

2. For the past few nights, after the dining room has closed, Susan's boss has asked her to go to his place for a drink. Although Susan has gone out with him and some friends once before, she is not interested in pursuing a relationship with him. When she tells him she is not interested, he tells her that a dining room supervisor job will be opening soon and that he could make sure she gets it if she takes him up on his invitation.

3. Barbara is a regular customer at a popular after-work bar where Bob works as a bartender. Barbara finds Bob to be a very good-looking man—so much so that she can't keep her eyes, or hands, off him. Bob doesn't like the attention Barbara gives him, but he feels he can't do much about it since she is the customer.

Such instances of sexual harassment can cost a company lost productive time, low morale, harm to its reputation, court costs, and punitive damages to harassment victims. In each of these three situations, there is an element of sexual harassment. While the second situation represents the typical exchange of sexual favors for employment opportunities, the first situation is an example of hostile environment sexual harassment in which the working environment was intimidating, hostile, or offensive due to physical, verbal, or visual (e.g., pornographic pictures) sexual harassment. The third situation represents third-party sexual harassment.

As a manager you are responsible for recognizing, confronting, and preventing the sexual harassment of both female and male employees by other employees or by nonemployees such as guests or people making deliveries. "An employer can be liable for customers who harass employees when the employer knew or should have known of the harassment and failed to prevent it." Both you and your employer will be considered guilty of sexual harassment if you knew about, or should have known about, such misconduct and failed to correct it. If you genuinely did not know that sexual harassment took place, liability can be averted if there is an adequate sexual harassment policy and the situation is corrected immediately.

Following are some specific actions that you can take to deal effectively with the issue of sexual harassment:

- Be familiar with your company's sexual harassment policy. Figure 12.9 is a sample policy. This policy should include disciplinary guidelines for people who are guilty of sexual harassment and guidelines for harassers who retaliate against those who turn them in. This policy might also include a formal complaint procedure for employees to use if they think they have been victims of sexual harassment, with provisions for immediate investigations and prompt disciplinary actions when appropriate.

- Educate your employees on how to recognize sexual harassment, how to report it when it occurs, and the steps that will be taken if an employee is guilty of sexual harassment.

- When an employee informs you of a possible case of sexual harassment, investigate the situation promptly according to your company policy. Your investigation is much the same as that done for any possible case of misconduct as just described. Don't assume that anyone is guilty or innocent.

- When you witness an example of sexual harassment, follow your policy and take appropriate and timely disciplinary action.

- Provide follow-up after instances of sexual harassment. Check with victims and witnesses that harassment has indeed stopped and that no retaliation is taking place.
- Prevent sexual harassment by being visible in your work areas, being a good role model, and taking all reported incidents seriously.

I. Policy

The policy of XYZ Hotels is that all of our employees should be able to enjoy a work environment free from all forms of discrimination, including sexual harassment. Sexual harassment is a form of misconduct that undermines the integrity of the employment relationship, debilitates morale, and therefore interferes with the work effectiveness of its victims and their coworkers. Sexual harassment is a violation of the law and will not be tolerated or condoned.

II. Definition of Sexual Harassment

Sexual harassment consists of unwelcome advances, requests for sexual favors, and other verbal or physical conduct of a sexual nature when:

1. Submission to such conduct is made either explicitly or implicitly a term or condition of an employee's employment, or
2. Submission to or rejection of such conduct by an employee is used as the basis for employment decisions, or
3. The conduct interferes substantially with an employee's work performance or creates an intimidating, hostile, or offensive work environment.

Sexual harassment is not limited to actions of hotel employees. Customers and clients may also be victims, or perpetrators, of sexual harassment.

Following are examples of sexual harassment.

- Unwelcome intentional touching or other unwelcome physical contact (such as pinching or patting).
- Unwelcome staring or whistling.
- Unwelcome sexually suggestive or flirtatious notes, gifts, electronic or voice mail.
- Offering an employment-related reward in exchange for sexual favors.
- Verbal abuse of a sexual nature.
- Unwelcome display of sexually suggestive objects or pictures such as pinups.
- Conduct or remarks that demean or are hostile to a person's gender.

III. Coverage: XYZ Hotels

XYZ Hotels prohibits sexual harassment during work hours or while on company property by all employees and by all nonemployees, such as customers and suppliers.

IV. Responsibilities

XYZ Hotels managers are responsible for preventing sexual harassment and educating employees about this subject. They are also responsible for setting a good example, taking every complaint seriously, investigating complaints fairly, and maintaining confidentiality.

XYZ Hotels requests that any employee with a complaint regarding sexual harassment make every effort to promptly present the complaint to their immediate supervisor or the human resource director. If the complaint involves the employee's immediate supervisor, or if the employee feels uncomfortable discussing the complaint with the immediate supervisor, the employee may speak to another supervisor.

V. Investigation Procedures and Disciplinary Action

Once a supervisor has received a complaint, he or she is to immediately contact the human resource department. After notification of the employee's complaint, a fair and confidential investigation will be initiated. The results of the investigation will be reviewed by the human resource director for possible disciplinary action.

If warranted, disciplinary action up to and including termination will be imposed. Retaliation against employees who file complaints or assist in investigating complaints may also result in discipline up to and including termination.

FIGURE 12.9: Sample sexual harassment policy.

❋ OTHER FORMS OF HARASSMENT

As noted earlier, *harassment* is defined as intimidating, hostile, or offensive behavior toward someone, or the creation of an intimidating, hostile, or offensive environment for someone, based on that person's national origin, race, color, religion, gender, disability, or age.

Examples of prohibited conduct include, but are not limited to:

1. Writing or displaying letters, notes, or e-mails that are derogatory toward any individual's race, color, marital status, sex, religion, national origin, disability, age, genetic information, sexual orientation, or military status.

2. Making comments, slurs, or jokes that are derogatory toward any individual's race, color, marital status, sex, religion, national origin, disability, age, genetic information, sexual orientation, or military status.

3. Unwelcome touching, impeding, or blocking movement based on any individual's race, color, marital status, sex, religion, national origin, disability, age, genetic information, sexual orientation, or military status.

4. Making gestures or displaying pictures, cartoons, posters, or magazines that are derogatory toward any individual's race, color, marital status, sex, religion, national origin, disability, age, genetic information, sexual orientation, or military status.

5. Continuing any of these conducts after being told or being otherwise made aware that the conduct is unwelcome.

6. Singling out or targeting an individual for different or adverse treatment with improper consideration of the individual's race, color, marital status, sex, religion, national origin, disability, age, genetic information, sexual orientation, or military status.

Intimidating behavior may involve threatening someone with harm of some type. Hostile behavior could include asking an employee to do something that is completely unrealistic, such as asking a pot-washer to be in charge of washing all dishes as well. Offensive behavior is generally ridiculing or taunting someone because of his or her color, for example. As a supervisor, you should be constantly on the lookout for intimidating, hostile, or offensive behavior because it has no place in the workplace.

In 2007, a jury ruled that a Subway franchise must pay $166,500 for a disability harassment lawsuit. The EEOC charged that a Subway owner and one of his managers subjected a former area supervisor to a disability-based hostile workplace, including teasing and name calling because she is hearing impaired and wears hearing aids. The supervisor was forced to resign her position after both owner and HR manager repeatedly mocked her privately and in front of employees. Taunts included, "Read my lips," "Can you hear me now?" and "You got your ears on?"[17]

The Supervisor's Key Role

LEARNING OBJECTIVE: Evaluate the role of supervisors in establishing and maintaining discipline.

The orderly and obedient carrying-out of the work of an enterprise depends almost entirely on the effectiveness of the first-line leader in establishing and maintaining discipline. It is the manager who transmits the rules and policies laid down by management. It is the leader who orients, trains, and provides continuous information to associates so that they know what to do, how to do it, to what standards, and what will happen if they don't. And if it comes to that, it is the leader who sees to it that what is supposed to happen does happen.

But the effective leader does not let it come to that. With prompt action, a teaching–helping approach to discipline, and sensitivity to people's motivations and feelings, leaders can usually keep incidents from developing into disciplinary problems. Leaders who are consistent and fair, who follow the rules themselves, who create and maintain a positive work climate with good communications and good person-to-person relations are usually able to maintain good discipline with a minimum of hassles, threats, and disciplinary actions.

By contrast, the leader who attempts to maintain discipline through threats and punishment is usually plagued with ongoing disciplinary problems because of the resentment and anger that such methods provoke. Such leaders often cop out by blaming workers for problems they have created themselves: "You just can't get good employees today." In this case, too, it is the leader who creates the prevailing condition of discipline.

Nobody ever claims that discipline is easy, and nobody has a foolproof prescription for success. It is supervisory leadership and example that set the climate and the direction, and it is the supervisor, acting one on one, who makes it all happen.

CASE STUDY: "They Like It the Way It Is"

Rita is head cocktail server at a high-volume singles bar that serves both food and drinks. She has responsibility for a large staff of part-timers, most of whom she worked with as a server before she was promoted. They are a lively bunch who regard themselves more as independent entrepreneurs doing business at this particular place than as loyal employees. Most of them pay little attention to rules, but they are all high-volume performers, and that, she figures, is what matters. She looks the other way and lets them get away with a lot.

Yesterday, her boss, Sam, who was recently hired to manage the entire operation, called Rita in for "a little talk." He offered her a cup of coffee, paused a moment, and then plunged in. "I want you to be aware that the discipline in your part of this operation does not measure up to standard and is causing a great deal of trouble," he said.

"The servers and kitchen staff are required to follow rules and are disciplined when rules are broken. They resent it when they see a cocktail server carry a drink into the employee lounge, have a cigarette with a customer, wear flashy jewelry, come in late and leave early, and take it for granted that the servers will cover for them. I'm sure you can understand how they feel."

"I don't see that their feelings are my problem," says Rita.

"I think they are," says Sam, "and I am asking you to begin enforcing the hours of work and the smoking and drinking rules, for starters. How you do it is pretty much up to you, although I will be glad to help you work things out. I suggest we meet again tomorrow to discuss your plan and set some improvement goals." Rita is astonished. "Listen, Sam," she says, "any one of my people could get a job anywhere else in town in five minutes, and I could, too. Improvement goals! They like it the way it is!"

"I know," says Sam. "But nobody else does. In fact, it has become a major problem that even customers have noticed, and its effect on the other employees could affect business. Think it over, and we'll talk again tomorrow."

Rita's first reaction is defiance and anger, but she senses that it won't do her any good. She would rather stay here than change jobs. She is proud of being a supervisor, the money here is the best in town, and it's a fun place. Her next reaction is panic. How in the world can she make her people toe the line?

Case Study Questions

1. What common mistake has Rita been making? What effects has it had?
2. Do you agree that high-volume sales are more important than enforcing rules? Defend your answer.
3. Is it workable to have different standards of discipline in different departments? Why or why not?
4. What is your opinion of Sam's approach to the problem? How well did he handle the interview? What risks is he taking?

KEY POINTS

1. *Discipline* refers to a condition or state of orderly conduct and compliance with rules, and also refers to action to ensure orderly conduct and compliance with rules. Both aspects of discipline are the responsibility of the leader, and both are essential to supervisory success.

2. The four essentials of successful discipline are a complete set of rules that everyone knows and understands; a clear statement of the consequences of failing to observe the rules; prompt, consistent, and impersonal action to enforce the rules; and appropriate recognition and reinforcement of employees' positive actions.

3. Negative discipline uses a fear-and-punishment approach and a four-stage progressive formula for disciplinary action: oral warning, written warning, punishment (such as suspension), and termination.

4. The positive approach to discipline is continuous education and corrective training whenever the rules and procedures are not being observed. The four-stage formula for disciplinary action includes oral reminder, written reminder, decision-making leave with pay, and termination (see Figure 12.4).

5. With positive discipline, the negative consequences of punishment do not fester their way through the work climate, and the boss and worker do not become adversaries. Positive discipline can result in reduced turnover, absenteeism, and disciplinary problems.

6. Some mistakes to avoid include starting off too easy about enforcing the rules, acting in anger, threatening to take any action that you do not carry out, putting somebody down in front of others, exceeding your authority, evading responsibility for taking action by shifting it to someone else, or disciplining unexpectedly.

7. The set of procedures to use when confronted with a serious infringement of rules includes collecting all the facts, discussing the incident with the employee, deciding on the appropriate action (if any is to be taken), documenting the process, and following up.

8. You can fire employees for just cause, meaning that the offense must affect the specific work the employee does or the operation as a whole in a detrimental way. In this chapter, we outlined 10 questions that need to be asked, such as, "Has the employee been treated as others have been in similar circumstances?" before deciding whether or not to terminate.

9. Figure 12.7 lists everything you need to think about before a termination interview. During the interview itself, you need to get right to the point, listen to the employee's response, state something positive about the employee, and then move on to severance arrangements and clearance procedures.

10. An employee assistance program provides confidential and professional counseling and referral service to employees with problems such as addiction and dependencies, family problems, stress, and financial problems.

KEY TERMS

decision-making leave with pay
dehire
discipline
due process
employee assistance programs (EAPs)
harassment
hostile work environment sexual harassment
intervention
just-cause termination

negative discipline
positive discipline
progressive discipline
quid pro quo
sexual harassment
third-party sexual harassment
uniform discipline system
work rules

REVIEW QUESTIONS

Answer each question in complete sentences. Read each question carefully and make sure that you answer all parts of the question. Organize your answer using more than one paragraph when appropriate.

1. Define *discipline*. Why is it necessary in a hospitality operation?
2. In determining disciplinary action, to what extent should you as a manager consider circumstances, intent, past history, seriousness of the offense, and consequences of the disciplinary action? If you vary the penalty according to such factors, how can you avoid making subjective and inconsistent judgments?
3. Describe the four essentials of successful discipline.
4. Compare and contrast positive and negative approaches to discipline.
5. Describe five mistakes commonly made by managers regarding discipline. Why is each one a mistake?
6. For the third day in a row, you have heard complaints about foods prepared by Jimmy, the grill cook. You had a similar problem just last month and the month before. Describe the steps you will go through.
7. Explain what a just-cause termination is.
8. Describe the kind of person a manager might refer to an EAP counselor.

ACTIVITIES AND APPLICATIONS

1. Discussion Questions
- What does discipline have to do with discrimination? To avoid discriminating against them, must you be more lenient with a woman, an older person, a minority person, or some other person protected by law?
- What relationships do you see between discipline and communication? Between discipline and performance standards? Between discipline and motivation? Between discipline and leadership?
- Which style of discipline do you support, positive or negative? Why?
- Discuss the state of discipline in a place where you work or have worked in the past. Was there a set of rules and consequences that most employees were familiar with? Was discipline prompt, consistent, and impersonal? Were employees recognized when they did things right?

2. Group Activity: D-O-C-U-M-E-N-T
Using the letters in the word *document*, think of how you should document employee performance, such as "descriptive" for the letter *d*.

3. Group Activity: Dos and Don'ts of Discipline
In groups, write up a list of 10 dos and 10 don'ts concerning discipline as discussed in this chapter.

ENDNOTES

1. Smallbusiness.findlaw.com. Retrieved July 27, 2007.
2. The original hot-stove model is generally attributed to Douglas McGregor of Theory X and Theory Y fame. It gave warning by being red hot. See Douglas McGregor, *The Professional Manager* (New York, McGraw-Hill, 1967).
3. Robert Bacal, "Five Sins of Discipline," Performance Management and Performance Help Center, performance-appraisals.org/Bacalsappraisalarticles/articles/sinsdisci.htm. Retrieved January 2005.
4. Shirley Rickel, director of human resources, Longboat Key Club & Resort, Sarasota, Florida, personal communication, October 4, 2007.
5. Bacal.

6. This section is adapted from Lawrence Steinmmetz, "The Unsatisfactory Performer: Salvage or Discharge?" *Personnel*, vol. 45, no. 3 (1968); and from Richard Martin, "Costly Liability Can Result from Botched Dismissals, Faulty Paperwork," *Nation's Restaurant News* (September 15, 2000).

7. James J. Lynott, "How to Discharge an Employee and Stay out of Court," *Restaurant Hospitality*, vol. 88, no. 7 (July 2004), p. 62.

8. This section is adapted from Richard Martin, "Costly Liability Can Result from Botched Dismissals, Faulty Paperwork," *Nation's Restaurant News* (September 15, 2000).

9. www.dol.gov/odep/documents/employeeassistance.pdf. Retrieved June 7, 2011.

10. www.eeoc.gov/policy/docs/harassment.html. Retrieved June 7, 2011.

11. Philip Deming, "Workplace Violence: Trend and Strategic Tools for Mitigating Risk," Society for Human Resource Management, White Paper, March 2006.

12. archive.eeoc.gov/types/sexual_harassment.html. Retrieved June 7, 2011.

13. Charles Robbins and David Grinberg, "Job Bias Charges Edged up in 2006, EEOC Reports," U.S. Equal Opportunity Commission, February 2007.

14. U.S. Equal Employment Opportunity Commission, "Caesars Palace to Pay $850,000 for Sexual Harassment and Retaliation," press release, August 20, 2007, www.eeoc.gov/eeoc/newsroom/release/8-20-07.cfm. Retrieved April 2011.

15. www.ohsu.edu/aaeo/investigation/hostile_environment.html. Retrieved June 7, 2011.

16. U.S. Equal Employment Opportunity Commission, "Charleston Temp Firm and Construction Company Sued by EEOC for Sexual Harassment," press release, August 7, 2009, www.eeoc.gov/eeoc/newsroom/release/8-10-09.cfm

17. U.S. Equal Employment Opportunity Commission, "Subway Franchise to Pay $166,500 for Disability Bias, Jury Rules in EEOC Lawsuit," press release, July 27, 2007, www.eeoc.gov/eeoc/newsroom/release/7-27-07.cfm. Retrieved April 2011.

Chapter 13

Decision Making and Control

M aking decisions is a built-in requirement of a supervisor's job. When associates run into problems, they bring them to you and you decide what they should do. When crises arise—equipment breakdowns, supplies that don't arrive, people calling in sick, hurting themselves, fighting, or walking off the job—you decide what should be done. When employees can't or won't do their jobs as they should, you decide whether to retrain them, motivate them, discipline them, or fire them.

For a supervisor in this fast-paced, time-pressured industry, decision making is your work, like it or not. It is not you but the people you supervise who make the products or deliver the services. You plan their work, hire and fire them, solve their problems, settle their arguments, grant or deny requests, deal with the unexpected, troubleshoot, and—by making countless small decisions you may not even think of as decisions—you see that the work gets done. How well the work gets done depends a great deal on how good your decisions are.

In this chapter, we explore the kinds of down-to-earth decision making that supervisors in the hospitality industry are faced with day after day. After completion of this chapter, you should be able to:

- Describe common approaches to decision-making.
- List the six steps for making good decisions.
- Analyze the pros and cons of participative problem solving.
- Identify ways to build strong decision making skills.
- Discuss the ways that supervisors control the work being done by employees.

Decision Making

LEARNING OBJECTIVE: Describe common approaches to decision making.

Human beings are constantly making decisions of one sort or another as they go about their daily lives. How do they do this, and what is special about a supervisor's decisions?

ELEMENTS IN A MANAGERIAL DECISION

decision
A choice among alternative courses of action directed toward a specific purpose.

Human resources and management decisions derive from the role and responsibility of being a manager. A manager's **decision** should be a choice among alternative courses of action directed toward a specific purpose. There are three key phrases in this definition that describe three essential elements in the decision-making process:

1. The first phrase is a *choice among alternatives*. If there is no alternative, if there is only one way to go, there is no decision to make: You do the only thing you can. Choice is the primary essential in decision making. You deliberately choose one course of action over others. You are not swept along by events or habit or the influence of others; you don't just go with the flow. You are making it happen, not just letting it happen.

2. The second essential of a managerial decision is a *specific purpose*. The decision has a goal or objective: to solve a specific problem, to accomplish a specific result. Like a performance-based objective, a decision has a *what* and a *how*. The specific purpose is the what.

A manager and a supervisor considering decision alternatives.

3. The third essential is a *course of action*. The decision is to do something or have something done in a particular way. This is the *how*. Making a decision requires seeing that the decision is carried out.

A good decision is one in which the course of action chosen meets the objective in the best possible way—usually, the one with the least risk and the most benefit to the enterprise.

❀ APPROACHES TO DECISION MAKING

Different people approach decision making in different ways. One way is to go deliberately through a series of logical steps based on the scientific method. You formulate your objectives and rank them, gather all the relevant facts, examine and weigh all the alternative courses of action and their consequences, and choose the one that best meets the objectives.

Management scientists recommend the **logical approach to decision making** (with variations). It takes time, something busy hospitality supervisors do not have, and it is probably foreign to their habits: Most people in operations tend to be doers rather than deliberators. It is better suited to weighty top management problems than to the day-in, day-out decisions of hotel and foodservice supervisors. Nevertheless, it can be adapted for solving their important problems, as we shall see.

At the opposite extreme is the **intuitive approach to decision making**: the hunch, the gut reaction, the decision that *feels* right. People who take this approach to decisions tend to be creative, intelligent people with strong egos and high aspirations. They draw on life experience to give them insight into what would work in another situation. Some entrepreneurs are like this: They may be driving around and suddenly they will say, "There's a perfect location for a restaurant," and they buy it, build a restaurant, and make it a big success.

But this approach does not work for everybody; much of the time it does not work for the people who practice it. When it does, there is bound to be a lot of knowledge, experience, and subconscious reasoning behind the hunches and the gut reactions. Most people would do about as well flipping a coin: You have a 50 percent chance of being right.

Your biases, habits, preconceptions, preferences, and self-interests are all at work right along with your knowledge, experience, and subconscious reasoning. It is almost impossible to see things clearly in a flash, even with extensive background and experience.

Many people have no particular approach to decision making. Some are *indecisive* and afraid of it: They worry a lot, procrastinate, ask other people's advice, and never quite come to the point of making up their minds. Others make *impulsive*, off-the-cuff decisions based on whim or the mood of the moment rather than on facts or even intuition.

Both the **indecisive** and **impulsive approaches to decision making** have trouble distinguishing important from unimportant decisions, and they confuse and frustrate their superiors and the people they supervise. Both types have a poor batting average in making good decisions and tend to be plagued with problems and frustrations.

✳ RATIONAL VERSUS EMOTIONAL DECISION MAKING

On occasion, all decision makers use too much emotion in making a decision. For example, consider a couple who spent a fortune in fancy decor when opening up a new restaurant. The rational part of decision making would question the amount being spent because it does not make sense if the restaurant cannot pay back the amount spent preopening. As managers it is important to weigh the consequences of our actions before we react purely on emotion. In the hospitality field, managers need to learn how to approach decision making using both emotion and rationality.

✳ KINDS OF DECISIONS

The decisions a hospitality supervisor is called on to make range from deciding how many gallons of cleaning compound to buy, up to solving problems that are affecting production, people, and profit. Some decisions are easy to make; others are more difficult.

Routine decisions are easier: what supplies to order and whether to take advantage of special prices; number of portions to prepare; weekly schedule assignments; assigning rooms to be made up; granting or denying employee requests for time off or schedule changes; what time to schedule an interview; whether to confirm a reservation in writing or by telephone; and number of banquet servers to bring in for a special event, for instance. All these decisions are simple to make because you have the historical data or the know-how on which to base your decision.

Sometimes you have a set routine or formula for finding answers to recurring decisions, and if there is nothing new to affect today's decision, you just follow the routine of what has been done before and you don't have anything to decide. This is really a standard operating procedure that may have been established by usage or may be spelled out in your policy and procedures manual.

Many decisions are less clear-cut and more complicated. They may affect many people. There may be many factors to consider. The wrong decision may have serious consequences. Among such decisions are hiring and firing, delegating, and making changes in the work environment or in the work itself, such as introducing a new menu, redecorating a dining room, and changing work procedures. Such decisions require time and thought.

Complicated time-pressure decisions arise when emergencies occur: from equipment breakdowns to accidents, fights, people not showing up, and food that does not meet your quality standards. Even if you have established routines for meeting emergencies, you must still make critical decisions, and one decision may require a whole string of other decisions to adjust to new circumstances.

If, for example, you pull a busperson off the floor to substitute for an injured kitchen worker, you have to provide for the busing, and when you ask servers to bus their own tables, you know that guest service will be affected and you have to handle that situation. A single incident thus demands one split-second decision after another.

This kind of decision making requires a clear grasp of what is going on plus quick thinking and quick action: **decisiveness**. These are qualities that you develop on the job. You develop them by knowing your operation and your people well, by watching your superiors handle emergencies and analyzing what they do, and by building your own self-confidence and skill as you make decisions on everyday problems.

Perhaps the most difficult decisions are those necessary to resolve problem situations. In a sense, all decision making is problem solving. But in some situations you cannot choose an appropriate course of action because you do not know what is causing the problem. For example, you discover that guest complaints about room service have increased during the last few weeks, but you do not know why.

The problem you are aware of (guest complaints) is really a symptom of a deeper problem, and you cannot choose a course of action until you know what that problem is and what is causing it. You need a quick study to isolate the problem: Is it in order-taking, preparation, or delivery

indecisive
A fear of making decisions, thus procrastinating or deferring the decision making to someone else.

impulsive approaches to decision making
Making off-the-cuff decisions based on a whim.

decisiveness
The ability to reach a firm conclusion.

where the system or procedure has failed? This is a far more complicated decision than, for example, deciding what to do about a stopped-up drain.

Some of the decisions that come up in your job are not yours to make. Sometimes the decision is made for you by company policy, such as prescribed penalties for absenteeism, improper food portioning, or smoking on the job. In such instances, the only thing you decide is whether the incident fits the policy specifications—was this the second or third time it happened, and does it deserve the penalty if the clock was fast and she thought it was closing time when she stepped outside to light up her cigarette?

You may have other problems that you do not have the authority to do anything about. If you supervise the dining room and your servers complain about one of the cooks, you do not have authority to discipline that cook. You can work with the head cook, who is on the same level you are, to resolve the problem. If this fails, you can send the problem up through channels to the food and beverage director. But you cannot give orders anywhere but in your own department. It is important never to make decisions or take action where you do not have authority and responsibility.

Since time is probably your biggest obstacle, it is essential to recognize which decisions are important and which are unimportant, which decisions you must make now and which can wait. As a rule of thumb, unimportant decisions are those that have little effect on the work or on people: Nothing serious will happen if you make the wrong choice.

Should you put tonight's specials on menu fliers or let the servers describe them when they take orders? What kind of centerpiece should you use on the chamber of commerce lunch table? What color paper should you have the new guest checks printed on? Such decisions are worth little more than a few seconds' thought: What is the problem? Who is involved? What are your choices? Is there a clear choice?

If there is no clear choice, make one anyway and move on. Getting hung up on unimportant decisions can be a disaster. It impedes the work, your boss, and your people lose confidence in you, and you lose your own perspective, your conceptual grasp of your job.

The opposite mistake—giving too little time to an important decision—can also have serious consequences. An important decision is one that has a pronounced impact on the work or the employees, such as a change in policy or procedures, what to do about a drop in productivity, or how to deal with friction between workers. In such situations the most serious mistake is not taking the time needed to consider the decision carefully. The wrong decision is hard to undo. It may cause serious consequences or increase the problem you are trying to solve.

In another interesting decision-making situation a restaurant owner might seriously consider forming an advisory board because it can enhance both the individual and collective performances of the leadership team. Over time, the board will potentially help perform four key functions:[1]

1. Develop and implement a strategic plan.
2. Steer clear of potential "avoidable" problems.
3. Create optimal solutions to issues and challenges as they arise.
4. Hold yourself accountable over time.

An advisory board does not replace the owner's primary role of leadership; rather, it is a mechanism to enhance results.

How to Make Good Decisions[2]

LEARNING OBJECTIVE: List the six steps for making good decisions.

When making decisions, six steps will lead to improved decision-making capabilities (see Figure 13.1). The following six steps are a simple version of the logical approach. The procedure is elastic and can be expanded or shortened to fit circumstances:

1. Define the problem and set objectives.
2. Analyze the problem: get the facts; the relevant who, what, when, where, how, and why.

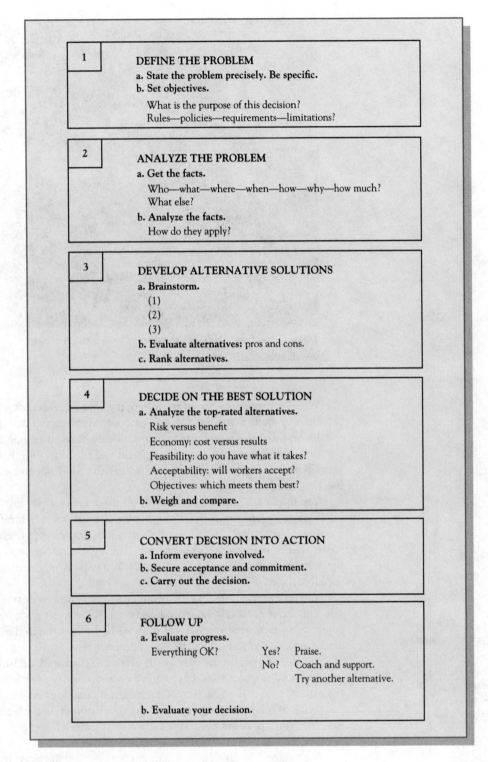

FIGURE 13.1: Pattern for making decisions.

3. Develop alternative solutions.

4. Decide on the best solution.

5. Convert the decision into action.

6. Follow up.

Making good decisions can sometimes take a team effort.
wavebreakmedia/Shutterstock

When you are out there on the job coping with an emergency, think about taking these steps. But chances are that you actually will without thinking, and this is as it should be. Suppose that the dishwashing machine breaks down suddenly at noon on one of the busiest days of the year. You don't think about defining the problem and the objective; they define themselves: you need enough plates, silverware, and glasses to get the food out to customers until you can get the machine fixed.

You might ask how many clean dishes there are—or you might not even stop for this—you know you are going to run out. You generate alternatives at lightning speed: Wash by hand in the bar sinks? Use the bar glasses? Ask the restaurant next door (a friendly competitor) for help (borrow dishes or share their dishwashing machine)? Use paper and plastic? You don't have time to figure out the pros and cons of each choice: You make a judgment on which is best, and you might use all of them.

You take action immediately by directing your employees and doing some of the work yourself. This requires more split-second decisions. Some won't work out, but you can work with what you have. In fact, what you did was to move logically through the six steps, but you also utilized one of the most critical success factors in decision making: **timing**. In many on-the-spot decisions of a supervisor's job, timing may be the overriding factor: A decision is necessary *now*.

The need to make hundreds of quick decisions every day might make you forget that there is any other way to make a decision. Quick decision making becomes a habit. Yet a quick decision is *bad* timing if it is something you need to think through. In the dishwashing machine breakdown example, you knew the necessary facts and the available alternatives, and you recognized the problem at once.

But in other circumstances you might not know all the facts. You might not recognize the real problem. You might jump in with the first solution that comes to mind. You may not realize the consequences it could have. If your decision affects your people and their work, it is worth taking the time to ensure the best possible outcome.

Let us take an example and run through the decision-making process using our six-step method. You have just hired a new server, Cindy, to replace Gary. She will work a split shift, 11:00 to 3:00 and 7:00 to 11:00 Tuesday through Saturday, which is Gary's last day. Cindy must be trained well enough to start work on Tuesday. She is an experienced server, having worked in a restaurant similar to yours. In the next five sections we follow this decision through to its solution.

timing
Selecting the time when taking action will be most effective; making a decision at the moment it is most needed.

❋ DEFINING THE PROBLEM

The first step has two parts. *State the problem precisely* and *formulate your objectives*: what you want to happen as a result of your decision. You start out by writing down *Who will train Cindy?* Then you realize that there is more to the problem than *who*; there are *when* and *what*: How much training does Cindy need? You restate your problem: *Give Cindy enough training to start on the noon shift on Tuesday.* This statement broadens your approach; you begin to include the training itself in your thinking. Figure 13.2 shows your worksheet as you develop your decisions.

Your objectives should restate your problem in terms of the results you expect. They should also include any rules or policies that apply, any requirements as to where, when, how, and so on, and any limiting factors such as time and money. These things are sometimes referred to collectively as **conditions and limitations**.

For your objective, then, you write down: *Cindy is to meet minimum performance standards by noon shift Tuesday. JIT method, on-site. Training expenses $50 maximum.*

❋ ANALYZING THE PROBLEM

The next step is to assemble the relevant data, but don't overdo it; keep the goal clear. It is easy to tell people more than they need to know, yourself included, so don't confuse yourself with too many facts. You can get more as you need them—and you probably will.

The easiest way to organize the **fact-finding** is to go through the who–what–when–where–how–why routine. The first question is who. After studying the weekly schedule, you put down: *Ashley or Karen (lunch), Charlie or Luis (dinner).* Some of these servers would make better trainers

conditions and limitations
Factors such as rules, policies, specific requirements, and limiting factors (e.g., time) that may apply when a problem is being defined.

fact-finding
The process of collecting all the facts about a certain situation.

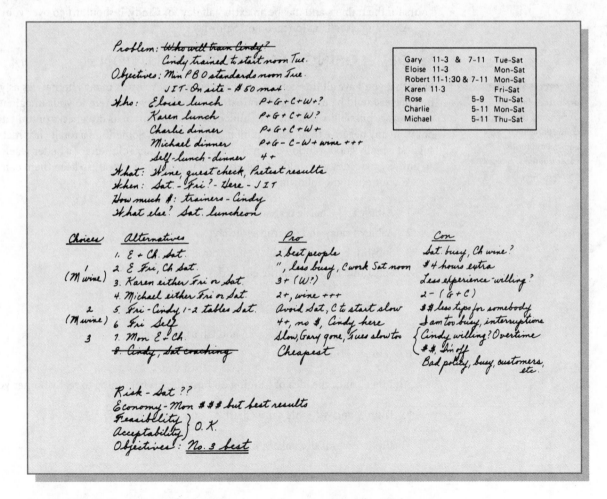

FIGURE 13.2: Decision making: manager's worksheet on training for Cindy.

than others, but you do not eliminate anyone at this point. You want to keep your mind open and your thinking positive. Besides, you need all the possibilities you can get because you might have to settle for a less than perfect solution. Then, reluctantly, you add *Self*. You really don't have the time.

The next question is *what*. You start to write *Server procedures*, and then you think, "Wait a minute, Cindy has been a waitress before; we don't need to train her from scratch." Then you think, "Just what does she know?" and right away, you pick up the phone and call her to come in on Friday at 9:00 A.M. for a pretest.

You know that she doesn't know wine service; they serve wine only by the glass where she worked before. She won't know your guest check system, either. But a lot of things will be the same. She's bright, and it shouldn't take long to train her. You make a few notes: *wine, guest check, pretest results*.

The next question is *when*. You put down *Saturday* and then you think, "Well, maybe part of Friday, too, since she's coming for the pretest anyway." And Saturday is so busy—you have that bridal luncheon for the mayor's daughter. You put down *Friday*?

The next question is *where*, and the answer is *here*, with your equipment and your setup, just as though she were working.

The next question is *how*. You write *JIT, show and tell*. You will have to coach your trainers on this, but since they were trained this way themselves and you have performance standards to go by, it shouldn't take long, maybe an hour on Thursday.

The last question is *why*, and you have answered that in your objective. Next you analyze the data you have gathered. Which server would make the best trainer? You rate them plus or minus on performance, guest relationships, communication, and willingness. What about costs? A few extra hours for the trainers and maybe an extra half day for Cindy. It shouldn't go over your $50 limit.

Anything else? That luncheon on Saturday.

❋ DEVELOPING ALTERNATIVE SOLUTIONS

brainstorming
Generating ideas without considering their drawbacks, limitations, or consequences; typically a group activity.

Now that you have all the facts, your next step is to develop as many alternatives as you can. The first stage should be uninhibited **brainstorming**: You give free rein to your imagination and put down every possibility that you can think of without regard to its drawbacks or limitations. You do not want any negative thoughts to inhibit your creativity. Sometimes a totally impractical idea will suggest a really good solution or a way of adapting another solution for a better result. Sometimes an entirely new idea will suddenly emerge. Ideas spark other ideas, so keep them coming.

You jot down the following:

1. Ashley and Charlie on Saturday
2. Ashley Friday and Charlie Saturday
3. Substitute Karen for Ashley either day
4. Substitute Luis for Charlie either day

After a moment's thought, you add:

5. Train Friday (Ashley or Karen and Charlie or Luis)
6. Yourself Friday

And then, since the idea of training on busy days is beginning to really bother you, you add:

7. Train Monday (Ashley and Charlie—Luis and Karen off)

Finally, in case all else fails, you add:

8. Let Cindy start working Saturday and assign someone to coach her

At this point you run dry. You think you have some good possibilities.

You now move to the second stage of developing alternatives: You weigh the pros and cons of each. You consider the good and bad points, keeping in mind how these would help or hinder the outcome and whether there would be side effects or bad consequences. It is very easy, in concentrating on achieving your objectives, to overlook other results that a course of action might have (the operation is a success but the patient dies).

The larger the problem, the more important this step is. In a major problem affecting production and people, thinking through the consequences is one of the most important steps of all. In our example, some alternatives might produce poor training quality, which could result in problems of service and cost. Or a personality clash might start a good server off on the wrong foot.

You start by listing the pros and cons. As you do this, you discover some things you hadn't thought of before. Here are your thoughts:

1. **Ashley and Charlie on Saturday**

   Pro: Ashley probably 4+ (willing?). Charlie definitely 4+ except on wine.

   Con: Very busy day, Charlie wine?

2. **Ashley Friday, Charlie Saturday**

   Pro: Best trainers. Friday not as busy, Cindy coming in Friday anyway. Cindy can work bridal luncheon Saturday.

   Con: Extra cost (Cindy 4 hours Friday)

3. **Substitute Karen for Ashley**

   Pro: 3+ trainer (willing?)

   Con: Less experienced than Ashley, probably less interested

4. **Substitute Luis for Charlie**

   Pro: 2+ (excellent performer, willing), superb on wine

   Con: 2+ (goes too fast, condescending)

5. **Train Friday, Cindy work one or two tables both shifts Saturday**

   Pro: Avoid training on Saturday. Break Cindy in on job gradually.

   Con: Extra cost, one or two fewer tables and tips for someone Saturday dinner—resentment?

6. **Yourself Friday**

   Pro: 4+ as trainer. No training cost. Cindy coming in anyway.

   Con: Important morning appointments. Off 2:00 to 5:00. Interruptions.

7. **Train Monday, Ashley and Charlie**

   Pro: Slow day. Best trainers. Gary gone. Tuesday good for first workday (slow dinner).

   Con: Will Cindy trade Saturday for Monday? Six-day week for Cindy. Your day off. Slightly over budget (overtime for Cindy).

8. **Cindy to work Saturday with someone coaching**

   Pro: Cheapest

   Con: Bad policy. Poor training, bad start, hidden costs. Too busy on Saturday. Guest confusion.

9. **Luis to do separate wine training before a shift with anyone who needs training in this skill (like Charlie)**

   Pro: Inexpensive

   Con: Need to make time—might need more than one time

The final step of this stage is to weigh the pros and cons and rank your alternatives. When you are making a very important decision with momentous consequences and you have plenty of time, you will rank all alternatives carefully. But if you have several feasible alternatives and limited decision time, you should pick out the top three or four alternatives and rank them. Don't throw the others out; you may have to come back to them.

Moving on, then, here is what you end up with:

- **First choice:** No. 2. Ashley train for lunch Friday, Charlie for dinner Saturday. Use Luis for wine training Saturday.
- **Second choice:** No. 5. Cindy to pretest and train Friday with Ashley and Charlie and work her regular hours Saturday. Luis to teach wine Saturday. Cindy to work bridal luncheon and one or two of Gary's dinner tables.
- **Third choice:** No. 7. Train Monday, Ashley and Charlie. (Must shave cost a bit further.)

You now have three alternatives that will meet your objectives reasonably well. Of the five remaining, you cross off Number 8: After weighing the pros and cons, you find it unacceptable. The rest are feasible alternatives, but you should not bother with them at this point. At least one of your three choices is bound to work.

❈ DECIDING ON THE BEST SOLUTION

Before making any decision of consequence, the decision maker should test the top-rated alternatives by asking five questions:

1. *Risk.* Which course of action provides the most benefit with the least risk?
2. *Economy.* Which course of action will give the best results with the least expenditure of time, money, and effort?
3. *Feasibility.* Is each course of action feasible? Do you have the people and resources to carry it out?
4. *Acceptability.* Will each course of action be acceptable to the people it will affect?
5. *Objectives.* Which course of action meets your objectives best?

These questions require you to do some more analysis. You must analyze benefits and risks and weigh one against the other. You must figure time, money, and effort and weigh cost against results. You must determine whether you have the people and resources needed in each case and whether it will be acceptable to the people concerned. Finally, you must weigh one course of action against another and decide which meets your objectives best.

You can run through this pretty quickly.

1. *Risk.* There isn't much risk anywhere. The biggest risk is probably in your first choice—Saturday being such a hectic day and leaving the dinner training until Saturday night. Charlie might come in late or things might get busy early. And there really isn't time for Luis to teach wine. So there *is* some risk in training quality that might make trouble later. Now, how about Monday? Monday is your day off; is there a risk there? You could stop in to see how things are going, and with your best trainers, little could go wrong and there's time to deal with it.
2. *Economy.* Monday is definitely the most expensive, but it would probably give the best results, and it's not *that* much more; you could shave it down to budget some way.
3. *Feasibility and acceptability.* Those are definitely the big questions; they may decide the whole thing. You pick up the phone and call Cindy. You find that Monday is okay with her; in fact, she'd like the overtime. She could also meet the other two schedules, but they sound a little confusing. You check with Ashley and Charlie. They can both meet all three arrangements and would like to do the training. They both think Monday sounds best.
4. *Objectives.* Which solution meets them best? You decide on Monday, largely on the basis of risk and, in the end, economy. Although it costs a bit more, you think the training quality will be the best, and Cindy will have a good start. It is a decision you are not likely to regret, and it will pay for itself in the long run.

✳ ACTION AND FOLLOW-UP

The next step in decision making is to hand the decision over to the people who will carry it out. At this point, good communication is the key to success. The people who will carry out the decision must fully understand it, and everyone affected must be informed. Every effort must be made to gain acceptance and commitment. If the people who must carry out a decision are involved in developing the alternatives, it usually pays off, but in our industry this often isn't practical.

You have already involved Cindy and Ashley and Charlie; you had to involve them before you could come to a decision. You know that they are committed. You pick up the phone again and tell the three of them that the plan for Monday is on. You give them the details of the Monday schedule and set a time on Thursday for training the trainers. You will give Cindy the pretest on Friday and pass along the necessary information to Ashley and Charlie. Luis will do a separate wine training before a shift for those who need it. You inform the assistant manager, who will be in charge on Monday. The decision is complete.

Altogether, this decision took you 15 minutes or less from beginning to end. Was it worth the time? You think definitely so. You will start Cindy off on the right foot. She will know what to do and how to do it, she will be confident instead of confused and anxious, and she will feel good about the trouble you have taken to provide her training.

Cindy will probably stay beyond the critical first seven days and turn into a good, productive worker. You have probably saved yourself a lot of grief by not making a snap decision for Saturday training.

You also have a lot of new insights on making decisions and on training. You might decide later to delegate training on a regular basis to Eloise and Charlie and give them more training than the quickie things that you can give them in an hour. The last step in decision making is to follow up—keep tabs on how things are going. It is an important step for several reasons. If problems develop, you can catch them early or even fall back on another alternative. If questions arise, you can answer them. Follow-up supports the people carrying out your decision: It reassures them and gives them confidence.

It also gives you a chance to evaluate your decision making. Is everything working out? Did you think of everything? Could you have done a better job in the time you had? Should you have given it more time? What have you learned from this decision? Such a review will help you develop skills and confidence.

Problem Solving

LEARNING OBJECTIVE: Analyze the pros and cons of participative problem solving.

problem solving
Using a logical process to identify causes and solutions to problems or to make decisions.

Problem solving is a special kind of decision making that involves more than a choice between courses of action: It involves identifying the cause of a problem and developing ways to correct or remove the cause. Usually you become aware of the problem through a symptom such as customer complaints, below-standard performance, or substandard food product or room cleanliness—some sort of gap between what is and what should be.

✳ PATTERN FOR SOLVING PROBLEMS

The chief difference between solving this kind of problem and simple decision making is that there are extra steps you must take before you can begin to generate alternative courses of action. The pattern goes like this:

1. Describe the problem.
2. Search out the cause (get the facts).
3. Define the real problem and set objectives.
4. Develop alternative solutions.

As a supervisor, it is important to have follow-up training for employees to see how they are progressing.

Goodluz/Shutterstock

5. Decide on the best solution.

6. Implement the decision.

7. Follow up.

As you can see, the last four steps are the same as in any decision-making process. The difference is in the first three steps. You do your digging for data before you attempt to define the problem and set objectives for solving it. If you try to develop decisions on the basis of your first impression of the problem, you might take the wrong action and the problem will recur. You will have mistaken a symptom for the real cause.

✳ PROBLEM-SOLVING EXAMPLE

To illustrate the problem-solving method, let us run quickly through an example. You are the manager of an independent restaurant and you have recently hired a new cook, a young graduate of a fine culinary school. This is her first full-time job. Your problem is that she is too slow. You think you will have to fire her; she is just too inexperienced. But the food has never been better, especially the daily "cook's specials" you asked her to develop. The customers are raving about the food but complaining about the service. What a dilemma! Where do you start?

The first step is to *describe the problem as you see it now*. This description should include not only what you see at this point as the primary problem but any other problem it is causing. You write down the following: *The cook is too slow; the food isn't ready at pickup time. Servers are angry because customers are blaming them for poor service. Customers are angry, too.*

The second step is to *search out the cause of the problem*. This involves getting all the relevant facts: the who, what, when, where, how, and why, the ongoing story of what is and is not happening, and any other relevant data.

It is obvious that you must get the cook's side of the story. When you first talk to her, she is very defensive. She says that the servers are harassing her and it makes her nervous and she can't work efficiently. You are sure there is more to it than that. That night you spend half an hour in the kitchen observing.

Naturally, the servers do not heckle while you are there, but the cook still has problems. You are surprised at how quickly she works. But she cooks the special at the last minute and she can't do that and plate and garnish everything at the same time. She gets farther and farther behind, so you step in and help her plate the orders.

You make a date to talk to her tomorrow. She is almost in tears and obviously thinks you are going to fire her. You reassure her of your desire to help and you compliment her on the food.

The food is fantastic. You talk to the guests and they rave about it, not only the special but also the regular menu items. It occurs to you that if you can solve this problem and keep this cook, the word will get around about the good food and your business will grow.

One of the ways to get at the basic problem is to ask a lot of comparison questions. What is being done that was not done before? What is different now, and what is the same? Let's see:

- The cook is different.
- The food is better.
- The menu is the same except for the specials.
- The cook is doing some special garnishes.
- The old cook cooked everything ahead except hamburgers and steaks. The new cook cooks the specials to order as well as the hamburgers and steaks.

The next day you talk to the new cook about the problem. You tell her how pleased you are with the food and you make it very clear that you want to solve this situation, which affects both of you. Can she help you analyze it? Finally, she says that she has been afraid to mention it, but the cook's special is a lot of extra work. She is really doing more work than your last cook. You acknowledge that this is true—but is it a piece of the problem?

Besides, she says, more specials are being ordered than any other item on the menu, so that makes still more work, especially at plating time. You ask her if she could cook the specials ahead and hold them, and she says "No, it's the sauce made in the pan at the last minute that makes the difference." There's your clue.

And the garnishes? She prepares them ahead, but it's true that you can't just plunk them onto the plate; you have to handle them carefully. You found this out last night—another clue.

You talk with various servers, and they verify that it seems to be the last-minute cooking that slows things up, and maybe the garnishes. Two waitresses confirm that some waiters are loud and nasty in their complaints while they wait for their orders and that the atmosphere gets very unpleasant.

As a stop-gap solution to this aspect of your basic problem, you speak to the offenders, pointing out that they are only making the problem worse. You believe that you have identified the real problem, and if you can resolve it, this side issue will disappear.

The third step in problem solving is to *define the real problem precisely and set objectives*. This corresponds to step 1 of the decision-making formula.

You have decided that you definitely want to keep this cook. But that does not solve any part of the problem. You write down what you now see as the real problem and state your objectives as follows:

Problem: Time between order and pickup is too long. Can't afford to lose this cook.

Objectives: To reduce the time between order and pickup specified in the cook performance standards, retain the present cook. Present menu must be retained for the next four months (since it was just printed). Cook's specials concept is to be retained if possible. Extra help may be hired if cost-effective. The cook must agree to the final decision.

You start to add: *Long-range objective: to expand business*. Then you come to your senses. Building a business around the skills of a particular cook is an entirely different ball game. You have to solve your present problem within its own frame of reference.

You are now in a position to generate and evaluate alternatives, decide on the best solution, and put it to work, as in the basic decision-making formula.

❋ PARTICIPATIVE PROBLEM SOLVING

In the case we have been following, it is logical for the manager at this point to consider bringing the cook into the next three steps of the problem-solving process. Is this a good idea in general? What are the pros and cons? In management theory, there is a school of thought with a strong following that believes in **group decision making**. They argue that many heads are better than one because:

> **group decision making**
> A process in which a group of people work together to come to a decision.

1. You get more information and expertise relevant to the decision.
2. You get more good ideas and can generate more and better alternatives.
3. You can arrive at better decisions because of the stimulation and interplay of different points of view.

They also argue that in practice:

4. People who have participated in making the decision are generally committed to carrying it out.
5. The coordination and communication necessary to carrying out the decision are simpler and better because everyone already understands what is happening.

This school of thought is associated with Theory Y management style. The experience of many managers who practice group decision making bears out that the theory can and does work. Other people take a dim view of group decision making and find the following problems with it:

1. It takes longer for a group to decide something than it does for one person to make the decision. Furthermore, it takes everybody away from his or her other work. (The decision may be better, but is it worth the total work hours required to make it?)

PROFILE Eric Walker

Courtesy of Eric Walker

Eric grew up in Kansas City and later moved to several sites in his career with the Ritz-Carlton, including Naples, Atlanta, and St. Thomas.

Currently, Eric is a catering and conference services manager at the Ritz-Carlton in Sarasota, Florida, where his general manager has high expectations of the newly created position. That's a challenge he likes. Having worked as a valet parker, doorperson, bellperson, and concierge, Eric has seen both sides of the supervision equation and knows that everything is not black and white. One of the secrets to his success is to empathize with people and to put himself in their shoes. He will do anything to make a guest happy so long as it will not harm the hotel.

Eric leads by example and would not ask any of his associates to do something he would not or could not do. An example of this is when he "jumps in" to help the setup team during the meeting room cleanups while the guests are hav-

ing a coffee break. He recognizes that the setup team is the most important group in his area and helps them whenever possible. As a supervisor, Eric also has high expectations and holds his associates accountable.

Eric's workday begins 6:00 A.M. when he goes over the day with his staff and he checks that breakfasts are ready. He then goes over the day with the client and monitors all the functions and meetings to ensure perfect guest service. In his role as the catering and conference services manager, Eric has to take care of all the meeting planners and their VIP's requests.

This means he communicates with virtually all departments. Eric is ranked number one, two, or three in all the guest surveys, with an average guest satisfaction score of 97 percent. Recently, Eric's general manager asked him which position he was eyeing next. Eric's reply was that he wanted to be the best in the company in his position before he moved to another position. Now that's commitment and dedication for you! No wonder the general manager and the executive committee, in view of his meritorious performance, nominated Eric for the J. W. Marriott employee of the year award.

2. Groups are often dominated by one person—usually the boss—because people want to please or are afraid to speak up or disagree, so there really is no advantage. (Here there is really no group decision.)

3. Group participants often get involved in winning arguments or showing off rather than in getting the best decision. (Groups often don't work the way they are supposed to.)

4. If consensus (general agreement of all participants) is required, people may go along with a decision they don't like just to get the meeting over with. (This is not a true consensus.)

5. Consensus leads to mediocre decisions that will appease everyone rather than the best decision. It can also lead to *groupthink* or conformity rather than to the creativity that is supposed to happen. (Groups may produce worse decisions.)

6. Self-seeking managers can use groups for their own purposes to shift blame in case of mistakes or to manipulate people into agreeing to a decision they do not want to carry out. (Here, again, there is no group decision.)

Clearly, group decision making is not a panacea. It works better in some types of organizations than it does in others, and it is more suitable to some problems, some leaders, and some groups than it is to others. Generally, groups work best when:

- Members are accustomed to working together as a team and have differing expertise and points of view but common goals.
- The leader is skillful at keeping meetings on target without dominating or manipulating (or allowing others to do so).
- The group is rewarded for making good decisions.

This combination of conditions is found more often at high corporate levels, but teams of hourly employees are becoming more involved in group decision making. There are times when including workers at some stages of problem solving makes a lot of sense.

This is especially true when the problem or decision involves specialized skills or experience that the supervisor does not have or when participation will motivate workers to accept the decision and carry it out. In our example, both conditions are present: The cook knows more than you

A group works at participative problem solving.

CandyBox Images/Shutterstock

do about cooking and about the particular problem, and you need her commitment to the decision.

The degree of participation in problem solving and decision making may also vary. This is similar to variations in leadership style running from autocratic to democratic (Chapter 2). Figure 13.3 illustrates such variations of participative problem solving. At one end of the scale the autocratic manager will make the decision alone and tell the workers how to carry it out. For example, a manager will decide on a menu item for tomorrow and order the cook to make it. There is no participation whatsoever.

Toward the middle of the scale, the manager will originate an idea and put it out for comment: "As a server, as a cook, what do you think of this menu item—will it sell?" The workers participate in the evaluation, but they do not take part in generating alternatives.

Farther along the scale, the manager says, "Give me some ideas for menu items and let's discuss them." Here the cook takes part in everything but the decision. Still farther to the right is the manager who says to the cook, "I want you to come up with a couple of new menu items by next Monday; anything you choose within reason is okay with me." Here, the manager delegates the decision with merely a precautionary restriction. At the extreme right is the group decision— clearly inappropriate to this simple problem.

Continuing our example, you as the manager invite the cook to help generate alternatives for solving the time-lag problem. The two of you come up with the following alternatives:

■ Develop specials that can be prepared ahead and simplify the garnishes.

■ Keep the current cooked-to-order specials the customers like so well and hire a part-timer to cook hamburgers and to plate and garnish each order.

■ Keep the current specials, simplify the garnishes, and have the servers plate and garnish the orders.

The cook thinks that she can handle any of the three. Although she will still be doing more work than your former cook, she likes the challenge of developing specials. She wants to work with you on the details of carrying out the decision.

Since your budget and the servers are involved, you will not include the cook in making the decision. You decide to have the servers participate in evaluating the third alternative. You hope they will see that the popularity of the current specials probably means increased business and higher tips in the future. But if they do not agree willingly to the extra work, this course of action will cause nothing but trouble. You need their commitment.

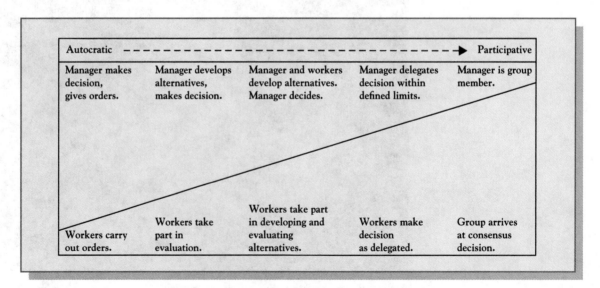

FIGURE 13.3: Range of participation in problem solving and decision making.

❋ SOLVING PEOPLE PROBLEMS[3]

Usually, your most difficult problems have to do with people. Problems about the work focus on products, procedures, schedules, time, costs, and other tangible things. Problems centered on people involve emotions, expectations, needs, motivations, and all the other intangibles associated with being human. The problem-solving steps are the same, but people problems require the sensitive practice of human skills.

Suppose that a cook comes in very late for the second time in a week and you slap a penalty on him, as your company procedure requires. He messes up everything all morning: scorches the soup, leaves the herbs and garlic out of the stew, drops an entire crate of eggs, and walks off the job. Two days later, you learn that his wife had walked out on him that day, leaving him with two little children to care for. You didn't get the facts. You didn't find the real problem. You made the wrong decision. You lost the worker.

A festering problem in many operations is the continuous antagonism between servers and cooks. Perhaps the underlying problem is unsolvable, since it probably has to do with self-image and professional jealousy. Often, each side looks down on the other.

Cooks are proud of their skills and their salaries, and they think of themselves as artists. They look down on serving tables, and they resent it when servers try to order them around. Servers, by contrast, look down on cooks for very similar reasons. In some instances, male-female rivalries are also involved. These are psychological conflicts that no manager has the skills to resolve.

Problems involving conflicts between people who work together surface quickly because most people don't hesitate to complain about each other. The usual problem-solving pattern is very appropriate here. But getting to the real problem might take time because of the number of people involved, their emotions, and probably their disagreement about the facts.

Yet identifying the real problem and solving it is more important than ever. The more people are affected, the more it affects the work. And if people's emotions are running high, they will carry them to the front desk, the lobby, the dining room, the kitchen, or the hospital bed. Guests and patients will not get the treatment they expect and deserve.

Guidelines in Solving Personal Problems

People problems flag you with symptoms—a drop in output, substandard quality, absenteeism, customer complaints—any gap between standards and performance. The first thing to do is get the facts and dig for the real problem. Don't make a hasty decision.

Personal problems are not yours to resolve, of course, but listening when people need to talk can help them to solve their own problems, or at least relieve their tensions enough to get on with their work. Advice is appropriate only if it helps to steer someone toward professional help. You might be able to help your distraught cook find a daycare center, or you might need to refer the person to an employee assistance program (see Chapter 12).

It is important to keep your own emotions out of your workers' problems and to maintain your supervisory role. Dependent people often try to manipulate the boss into telling them how to live their lives. Active but neutral listening, as described in Chapter 4, is the best approach to such problems. The time this takes is appropriate only if the problem is interfering with the work.

There are, however, ways to eliminate the friction. Sometimes the best decision is to choose not to solve the real problem but to bypass it. Some managers have made the decision to use a food expediter to receive the orders from the servers and transmit them to the cooks, so the rivals have no contact at all.

Win/Win Problem Solving[4]

For dealing with problems involving one person, an interesting participative approach includes the worker from the beginning to the end of the problem-solving process. It is known as **win/win problem solving** because everybody wins. People who have used it say that it solves the problem 75 percent of the time.

The win/win concept is difficult for many supervisors to accept. When a supervisor is dealing with a worker who is causing a problem, it is very natural to think of the situation as a contest that the supervisor must win. Win/lose is a competitive concept that pervades our culture: ball game,

win/win problem solving
A method of solving problems in which supervisor and employee discuss a problem together and arrive at a mutually acceptable solution.

tennis match, arm wrestling, election, war—whatever the contest, there is a winner and a loser; that's what it's all about.

In win-or-lose terms, you as a supervisor have four possible ways of approaching the problem solving. The first is a *win/lose* stance: You say to the worker, "You've gotta shape up or else; if you don't shape up, you're fired." You win, the employee loses. Of course, you win the battle, but you lose the war: You either have to hire a new employee or, with a different penalty, put up with a continuation of the conflict on the guerrilla level.

The second approach is a *lose/win* posture: retreat and appeasement. You don't take a stand, you let the worker get away with things, and you back away from any decision. You lose and the employee wins. And soon you lose not only the battle but also your job.

The third approach is *lose/lose*: compromise. You give up something in exchange for the employee's giving up something, and each of you has less than before. You both lose. Neither of you is satisfied and the problem is likely to reappear, perhaps in another form.

The fourth approach is *win/win*: You collaborate to find a solution that satisfies both of you. You include the employee from the beginning of the problem-solving process, and you go through the following steps:

1. Together, you establish the facts and identify and define the problem. As the supervisor you make it clear that both you and the employee will benefit from getting the problem solved. You pull out all your interviewing skills; you listen, encourage, and let the worker vent feelings and complaints. Finally, you agree on the definition of the problem.

2. Together, you generate all possible alternative solutions—no vetos at this point. You keep going until neither of you can think of any more.

3. Together, you evaluate the alternatives and pick the one that is best for both of you.

4. Together, you carry out the agreement. You follow up at intervals to see how the solution is working.

Suppose, for example, that you have a desk clerk who does not get to work on time. After considerable discussion you agree that the problem is that her starting hour of 7:00 A.M. is incompatible with her home situation. She has two young children to get ready for school.

You generate alternatives: Let her husband deal with the children. Have her pay someone to come. Put her on a different shift. Have her work 9:00 to 5:00 instead of 7:00 to 3:00. Terminate her. Put her in a different job: the office, payroll clerk. And so on.

You go over the alternatives and finally agree that she will work the evening shift starting an hour late. She is happy that she can handle both ends of the school day and still make almost as much money. You are happy that you will not have to hire and train a new desk clerk. You already have someone on the evening shift who would like to trade shifts. You both win.

For many supervisors, the win/win approach represents a major shift in attitude. It denies the traditional assumption that problem employees are adversaries in a contest, replacing it with the far healthier assumption that both parties to the problem are in it together.

It goes right along with the Theory Y idea that jobs can be structured to fulfill personal goals and company goals at the same time. And it fits perfectly with the humanistic approach to management that seeks to build a positive work climate and an atmosphere of cooperation and trust. For supervisors who can make that shift in attitude, it is certainly another string to one's bow.

Building Decision-Making Skills

LEARNING OBJECTIVE: Identify ways to build strong decision-making skills.

The ability to make good, sound, timely decisions is one of the most important qualities on which a manager is judged. It is essential to running a tight ship and to being a good manager of people. You can learn a great deal about making decisions from books and from observing people who are good decision makers. But the only way to build a skill is to practice it. Here are some guidelines to help you along the way:

- Make sure that the decision is yours to make, that you have both the authority and the responsibility. Make each decision in the best interest of your employer, not your own interests.

- Accept your responsibility fully: Face decisions promptly. Be ready to take unpopular stands when they are necessary.

- Sort out the important decisions from the inconsequential ones. Make minor decisions quickly. Make major decisions deliberately, seeking out the basic problem and considering consequences before you take action.

- Calculate the risks, and do not be afraid to take them if they are worth the benefits.

- Timing is important; often, it is everything. Adapt your decision making to this overriding requirement.

- Be alert to signs of problems. If you let a situation become a crisis, it may be too late for a good decision.

- Keep an open mind when investigating a problem. Avoid jumping to conclusions, and stay away from your own biases, prejudices, and self-interests. Remember, too, that the easiest solution is not necessarily the best.

- Do consult your supervisor when a problem is truly beyond your ability or experience.

- Make sure that you are not part of the problem yourself.

- You will make some bad decisions along the way—everyone does. Don't brood over them; learn from them.

- Follow up on your important decisions to see how they are working out. Were they good decisions? What would you do differently next time?

- Look at each situation from its own unique perspective. When presented with a situation similar to an earlier one, some supervisors will make the same decision as that used earlier. Although this sometimes works, it is best to examine each situation individually.

Controlling

LEARNING OBJECTIVE: Discuss the ways that supervisors control the work being done by employees.

controlling
Measuring and evaluating results to goals and standards previously agreed upon, such as performance and quality standards, and taking corrective action when necessary to stay on course.

Controlling is a process by which managers measure, evaluate, and compare results to goals and standards previously agreed upon, such as performance standards, and take corrective action when necessary to stay on course. There are visible controls throughout your workplace: door locks, time clocks, the bartender's measuring device, keys to the cash register.

❋ THE CONTROL PROCESS

A major area of control for hospitality supervisors is cost: food cost, beverage cost, labor cost, supply cost, energy cost. Cost control is a process by which managers try to regulate costs and guard against excessive costs in order to have a profitable business. It is an ongoing process and involves every step in the chain of purchasing, receiving, storing, issuing, and preparing food and beverages for sale, scheduling employees, and using supplies and energy.

On a daily basis you will be involved in cost-control techniques such as checking vendors' invoices and counting cash. The exact methods you use will vary from place to place, depending in part on the nature and scope of your business, but the principle of sparing your employer excessive costs remains the same.

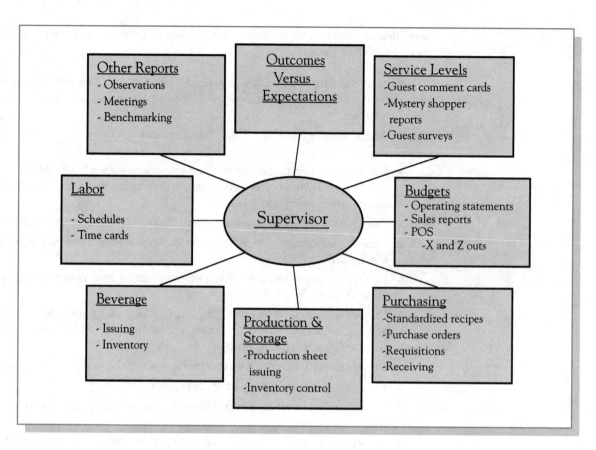

FIGURE 13.4: Examples of control tools for various departments.

In the hospitality industry, excessive costs are often due to inefficiency, theft, and waste—factors that you can influence. For instance, you may check production records against recorded sales to ensure that all quantities produced are accounted for. Figure 13.4 displays various control tools that can be used, depending on the department (e.g., comment cards to access quality of service, purchase orders to control purchasing expenditures). As a supervisor, you can use the following control techniques:

- Require records and reports (such as production reports; food, beverage, and labor costs; and income statements).
- Develop and enforce performance standards.
- Develop and enforce productivity standards.
- Develop and enforce departmental policies and procedures.
- Observe and correct employee actions.
- Train and retrain employees.
- Discipline employees when appropriate.
- Be a good role model.

productivity standards
A definition of the acceptable quantity of work that an employee is expected to do.

As discussed in Chapter 7, performance standards define how well a job is to be done (the quality aspect); productivity standards define the acceptable quantity of work an employee is expected to do (following performance standards, of course). For example, you may determine that it is a reasonable expectation for a housekeeping employee to complete cleaning a guest room according to your performance standards in 30 minutes, for a server to be able to serve four tables at a time, or for a hospital tray line to produce three trays per minute.

Control in the hospitality industry means providing information to management for decision-making purposes. A good example is the food cost percentage. Let's say we have a casual Italian restaurant and its food cost percentage is budgeted to be 28 percent, but the inventory just taken and calculated shows it to be 33 percent. Then at least you know it and can begin the fun part of figuring out what is going on!

An old Swiss chef once said, "Before you can stop someone stealing the chicken, you first must know how to steal the chicken." Control in a situation like a difference in the actual result from what was expected looks at each step in the process—ordering, receiving, storaging, issuing, prepping, cooking, holding, serving, and billing.

Taking another common situation in hospitality, if kitchen labor costs are budgeted to be 14 percent of sales, but turn out to be 18 percent, then we know we must impose a better control and not wait for the end of the month for the result. We would need to monitor labor costs on a daily basis, so as to avoid a repeat situation.

productivity
How effectively an operation converts an input (e.g., food) into an output (e.g., a meal).

Productivity is the output of employees' services and products in a given time period. In a restaurant, inputs such as food and service are used to produce outputs such as meals. Productivity can generally be thought of as a ratio:

$$\text{Productivity} = \frac{\text{Outputs (products or services: quantity and quality)}}{\text{Inputs (resources to produce outputs such as food, labor, and supply costs)}}$$

Productivity measurements tell much about an operation's efficiency and performance and are the common denominator for comparing your operation with others. They also help when developing budgets and allocating resources. Productivity in the hospitality industry is as much a matter of quantity as it is quality.

For example, a server might be able to handle six tables in 20 minutes, but the quality of the service is inferior to that of another server who takes care of fewer guests in the same period of time. Figure 13.5 lists common foodservice productivity measures.

It's amazing to think that a high proportion of employees will likely steal if given the chance. The trick is to prevent that from happening. It begins with the Swiss chef's words: "Before you can stop someone stealing the chicken, you must first know how to steal the chicken."

This was borne out when Julian E. Montoya, founder of the Burrito King chain of restaurants, saw his chain go from success to near disaster in just a few years due to increasing competition and employee theft. Montoya recalls the difficulty of monitoring all of his stores at the height of expansion.

Productivity Measures

Sales Per Employee: $\dfrac{\text{Sales}}{\text{Number of Employees}}$

Labor Cost Percentage: $\dfrac{\text{Labor Costs \$}}{\text{Sales}}$

Departmental Labor Cost Percentage: $\dfrac{\text{Department Labor Cost}}{\text{Departmental Sales}}$

FIGURE 13.5: Common foodservice productivity measures.

One night he went to one of his restaurants unannounced, donned an apron, and went up front to cover the cash register. "The cook took me aside and said, 'Here's how it works: You take the order; you take the cash and don't put it in the register. At the end of the night, we split it.'"[5]

✳ ADDITIONAL TYPES OF CONTROL

feedforward control
Control measures designed to prevent problems before the activity begins.

Other types of control are feedforward control, concurrent control, and feedback. With *feedforward* control, the idea is to prevent problems before the activity. When McDonald's opened in Moscow, it sent a team of experts to show the Russians how to grow potatoes, how to make and bake the buns, and how to ensure the quality of the beef so it would make a burger that tastes the same as the ones in the United States.

concurrent control
Control measures conducted during the activity.

Concurrent control is control that is conducted during the activity. Supervisors practice concurrent control by walking around and observing the work in their areas. This is known as *management by walking around*. The manager can immediately see if something is being done wrong and can show the associate how to do it the correct way.

Feedback is the most-used control. It provides accurate information after the fact. Management uses such information in a variety of ways. Daily reports on operations provide feedback on how well the operation performed. The information provided on the daily report can help managers and supervisors compare the results with what was forecast.

feedback
Giving information about the performance of an individual or group back to them during or after performing a task or job.

Feedback can, when shared with associates, motivate them to even better performance. A guest survey or a positive write-up in a newspaper or popular blog encourages most associates to strive for even better results. People generally want and welcome feedback on their performance, and this form of control provides it.

Finally, control provides information necessary for management to make informed decisions. Control is the last element of the management functions and should interface with planning to make for a continuous process.

✳ CASE STUDY: Who's the Boss?

Leon has been head chef at the Elite Café since it opened 25 years ago. The little restaurant has been a landmark in a small seaside resort town and up to a year or so ago had always been crowded with customers who came back to enjoy the same fresh seafood dishes they remembered from years before.

In the past year, however, there has been a noticeable drop in its business, owing to competition from several new restaurants that feature nouvelle cuisine, gluten-free offerings, and ethnic cuisines.

Leon's boss, Dennis, the restaurant manager, is an eager young man fresh out of a college hospitality program. He sees what is happening and wants Leon to change the menu, but Leon flatly refuses. He says that the food is as good as it ever was—the best food in town—and that Dennis simply isn't promoting it properly and is probably making a lot of other mistakes, too.

Leon makes it clear that he has no respect for college graduates who haven't paid their dues and "gotten their hands dirty"—a figure of speech that is all too appropriate for Leon, whose sanitation practices are old-fashioned, too.

The other employees are aware of this ongoing situation between Leon and Dennis and are beginning to take sides. Dennis is aware that he must do something quickly. But what? Dennis sees his main problem as regaining the café's share of the market and putting it out front, where it has always been. He can see only the following alternatives:

1. Fire Leon for insubordination. This is what he would like to do. But Leon is an excellent cook and no one on his staff can duplicate his chowder, his lobster bisque, and some of the other classics, and there are no recipes to follow.

2. Try it Leon's way—a marketing program emphasizing an old-timey image and ambience—the good old days, tradition. Dennis's heart is not in this approach—he does not believe that Leon is right and knows the other employees will protest.

3. Discuss the problem with his boss, the owner of the café. She is an older woman who really doesn't understand the restaurant business—and besides, Dennis doesn't want to admit to her that he has a problem.

4. Get some expert advice on market trends and how to make a market study: Hire a consultant or pay a visit to his favorite professor at the hospitality Institute.

Case Study Questions

1. What do you think of Dennis's four alternatives? What are the pros and cons of each? What are the consequences?
2. What do you think is the real problem? How would you define it?
3. What should Dennis's objectives be?
4. Is Dennis himself part of the problem? If so, does this make it harder or easier to solve?
5. Are there other alternatives besides those Dennis has listed? Suggest as many as you can, and give pros and cons for each.
6. Who do you think is right about the menu: Dennis or Leon?
7. Is it possible for Dennis to change Leon's opinion of him? If so, how?
8. Do you think Dennis and Leon might ever get together using the win/win problem-solving method? Would it be appropriate in this situation?

KEY POINTS

1. A supervisor's or manager's decision should be a conscious choice among alternative courses of action directed toward a specific purpose.
2. Different people approach decision making in different ways. Examples include logical, intuitive, indecisive, and impulsive approaches.
3. The decisions that a hospitality supervisor is called on to make range from those that are easy to make to complicated time-pressure decisions to problem solving.
4. It is essential to recognize which decisions are important and which are unimportant, which decisions you must make now and which can wait.
5. The following six steps are a simple version of the logical approach to decision making: Define the problem and set objectives; get the facts (who–what–when–where–how–why); develop and rank alternative solutions; decide on the best solution by examining risk, economy, feasibility, acceptability, and objectives; convert the decision into action; and follow up.
6. When making decisions, your timing can be very important.
7. Problem solving is a special kind of decision making that involves more than a choice between courses of action. It involves identifying the cause of a problem and developing ways to correct or remove the cause.
8. The chief difference between problem solving and simple decision making is that there are extra steps that you must take before you can begin to generate alternative courses of action. The pattern goes like this: describe the problem, search out the cause, define the real problem and set objectives, develop alternative solutions, decide on the best solution, implement the decision, and follow up.
9. Group decision making is advantageous because you get more information relevant to the decision as well as more ideas. People thinking together can arrive at better decisions, and people who have participated in making the decision are generally committed to carrying it out. Critics of group decision making say that the process takes too much time and tends to be dominated by one person (usually, the boss). If consensus is required, critics say that it leads to mediocre decisions that will appease everyone rather than the best decision.
10. Group decision making is not a panacea. It works best when members are accustomed to working together as a team and have differing expertise and points of view but common goals, when the leader is skillful at keeping meetings on target without dominating or manipulating, and when the group is rewarded for making good decisions.

11. The degree of participation in problem solving and decision making may also vary.

12. For dealing with problems involving one person, an interesting participative approach, win/win problem solving, means that you find a solution that satisfies both of you. You include the worker from the beginning of the problem-solving process, from defining the problem through to carrying out the agreement.

13. Some important decision-making skills include make sure the decision is yours to make, face decisions promptly, sort out the important decisions from the inconsequential ones, calculate the risks, think about timing, be alert to signs of problems, keep an open mind when investigating a problem, consult your supervisor when necessary, make sure that you are not part of the problem, learn from your decisions, and follow up on your decisions.

14. Controlling is a process by which supervisors measure, evaluate, and compare results to goals and standards previously agreed upon, and take corrective action when necessary to stay on course. Figure 13.5 gives examples of controls commonly found in the hospitality industry.

KEY TERMS

brainstorming
concurrent control
conditions and limitations
controlling
decision
decisiveness
fact-finding
feedback
feedforward control
group decision making

impulsive approach to decision making
indecisive approach to decision making
intuitive approach to decision making
logical approach to decision making
problem solving
productivity
productivity standards
timing
win/win problem solving

REVIEW QUESTIONS

Answer each question in complete sentences. Read each question carefully and make sure that you answer all parts of the question. Organize your answer using more than one paragraph when appropriate.

1. What are the three essential elements in a managerial decision?
2. Describe four different approaches to decision making.
3. What is decisiveness?
4. Describe the six steps in decision making.
5. What is *brainstorming*, and what are its pros and cons?
6. How can you test which alternative is best?
7. Explain the relationship between decision making and problem solving. Why would you group them together? Why would you consider them separately?
8. Describe the pattern for solving problems.
9. What are the pros and cons of group decision making?
10. What are the steps involved in win/win problem solving?
11. List 10 tips for making good decisions.
12. Give five examples of controls commonly found in hospitality operations.

ACTIVITIES AND APPLICATIONS

1. Discussion Questions
- What is the difference between decisiveness and decision making? What is their relationship?
- What kind of decision maker are you: logical, intuitive, impulsive, or indecisive? Do your decisions usually turn out well? How can you improve them?

- What relationships do you see between decision making and responsibility, authority, and accountability? What supervisory responsibilities discussed earlier in this book involve decision making? Give examples of the kinds of decisions required.
- What situations do you see in the hospitality industry where participative decision making would be useful? Explain. Where would it be detrimental or impossible? Explain. If possible, give instances from your own experience where workers participated in some phase of decision making, and comment on the process and outcome.

2. Group Activity: Decision Making

In groups of two, each person presents a work or school problem to solve. Using the pattern in Figure 13.1, go through the steps for each problem. The person who brought up the problem should write down the results of the discussion on his or her problem.

3. Brainstorming

Brainstorming is a technique that can be useful in a variety of training and other supervisory situations. It requires a few ground rules to promote participation: All ideas must be accepted, the pace needs to be fast, ideas need to be recorded, and some of the ideas must be used.

To learn more about brainstorming, take part in a brainstorming session in which you need ideas on the advantages of participative problem solving and the disadvantages of participative problem solving. Were the ground rules met?

 # ENDNOTES

1. Bradley S. Schneider, *Restaurant Hospitality Cleveland,* vol. 91, no. 2 (February 2007), p. 62.
2. This section is adapted from Michael Tsonton, "New Line Order," *Restaurants and Institutions,* vol. 110, no. 12 (May 1, 2000).
3. This section (including the guidelines but not the win/win strategies) is adapted from John Walsh, "Reservations Manager Maintains Steady Business Stream," *Hotel and Motel Management,* vol. 215, no. 19 (November 6, 2000).
4. This section is adapted from Barbara Young, "Profitable Partnership," *National Provisioner,* vol. 214, no. 10 (October 2000).
5. Stephen P. Robbins and Mary Coulter, *Management,* 8th ed. (Upper Saddle River, NJ: Prentice Hall, 2005), p. 477.

Chapter 14

Delegating

You often hear managers in the hospitality industry, especially restaurant managers, talk about the 60, 70, even 80 hours a week they put in just to keep on top of their jobs. They tell you about the constant pressures of the job and how hiring good employees is more difficult than ever.

There is no doubt that a manager is in a high-pressure position and that the industry is plagued with people problems. But does it have to be a constant, never-ending race between the work to be done and the time there is to do it? In this chapter, we discuss one management tool for alleviating the problem: delegation.

Management experts always recommend delegation, yet busy supervisors and managers in our industry seldom delegate.

In this chapter, we examine the delegation process and suggest how to put it to work successfully in a hospitality enterprise. After completing this chapter, you should be able to:

- ■ Explain why delegation is necessary for a supervisor.
- ■ Describe supervisor accountability in delegation.
- ■ List the benefits of delegation.
- ■ Discuss common reasons why both supervisors and employees might resist delegation.
- ■ Discuss the importance of each step in successful delegation.

What Delegation Means

LEARNING OBJECTIVE: Explain why delegation is necessary for a supervisor.

delegation
Giving a portion of one's responsibility and authority to a subordinate.

In a nutshell, **delegation** is a skill of which we have all heard but which few understand. It can be used either as an excuse for dumping failure onto the shoulders of subordinates or as a dynamic tool for motivating and training your team to realize their full potential.[1]

Since you are responsible for the entire output of your unit or department, you delegate responsibility for certain parts of the work to people you hire to do certain jobs—you delegate cooking to the cooks, front desk work to the front desk clerks, and so on. Certain responsibilities you keep for yourself: hiring, keeping track of labor and material costs, making key reports, and so on.

Usually, we do not think of giving people jobs to do as delegating responsibility for the work, but it is—or it should be. Supervisors who are on people's backs all the time—telling them what to do, telling them what they are doing wrong, directing them at every turn—have delegated little or nothing. Supervisors who train their people and then trust them to carry out the job have delegated the responsibility for doing the work. Which supervisor has more time and fewer hassles?

In nearly every job there are variations in the degrees of responsibility attached to that job. If the dishwasher discovers that the gauges are registering in the red zone, whose responsibility is it to correct the water temperature? If the manager has delegated this responsibility to the dishwater and has given the proper training, it is the dishwasher's responsibility. If not, the dishwasher must report the reading to the manager, and the manager must fix it.

Who orders the supplies for the kitchen, the cook or the manager? Who receives the supplies? Who stocks the bar? Who closes the cash register? The manager who feels it necessary to attend to every last one of these things personally is the one who works 80 hours a week. In addition, things probably don't run very well and there are constant crises because the manager cannot be everywhere at once. In short, that manager is not delegating. That manager is trying to do all the work. That manager is not managing.

Essentials of Delegation

LEARNING OBJECTIVE: Describe supervisor accountability in delegation.

There are three aspects of delegation: (1) responsibility, (2) authority, and (3) accountability.

As a supervisor, you have been given **responsibility** for certain activities and the results they are expected to produce. That is your job, your ultimate responsibility. Your boss delegated this responsibility to you when you took over the job. When you delegate, you give a portion of this responsibility to one or another of your employees—you pass along responsibility for certain activities and the results you expect them to produce. However, you maintain ultimate responsibility.

When you took over the job of supervisor, you were given the **authority** you need to carry out your responsibilities, the rights and powers to make the necessary decisions, and take the necessary actions to get the job done. When you delegate a portion of your responsibilities, you in turn must give the person assuming these responsibilities the authority to carry them out, carefully defining the terms.

If you delegate responsibility without such authority, you make it impossible to fulfill the responsibility. Suppose, for example, that you give Tom responsibility for stocking the bar but you do not give him the authority to sign requisition slips. The storeroom is not going to release the liquor because Tom's signature has not been authorized, so Tom cannot fulfill his responsibility, and there is not going to be enough gin for the martinis when customers come in.

As a supervisor, you are accountable to your boss for the results expected of you. **Accountability** means that you are under obligation to your boss to produce these results. People to whom you delegate are accountable to you for the results you expect. Accountability goes automatically with the responsibility delegated; it is the other side of the coin. Delegating responsibility does not relieve you of either responsibility or accountability. If your employee does not come through for you, you must find another way to achieve the results. You cannot shift the blame even though your employee is at fault. The ultimate responsibility is always yours.

The lines of responsibility and authority in an organization provide the anatomy of its organization chart, its **chain of command**. They are the lines along which responsibility is delegated from the top down to the least member of the organization. The chief executive officer delegates

responsibility
The duties and activities assigned to a given job or person, along with an obligation to carry them out.

authority
Possessing the rights and powers needed to make the decisions and to take the necessary actions to get a job done.

accountability
An employee's obligation to a supervisor to carry out the responsibilities delegated and to produce the results expected.

chain of command
Lines along which responsibility and authority are delegated from top to bottom of an organization.

A supervisor delegating.
© Wavebreak Media Ltd./Corbis

responsibility and authority to senior vice presidents, who delegate to the managers who work for them, who, in turn, delegate to the people who report to them, and so on, right down to the night cleaner or the person who does nothing but spread mayonnaise on bread.

At each level, the person delegating has responsibility for the results expected from all those on down the line at whatever level, and the chief executive officer has responsibility for the entire operation.

Accountability moves right beside responsibility but in the opposite direction (see Figure 14.1). All employees are accountable to whoever delegated responsibility to them, so the accountability moves right up to the top along the same lines on which authority and responsibility move downward, and ultimately everyone is accountable to the chief executive officer, who is accountable to the owners or a board of directors.

This organizational anatomy tells who has responsibility at each level for everything that happens or fails to happen. It determines whose head will roll when someone fails to deliver the results expected. If the failure has dire consequences, it may not only be the head of the worker who failed to deliver but the head of the worker's supervisor who let it happen.

The lines of responsibility and authority are also the **channels of communication** from level to level up and down the organizational ladder. *Going through channels* means that when you send information, requests, or instructions to people on levels above or below you, you go one level up

channels of communication
The organizational lines (corresponding to the chain of command) along which messages are passed from one level to another.

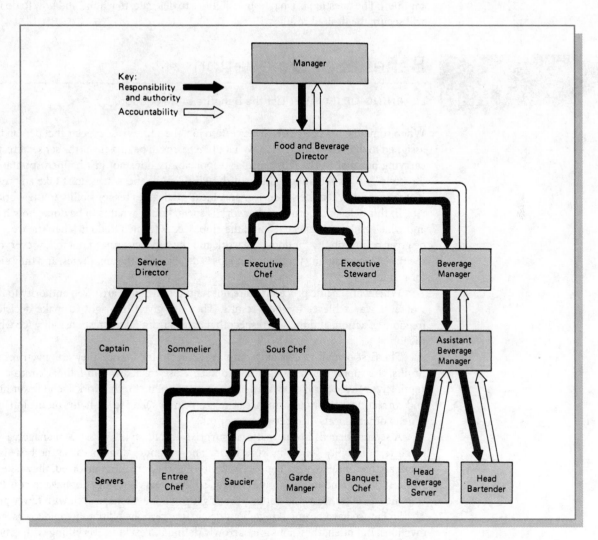

FIGURE 14.1: Anatomy of an organization: the lines of responsibility, authority, and accountability; the chain of command; the channels of communication.

or down your own channel. You do not ask the chief executive officer for authority to spend company money even if he is your own father; you ask your immediate superior. You do not give the head cook's second assistant instructions about the dinner; you pass them through the head cook.

This keeps everyone informed of what is going on and keeps the lines of responsibility and authority straight. If you violate the chain of command, there is bound to be somebody who doesn't know about the delegation and that person's immediate reaction to someone new giving orders is: "Hey, you're not my boss, you're not supposed to do that!"

It is especially bad to cross channels. You do not give orders to someone on another channel because you do not have responsibility for their work and they are not accountable to you. The dining room supervisor cannot give anyone in the kitchen something to do; it has to go up through channels to the manager of the restaurant or the food and beverage director of the hotel and then down through the head cook or the executive chef to the right person. That way everybody knows what is going on. When you delegate responsibility and authority, you keep your own supervisor informed because your boss also has responsibility for what you delegate and is accountable for the results.

Delegation, then, is a managerial tool by which responsibility for the work is divided among people, level by level, throughout the organization. Supervisors are concerned with getting the work done that has been delegated to them and achieving the results for which they are accountable. They have the authority to delegate portions of their responsibility to the people who work for them. The question is what responsibilities to delegate, to whom, and how to do it in a way that will secure the desired results.

Benefits of Delegation

LEARNING OBJECTIVE: List the benefits of delegation.

When someone has been hired for a certain job, the supervisor expects that person to do the work assigned to that job classification. What often seems to be missing is the sense of responsibility for carrying out that work. This sense of responsibility does not just happen spontaneously, as the old-style Theory X manager thinks it should ("People these days don't take any responsibility"). It comes about when the supervisor specifically delegates responsibility to the worker.

In this delegation process, the work is spelled out—what is to be done, how it is to be done, and what results are expected—and the employee is trained to do it. Then the worker can indeed be given responsibility for doing the work and the supervisor need not follow everyone around all the time. Responsibility for their own jobs is delegated to the employees, and they are accountable for results.

Wherever specific procedures are not required, workers are given authority to do the work in whatever way achieves the best results. They are given the right to make decisions, a certain freedom of action, and the self-respect that comes from taking responsibility for what they do on the job.

The first benefit is that once employees are trained, once they are given responsibility for results, the supervisor no longer has to keep close track and can fall back into a coaching and supportive role. Supervisors who delegate responsibility for the work spend fewer hours watching and correcting worker performance. They can either spend fewer hours on the job or devote more time to other aspects of supervision, or both.

A second benefit is one that may surprise an old-style Theory X manager: People who are given responsibility generally work better and get more done because the boss is not on their backs all the time. They are happier in their jobs, they are more involved, they take pride in their work, and they tend to stay around longer, so the supervisor does not have to constantly hire and train new people. More and better work is getting done, perhaps even with fewer people.

So far, we have been talking about delegating responsibility to people for carrying out their own jobs. But often, these jobs are narrowly defined to exclude everything with even a whisper of a risk. Many jobs could be broadened to include related responsibilities. The dishwasher job could include responsibility for correcting the water temperature. The bartender job should include stocking the bar. The cashier should close down the register. These things make more sense.

Job descriptions and performance standards can be broadened to include these duties so that all workers in this job classification will be trained and given responsibility for them. The added responsibilities will make the employees in these jobs feel more important and, at the same time, free the supervisor of still more detail and interruption.

The supervisor might then see the possibility of giving the more promising people still more responsibility by delegating some of the routine management duties or even some duties that are not totally routine. It would mean careful planning, more training, and more follow-up, but it would ultimately relieve the supervisor of still more time-consuming detail, and it would develop a promising employee in a new direction: more responsibility, new skills, a new interest, a new way of thinking that might produce fresh ideas, and better ways to do the delegated task. This is a third benefit of delegation: *developing your people and multiplying their contribution.*

Developing people is part of a supervisor's job. It is a way of putting people's capacities and potential to work for the benefit of the operation. Many people in our industry are underemployed, and these employees are a valuable, untapped resource of ability and intelligence. To such high-potential employees you can delegate small units of your own job (e.g., a daily routine, a weekly report, a troubleshooting task).

By training them to take over such tasks, you are increasing their skills and opening up their future while giving yourself more time to manage. People given new responsibilities and the opportunity to learn new skills become more motivated and more committed to their work, and they usually do it well, often with imagination and creativity.

In this way you can gradually expand such people's experience and prepare them for promotion. What if you lose them through promotion? You might be promoted yourself precisely because you are doing this kind of thing.

Delegation is a conceptual skill. It requires you to see your own job as a whole and find what parts of it can be delegated. Far from lessening your control over the work of your department, it actually tightens up the operation, leading to greater efficiency.

Herein lies a fourth benefit: Greater efficiency means less waste and confusion, lower costs, less conflict, higher morale, and less turnover in personnel. Greater efficiency makes everybody happier, including the guests.

Finally, delegation will sharpen your leadership skills, both conceptual and human. The essence of supervision is getting things done through people. Learning to delegate is not easy, but it will make you grow, both in your job and as a person. Success in delegating will increase your own confidence and your satisfaction in your job, and it will prepare you for advancement.

Delegation not only benefits the supervisor, but also the people to whom one delegates. Studies indicate that most people want more responsibility, and they want the opportunity to grow and develop. The ways that people to whom you delegate can benefit are:

- They become more productive and valuable to the organization and team.
- By learning new things, they improve their self-esteem.
- They become resources for people who need help and function as backups when needed.

> If you've got the right team and you're not delegating, they will see this as a lack of trust and, more important, a lack of leadership and judgment. Remember that what you consider boring might be good development for someone else.[2]

Why People Resist Delegation

LEARNING OBJECTIVE: Discuss common reasons why both supervisors and employees might resist delegation.

If delegation has so many benefits, why is it so rarely practiced in the hospitality industry? There are two sides to the answer. On the one hand, it is very difficult for many supervisors to delegate, or even to believe that it will work. On the other hand, many workers do not want to assume responsibilities. Sometimes the supervisor's reasons and the worker's reasons feed on one another and make it even more difficult to initiate the process.

❋ WHY MANAGERS HAVE TROUBLE DELEGATING

The Theory X manager—and there are still many of them in the hospitality industry—simply does not believe in delegation. Since this type of manager believes that people are by nature lazy and avoid responsibility and must be coerced, controlled, and threatened with punishment to get anything done, the matter ends right there. *They do not believe that delegation, properly carried out, can work and that at least certain kinds of people will take responsibility.* They will not even try it. If they were to try it, they would not do it right, they would not trust their people, they would expect it to fail, and it would.

Many supervisors are afraid that if they let go of the work—if they delegate—the work will not be done right. They, too, do not trust their employees. This is why they are on their employees' backs all the time, overseeing, correcting, and looking for mistakes. This might be their idea of on-the-job training, but it breeds resentment and causes people to leave. It is true that work delegated may not always be done right; you have to train people carefully, trust them, and expect some mistakes at first. If you don't, delegation will not work for you, either.

Some managers believe that their constant presence and their personal control of every last detail are indispensable to the success of the operation. This is an ego problem: They have to feel that without them everything will fall apart, that something terrible will happen if they are not there. Perhaps secretly, even unconsciously, some are afraid that nothing will happen and that things will move right along without them, and they do not want to find this out. For some, this might be the most compelling reason of all for not delegating anything. For such people, power, authority, and tight control are essential to their own security.

Supervisors who are not confident in their own jobs may be afraid to delegate because employees may turn out to do the work better than they did themselves. This is a very threatening idea. How could they handle this—how could they save face? Would the workers want more money, or even take over their job? Would they lose these good workers through promotion? Such fears are powerful inhibitors.

PROFILE   Jerry Ansell

Courtesy of Jerry Ansell

As a supervisor, I delegate every day. Once I have formulated the shift's "battle plan" of what needs to be done by whom (including myself—I cannot delegate everything) and by when, I share that information with each of my team members. We are a high-volume restaurant so I select the best person for each job. We have a brief discussion on how we will meet the goals for the shift, and then we go to it with our battle cry!

Clear communication is critical in delegation—it can be from carefully worded recipes, in Spanish and English, and instructions to verbally tell a team member duties and responsibilities. Always double-check to see if the team member's got it straight. It is also important to give the appropriate authority along with responsibility. For example, to what extent can a kitchen supervisor, who is in charge of a shift, discipline a cook? Both the supervisor and the cooks need to know so there are no unfortunate situations.

Delegating can help develop employees, raising them to the next level. As the employees accomplish more, they can accept more responsibility. After delegating a job, I periodically check to see how the progress is going. What I find important is for the employees to be able to see their progress by being able to check their progress themselves. For example, if a cook starts making a dish at 2:00 P.M., then by 2:45 she should be at stage 3 in the production schedule of the menu.

Many supervisors do not want to take responsibility for the mistakes of others. They may be afraid to be dependent on others. They may worry about what will happen to their own job if they delegate responsibilities and their people do not come through with the results. They may be afraid of what their boss will do, and in some cases, this might be a very legitimate fear.

Although fear of one sort or another is a major reason for not delegating, in other cases the reason might be habit or momentum. Some people simply cannot delegate. They have always done things themselves. They have gotten where they are by doing, not by letting others do, and they cannot let go. You see this sometimes in family corporations or companies where the president is 97 years old and is still running the business. These people cannot let go of the reins. They do not know how to do it any other way.

Many supervisors who are newly promoted from hourly jobs also have trouble shifting from doing to managing. We have talked about this boomerang type of management before: Supervisors slip back into doing the work themselves because it is easier and more comfortable than getting others to do it. This may very well be the most common reason for failure to delegate. When you are not at home with your new responsibilities, delegating them can be a scary and painful prospect.

Sometimes, the momentum of the operation takes over common sense. *Many supervisors say that it is quicker to do something yourself than to train someone else to do it.* A manager will tell you, "I can make coffee in a 5-gallon urn in five minutes, but it will take me half an hour to train an employee to do it." The supervisor never has half an hour to spare, so the supervisor makes the coffee and it becomes part of the supervisor's job. But training an employee, a one-time expenditure of 30 minutes, would save some 60 hours a year of the supervisor's time. Hundreds of such decisions are made in hospitality enterprises because short-term pressures override the long-term gains of delegation, or because the supervisor cannot see beyond the next five minutes, or has the habit of doing rather than delegating.

Sometimes, tasks that could easily be delegated to a promising worker might involve important people in the organization, or perhaps information the manager might not want to share with any of the workers. Or they might be detailed tasks that the manager really enjoys doing. In such cases, a manager might decide that the personal gains in hanging onto these tasks outweigh the time saved or other benefit to the organization. This is a decision of questionable wisdom but not an uncommon one.

Sometimes, for both good and bad reasons, supervisors resist delegation simply because they do not want to lose touch with what is going on.

Sometimes, there are reasons more substantial than fears or habits or self-interest that keep supervisors from delegating. *There may be no workers who are qualified and willing to take on work the supervisor would really like to delegate.* The ability and willingness of the employees are of critical importance to the success of delegation, so let us see why employees do not want the responsibilities that delegation entails. You create barriers that prevent you from delegating when you:

- Prefer to do the work yourself or think no one else can do it as well.
- Feel a strong need to work at tasks with which you are familiar.
- Feel threatened by the possibility that someone else might not complete a task for which you are responsible.
- Fear the loss of power.
- Delegate without planning. It is very important to set deadlines, explain the task's objectives, and transfer authority.

Why Some Employees Won't Accept Responsibility

LEARNING OBJECTIVE: Discuss common reasons why both supervisors and employees avoid delegation.

Some employees in the hospitality industry are barely able to do their jobs at the minimum level of acceptability. Others are very dependent people who want to be told what to do all the time and are afraid to make decisions. A few are hostile types who are just waiting for a chance to get hold of the ball and run with it in the opposite direction; they don't trust you and you don't trust them. None of these are good candidates for delegation beyond the specific barebones tasks of their jobs. Even if they were willing, they are not able to assume additional responsibilities.

In delegation, fear plays a part for many employees, just as it does for managers. *Fear of failure* is common among people who lack self-confidence; they doubt their own capabilities to carry out new tasks. They do not trust themselves.

Others fear the consequences of the mistakes they may make in a new assignment. They might be afraid of the boss's criticism or anger. Their relationship with the boss might not be good enough to make them willing to risk the mistakes.

Sometimes an employee who is offered an extra responsibility may be afraid of rejection by other employees. If others are jealous, or see the new assignment as a defection to the management side, or think it is unfair to themselves, they might give the employee in question a hard time—or at least the employee might think they will. Getting along with one's peers, being part of a group—that feeling of belonging—is often more important than having responsibilities and rewards.

Many employees will refuse added responsibilities if they see them simply as meaningless extra work that they have to do. If there is nothing in it for them—no interest, no reward, no extra pay, no recognition or independence or challenge or opportunity for growth, just more drudgery—they will perceive the added work as an imposition, and they will resent it. They will refuse it outright or find ways not to do it, and the attempt to delegate will backfire. Adding more work without adding interest, challenge, or reward, known as job loading, should be avoided at all costs.

Finally, there are highly capable workers who are satisfied with what they are doing and *simply do not want to be given more responsibility or to be developed and pushed up the corporate ladder.* Not every cook aspires to be a food and beverage director; some people just love to cook. Most people who are "only working until" are not interested in taking more responsibility; they are only marking time.

Some people want a routine job that makes no demands on the mind because they are writing the great American novel. The professional dishwasher we mentioned several chapters back was utterly happy as a dishwasher and refused all offers of advancement. The dishroom was his empire, he was in charge, he was proud of it, nobody bothered him, it was where he belonged, and it fulfilled all his needs.

Delegation, then, is sometimes a relationship between two fearful and reluctant parties. How can one avoid its fears and follies and reap its benefits and rewards?

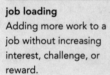

job loading
Adding more work to a job without increasing interest, challenge, or reward.

How to Delegate Successfully

LEARNING OBJECTIVE: Discuss the importance of each step in successful delegation.

Certain conditions are essential to successful delegation. You have met them in earlier chapters.

❁ CONDITIONS FOR SUCCESS

One condition is *advance planning.* This should include an overall review of who is responsible for what in your department at this time, what further responsibilities could be delegated, who is qualified to assume greater responsibilities, what training would be necessary, how various shifts in responsibilities would affect others, and when these shifts would appropriately take place. Delegation involves rearranging things, and it brings your conceptual skills into play.

One of the conditions of successful delegation is to have a positive attitude toward your associates.
CandyBox Images/Shutterstock.

You have to look beyond the daily operational detail to the larger picture and get it all into focus. In addition to general overall planning, delegation requires a specific plan for each instance of delegation so that everything is clear to everyone concerned and the groundwork is prepared properly. We say more about this in the next section.

A second condition for successful delegation is a *positive attitude toward your people*. You cannot have Theory X beliefs about your people and expect delegation to work. You don't have to be an all-out Theory Y manager, but you must have good relationships with your people, know their interests and their capabilities, and be sensitive to their needs and their potential. You must respect them as individuals and be interested in developing them, for both your sake and theirs. You need to develop the kind of leadership skill that gives people belief in themselves and makes them want to come through for you.

A third condition is *trust*. There has to be trust between you and the people to whom you delegate: You trust them enough to share your responsibility with them, and they trust you not to put something over on them or get them in over their heads. Only if you both have this trust can you get the commitment necessary to make delegation work.

A fourth condition of successful delegation is the *ability to let go and take risks*, to let your workers make some mistakes and to give yourself the same privilege. Each time you delegate a responsibility, it is going to be new for you and new for the other person. Some mistakes are bound to be made; it is not going to be perfect from day one. But when a mistake happens, don't panic and don't jump on the other person. Take a coaching approach. The employee learns under your leadership, and you improve your leadership skills. That is what true on-the-job training is all about. You can both learn something from every mistake, and that is how you both grow.

A fifth condition of successful delegation is *good communications*. You must keep the channels open and use them freely, send clear messages, and keep everyone informed who is affected by the delegation: the person to whom you delegate, the employees in that area, and your boss.

Make sure that the people to whom you delegate know the terms of their authority and the extent of their responsibility. The more you delegate—the more people there are who share your responsibility—the more important good communications become.

The sixth condition is *commitment*. If you can involve your people in the planning and goal setting for their new tasks, they will become committed to achieving the results. You, in turn, must be committed to train, coach, and support as needed. Don't just dump the job on somebody else and abdicate.

❄ STEPS IN DELEGATION

The first step in delegation is to *plan*. You need to identify tasks that can be assigned to someone else, and you need to figure out which of your people are able and willing to take them on. Begin by listing all the things you do. You might find it useful to keep a chart for several days on which you note absolutely everything you do in each quarter-hour. Then you sort out your activities and responsibilities into groups, as shown in Figure 14.2. After you have done this, you can arrange things that should and could be delegated in some kind of order: order of importance, or ease of delegating, or time saved, and choose which one or two you will tackle first.

Next, you must look at your people. Choosing the right person for the responsibility is a key ingredient in successful delegation. Figure 14.3 can be useful to you here. Motivation and ability are both essential to success. Who among your people is both able and willing? If there is no one with both qualities, is there someone you can train, or someone who would be willing if you could overcome their fears or make the content of the task more attractive or offer appropriate recognition and reward?

Once you have identified the task and the person you want to have perform it, the second step is to *develop the task in detail* as a responsibility to be delegated. Define the area of responsibility, the activities that must be carried out, the results you expect, and the authority necessary to fulfill the responsibility. This is all very similar to the procedure you use in developing performance standards.

In fact, a system of performance standards is an excellent tool for use in delegating responsibility. You can turn people loose in their jobs because you have told them exactly what you want, have set the achievement goals, and have trained them in the skills needed. They can take the responsibility from there and leave you free to manage.

In any delegation you do the same thing. You spell out the essential content and detailed requirements of the task, define the limitations, and specify the results expected. Within these limits, people will be free to do the job in their own fashion. You will also spell out the specific authority that goes with the responsibility delegated: what kinds of decisions can be made without checking with the boss, what money can be spent, what actions they are authorized to take on behalf of the boss or the enterprise, and so on. Figure all this out ahead of time, and then take the third step—delegate.

The third step has three parts: *Delegate responsibility* for the task and the results expected, *delegate the authority* necessary to carry it out, and establish accountability. As we have seen, these are the three interlocking parts of delegation, and they must be spelled out clearly.

When you delegate, you meet with the chosen employee in a private interview in which you describe the task, the results you expect, and the responsibility and authority it entails. It should be an informal person-to-person discussion. You should present the new assignment in a way that will stimulate interest and involvement: ask for ideas, make it a challenge, mention its present and future benefits, offer rewards if appropriate, and express confidence.

Take a "we" approach, indicating your availability for support and your continuing interest in their success. Promise training if it is needed. However, do not put pressure on by ordering, threatening, or making it impossible to refuse. *There must be agreement on the employee's part to accept the delegation.*

Delegation is a **contract**. You cannot just give responsibility to people; they must accept the responsibility. They must also accept the accountability that goes with the responsibility. Unless you have fully given responsibility and authority and the other person has fully accepted responsibility and accountability, true delegation has not taken place.

contract
An agreement between two parties that is fully understood and accepted by both.

Things you must do yourself (e.g., hiring, evaluation, rewards, discipline, termination)

Things you should do yourself that someone could help you with and could do when you cannot be there (e.g., scheduling, data for cost percentages)

Things you now do yourself that someone else could be trained to do (e.g., training new personnel, requisitioning, ordering, receiving)

Things you should have others do with help and guidance from you (e.g., customer counts, time sheets, guest room counts, occupancy rates)

Things others must do (e.g., their own jobs)

Things no one is doing that others could do (e.g., new menu items, specialty drinks, promotion ideas)

FIGURE 14.2: A way of organizing tasks that can be delegated.

It is important for your employees to know that you are sharing your responsibility with them; you are not dumping it on them and abandoning them. You, too, are accountable for the results. Give them plenty of chances for questions and plenty of reassurance for lingering doubts.

If you have matched the right person with the right assignment and have communicated it in the right way, they will be interested, pleased, motivated, challenged, and glad to have more responsibility. If you include them in setting goals for the project, you will gain their commitment to achieving them.

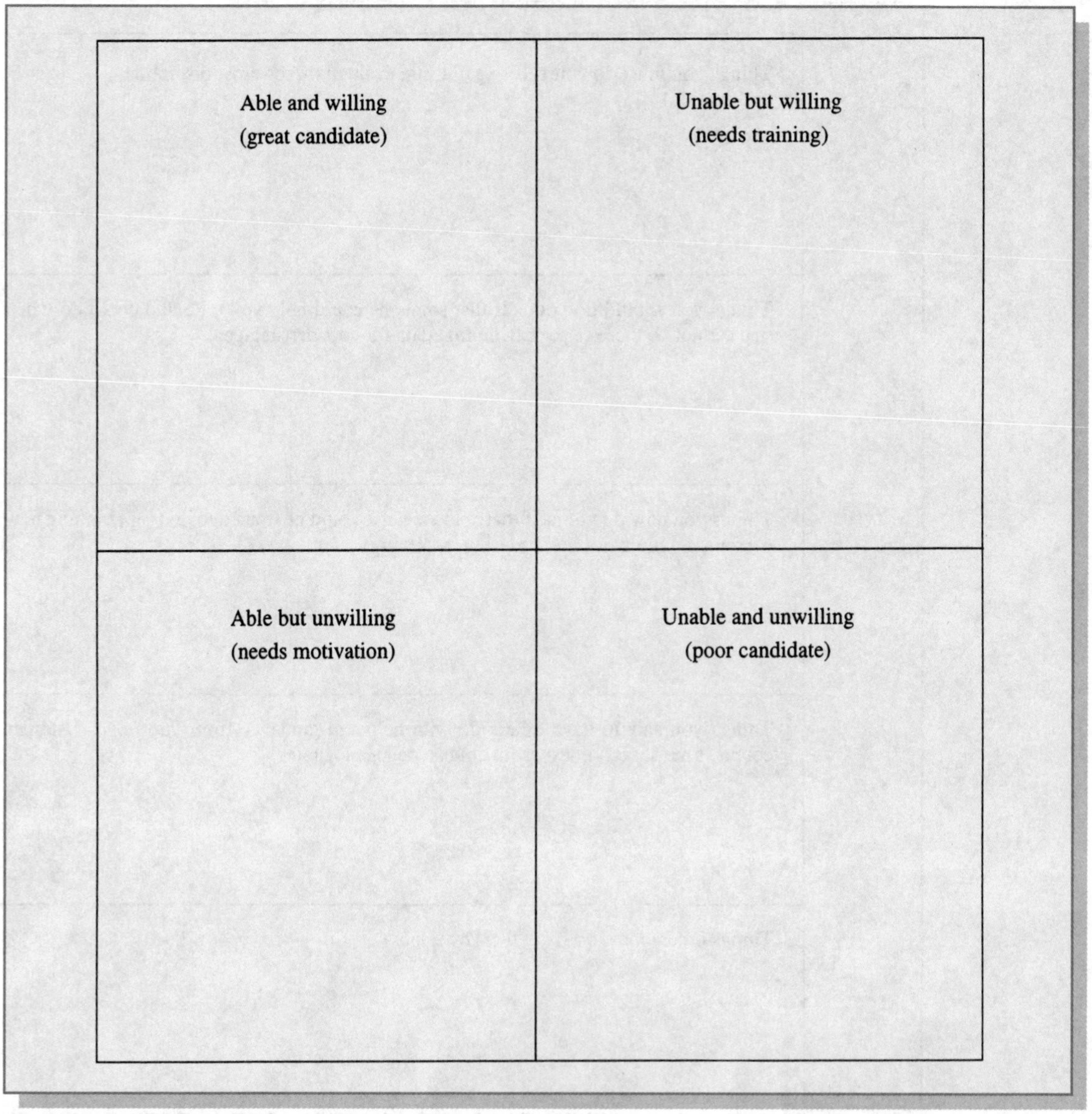

Able and willing (great candidate)	Unable but willing (needs training)
Able but unwilling (needs motivation)	Unable and unwilling (poor candidate)

FIGURE 14.3: A way of planning the assignment of tasks to be delegated.

Set checkpoints along the way for following progress. They give you the means of keeping the employee and the assignment on target. You can modify or adjust the assignment, correct mistakes, and give advice at critical points without taking back the entire job. Checkpoints are your controls. If you can't set up controls, either don't delegate the job or redesign it so that you have some other means of tracking performance.

The fourth step in delegation is to *follow up.* Train your employees as needed. This is something they have never done before, so you go through the whole story: what you want done, how you want it done, to what standard. If you don't, they will take the easiest way to do it. When they are ready to go, communicate the new status to everyone concerned, follow channels, and make good on immediate rewards promised, such as relieving them from other duties to make time for the new ones.

Then slip into the coaching role. Stay off their backs: Don't oversupervise and overcontrol; let them work out their own problems if they can. If they have trouble making decisions and keep asking you what to do, turn the questions back to them—ask them what *they* think. Encourage them to go it on their own. Don't let the responsibility you have given them dribble back to you.

When employees try to dump their assignment back on you, it is called reverse delegation. It might occur because the employee lacks confidence, doesn't really know enough to do the job, is afraid of making a mistake, or simply does not want the added responsibility. You need to listen to the employee and discuss the impasse, but make it perfectly clear that the task is still the employee's responsibility to complete. If you take back incomplete work, you will support the employee's dependence on you. The best way to handle reverse delegation can be stated as: "Don't bring me problems, bring me solutions."

Observe the checkpoints, assess progress, give feedback, and help them reach independence in their new assignments. Then congratulate yourself on two things: You are learning how to delegate successfully, and you are developing your promising employees. This is genuine on-the-job training (not the magic-apron type), and you are developing genuine management skills.

❀ COMMON MISTAKES IN DELEGATION

When you have had no experience in delegating, it is easy to make mistakes. Perhaps one of the most common is *not communicating clearly*. Employees must understand what you want done, how you want it done, what results they are accountable for, and what the goals and standards are. They must understand the area, extent, and limits of their responsibility. They must know what authority they have and its limitations—what they are empowered to do, what decisions they can make on their own, and what decisions they must refer to you.

If you have done your homework carefully before you meet with them, you can communicate clearly. Make sure they have understood by asking them to summarize for you the essentials of the agreement. In many cases it may be wise to put things in writing.

But often in this time-pressure industry it is easy to skip the planning stage, and it is easy to crowd the delegation of a responsibility into one of those 48-second interchanges of which your day is made. This is taking a big risk. You cannot communicate the details of the assignment itself and the implications of responsibility, authority, and accountability in that length of time. After you have delegated for a while to people who have become experienced in sharing your responsibilities, 48 seconds might be enough. But the first time you delegate to first-time delegates, make it a big deal. Communicate everything clearly, and check to see that everything is understood between you.

Another mistake it is easy for a first-timer to make is to *oversupervise*, simply because you are nervous about the whole thing. In this case you soon revert to being the boss and taking back the responsibility you have delegated. You have to remember that it won't all go perfectly, that you have picked someone you trust, that you do have checkpoints and controls, and that the only way to learn to delegate is to stop being bossy. If you jump in and correct small mistakes all the time, they are not going to come to you for help when they have a real problem.

It is also easy to make the opposite mistake—*not taking time enough to train* employees in their new responsibilities and *not giving them enough support*. In this case, they might become discouraged and lose their enthusiasm. They may do a poor or mediocre job, or they may leave because things are not going well for them and they are discouraged about themselves. You must take time to give them the training they need, and they must always have the feeling that you are supporting them and that they can come to you to discuss problems. Furthermore, they must experience success and build confidence if the delegation is to prove fruitful.

Delegating without setting up controls—built-in ways to monitor performance—is another common mistake. If you have not had much experience with delegation, you might overlook this essential. You need checkpoints—periodic reports, reviews, conferences—so that you can keep track of things without being involved in the work but can intervene if necessary to keep things from getting out of hand.

Still another common mistake is *job loading*, mentioned earlier—increasing the workload without adding any new responsibility, interest, or challenge. Suppose that you raise the number of rooms that must be cleaned in a given time period. This is not delegation. It is bound to cause resentment and will complicate rather than simplify your life. It demotivates the worker. In contrast, a task that includes new responsibilities can be a motivator even when extra work is involved, as long as the worker welcomes the responsibility and as long as the work can be done in the time there is to do it.

A similar mistake is to *assign dead-end, meaningless, boring, unchallenging tasks without offering any kind of incentive or reward*. In this case, an increase in responsibility is not enough—it is just an extra burden, and there is nothing in it for them. You have to make it worth their while: extra money, more status, shorter hours, a promise of something better at a specific time, whatever will cause a worker to accept the responsibility willingly.

Delegating to the wrong person is another common mistake. If you know your people well and plan the delegation carefully, this will not happen to you.

A few supervisors make the mistake of *delegating unpleasant parts of their job that involve the boss–subordinate relationship, such as discipline or termination*. This simply amounts to abdicating the role of boss. The employee cannot handle it and everyone loses respect for the boss who passes the buck. Follow the advice of President Truman: "The buck stops here."

Setting up overlapping responsibilities is another mistake that people sometimes make. You may carelessly give the same task to two different people, or give someone an assignment that involves someone else's department without clearing it with that department head. If your own boss is a disorganized person and the lines of authority and responsibility are not too clear, you may even find yourself delegating a responsibility that does not belong to you. Be sure that you know where you stand in the organization and what you are doing, and be careful to keep everyone informed.

You can avoid these mistakes if you plan carefully, know your people and your own responsibilities, keep in touch with what is going on, and keep your overall goals in sight—to manage your people to produce a smooth-running operation with everyone contributing the best of which each is capable.

✳ ADAPTING DELEGATION TO YOUR SITUATION

There are few universal rules about what tasks you should delegate and what you should keep for yourself. Generally, you should not delegate responsibility that involves your relationship with subordinates, such as hiring, evaluating, disciplining, and terminating. You should not delegate tasks that require technical expertise that only you have, or tasks that involve confidential information, and you should not dump unpleasant tasks on people who don't want them by passing them off as "delegating responsibility." Other than these, there are few tasks that you should avoid delegating if the delegation makes sense.

Delegating time-consuming and routine tasks frees your time and attention for managing.

Delegating time-consuming and routine tasks frees your time and attention for managing.
Lucky Business/Shutterstock.

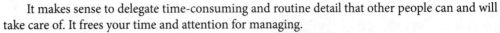

It makes sense to delegate time-consuming and routine detail that other people can and will take care of. It frees your time and attention for managing.

It makes sense to train others to take over tasks and responsibilities that must continue when you are not there. You must provide for emergencies and for your off-hours and vacations. You must have people who can assume your day-to-day responsibilities when necessary. If your unit or department cannot run without you for a while, you are not doing your job.

It makes sense to delegate tasks and responsibilities that motivate and develop your people. If you know your people and their interests, talents, and shortcomings, you can match the responsibility to the person. You can give them work that interests them, challenges them, makes them feel important and valued, gives them the satisfaction of achievement, and helps them grow.

It makes sense to plan such growth for people of high potential, to add further responsibilities over a period of time, to groom them to take your place someday or climb your company's career ladder or move to a better position somewhere else. Although you might lose them in the end, they will more than repay you in what they contribute to you and your operation as they grow.

You are the only person who can decide what makes sense in your area of supervision. You are the only one who knows the tasks and the people. Taking the first steps of delegating can be scary and even painful, but once you have done it, you are on your way to being a manager in every sense of the word. You do not have to do it all at once; there are degrees and stages of delegation. Take it one task at a time, one step at a time, and start with a task and a person you are pretty sure are made for each other.

Delegating responsibilities, making jobs more interesting and challenging, and helping people grow multiplies your own effectiveness many times over—far, far beyond anything you could do by keeping all your responsibilities to yourself.

 ## CASE STUDY: Too Much, Too Fast?

Joanne is manager of an in-plant, self-service cafeteria for an insurance company headquarters with 1,000 employees, most of whom eat breakfast and lunch there. In addition to managing the cafeteria, she is responsible for stocking sandwich and dessert vending machines. She has been supervising all her employees directly but has decided that it would be better if she delegated the major food-preparation responsibilities to her three best employees in order to devote more time to customer relations.

After lunch on Wednesday, she calls the three employees together and explains her plan.

"I am going to delegate to each of you responsibility for preparing the food in your department and keeping the counters and steam table stocked during the serving period. Jasmine, you will be in charge of salad and sandwich preparation. Michelle, you will do the desserts and baked goods. You two will also prepare the food for vending. Robert, you will be responsible for all the hot food: soups, entrées, vegetables, and so on.

"Your co-employees—you each have two—will become your assistants, and you will direct their work. I will be on hand at all times, but I will be talking with guests and supervising the rest of the staff: the breakfast cooks, cashiers, cleanup crew, dishwashers, and so on. I will also continue to do the ordering, receiving, staffing, and so forth.

"Now, you all have seen me in action in your departments, and you know what my methods and standards are. Make the usual menu in the usual quantities. Just do everything as I would do it, and come to me with questions. We will start tomorrow."

The first day of the new regime is a near-disaster. No one makes the beverages and no one stocks the vending machines, although the food is prepared for them as usual. Both Michelle and Jasmine prepare the cantaloupe and the fruit/cheese plates. One of Robert's assistants does not show up, and instead of asking Joanne to get a substitute cook, he and his other assistant try to keep up with the demand. The result is a large and growing crowd of complaining guests waiting for the hot food. Jasmine's two assistants refuse to take orders from her and go to Joanne saying, "Hey, she's not our boss. Who does she think she is, telling us what to do?" One of Michelle's assistants resigns in a huff in the middle of lunch because she thinks she should have had the job instead of Michelle, and Michelle is snapping at her. The other complains to Joanne about Michelle after the serving period is over.

Joanne spends the entire day putting out fires (some of them are still burning), dealing with complaining guests, and trying to find a replacement for the worker who resigned. She ends the day harassed and embarrassed. She is pretty sure that all her employees except those who are mad are laughing at her, and she will probably have trouble with everyone for several days, including the guests. She hopes that her boss at the catering company she works for does not hear about this.

Case Study Questions

1. What basic mistakes did Joanne make?
2. Why do you think she did not foresee what happened?
3. How could she have avoided the reaction of Jasmine's and Michelle's assistants? How could she have avoided the reaction of Robert, who tried to work shorthanded?
4. What should she do now? Should she withdraw the delegation or try to make it work? If the latter, then make a detailed plan for her to carry out.
5. How will she handle all her other workers tomorrow to keep their respect?
6. What should she do about pacifying customers?
7. Should her boss at the catering company headquarters be involved in any way? Does her boss share the responsibility for what happened?

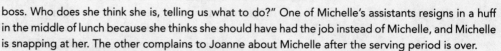

KEY POINTS

1. Delegation is a managerial tool by which responsibility for the work is divided among people, level by level, throughout the organization.
2. There are three aspects of delegation: responsibility, authority, and accountability. As a supervisor, you have been given the responsibility for certain activities and the results they are expected to produce, the authority (or rights and powers) to carry out your responsibilities, and the accountability (or obligation to your boss) to produce these results.
3. The lines of responsibility and authority in an organization provide the chain of command. Accountability moves right beside responsibility but in the opposite direction. All employees are accountable to their boss, so the accountability moves right up to the top along the same lines on which authority and responsibility move downward.
4. The lines of responsibility and authority are also the channels of communication.
5. The benefits of delegation include the fact that the supervisor can spend more time coaching and performing other duties instead of watching and correcting performance; the employees can usually work better; the supervisor can develop employees; and greater efficiency results, which means less waste and confusion, lower costs, less conflict, higher morale, less turnover, and happier customers.
6. The Theory X manager does not believe in delegation because he or she believes that employees will not take responsibility and/or the job will not be done right. Some managers believe that their constant presence and personal control of every last detail are indispensable. Other managers don't delegate because they are afraid the workers might do the work better than they do it themselves, or simply out of habit. Sometimes, supervisors resist delegation simply because they do not want to lose touch with what is going on.
7. Employees at times may not accept new responsibilities because of fear of failure or fear of being rejected by coworkers, dislike of meaningless extra work, or simply a lack of desire to be pushed up the career ladder.
8. Conditions for delegating successfully include advance planning, a positive attitude toward your people, trust, the ability to let go and take risks, good communications, and commitment.
9. The steps in delegation include planning, developing the task in detail, delegating responsibility for the task and results expected, delegating authority and establishing accountability, setting checkpoints along the way for following progress, and follow-up.

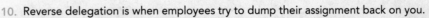

10. Reverse delegation is when employees try to dump their assignment back on you.
11. Common mistakes in delegating include not communicating clearly, oversupervising, not taking time enough to train and give support, not setting up controls, job loading, assigning dead-end work without any reward, delegating to the wrong person, delegating unpleasant parts of the job that involve the boss–subordinate relationship, and setting up overlapping responsibilities.

KEY TERMS

accountability
authority
chain of command
channels of communication
contract

delegation
job loading
responsibility
reverse delegation

REVIEW QUESTIONS

Answer each question in complete sentences. Read each question carefully and make sure that you answer all parts of the question. Organize your answer using more than one paragraph when appropriate.

1. The ability to delegate has been called one of the hallmarks of a good manager. Explain its importance to management.
2. Discuss in detail the three aspects of delegation.
3. Why do some supervisors resist delegation? Why do some employees resist delegation?
4. What conditions are necessary for delegation?
5. Describe the steps in delegating tasks.
6. What is reverse delegation?
7. Discuss common mistakes in delegating.

ACTIVITIES AND APPLICATIONS

1. Discussion Questions
■ How can a supervisor delegate responsibility, yet retain it at the same time? How does this principle work out in practice? Give examples.
■ In what ways do the concepts of a performance standard system apply to the delegation process? How does delegation draw on communication and training skills?
■ With what management styles is delegation compatible? Why doesn't delegation work for a Theory X manager?
■ Why does delegation so often involve fear, and why does it require courage?
■ Have you ever been delegated a task while working? How did your supervisor handle it? How did you feel about it?

2. Group Activity: Delegation Role-Play
In a group of four, think of a situation in which a hospitality supervisor delegates a job to an employee. While two students role-play, the supervisor telling the employee about the delegated task, the other

two students can be observers. When the role-play is done, the observers can take a turn being supervisor and employee.

 ## ENDNOTES

1. Gerard M. Blair, "The Art of Delegation," www.see.ed.ac.uk/~gerard/Management/art5.html. Retrieved February 8, 2001.
2. Patrick Dunne, "Time to Delegate," *Management Today* (June 2003), p. 80.

Glossary

401(k) Plan A defined contribution that employees can elect to defer from their salary before taxes—some employers match these contributions—there are limits on the dollar amount that can be deferred each year.

Accountability A worker's obligation to a supervisor to carry out the responsibilities delegated and to produce the results expected.

Active listening Encouraging a speaker to continue talking by giving interested but neutral responses that show that you understand the speaker's meaning and feelings.

Adult learning theory A field of research that examines how adults learn.

Age Discrimination in Employment Act of 1967 (ADEA) An act that makes it unlawful to discriminate in compensation, terms, or conditions of employment based on a person's age. The ADEA applies to everyone 40 years of age or older.

Agenda A written statement of topics to be discussed at a meeting.

Alternative dispute resolution (ADR) Problem solving and grievance resolution approaches to address disputes.

Americans with Disabilities Act (ADA) An act that makes it unlawful to discriminate in employment matters against the estimated 43 million Americans who have a disability.

Anaphylactic shock An allergic reaction in which the throat swells up to the point that the victim cannot get air into the lungs. Without proper treatment, it can be fatal.

Appraisal See **Performance evaluation**.

Appraisal interview, appraisal review, evaluation interview An interview in which a supervisor and an employee discuss the supervisor's evaluation of the employee's performance.

Approaches to decision making **impulsive** (making an off-the-cuff decision), **indecisive** (never quite making up one's mind), **intuitive** (making decisions that feel right), **logical** (making a deliberate stepwise process to make decisions).

Authority Possessing the rights and powers needed to make the decisions and take the requisite actions to get a job done.

Authority, formal The authority granted by virtue of a person's position within an organization.

Authority, real or conferred The authority that employees grant a supervisor to make the necessary decisions and carry them out.

Autocratic Behaving in an authoritarian or domineering manner.

Base compensation The base amount for a particular job that may be incrementally increased.

Behavior modification Effecting behavioral change by providing positive reinforcement (reward, praise) for the behavior desired.

Big Brother/Big Sister training method See **Buddy system**.

Body language Expression of attitudes and feelings through body movements, positions, and gestures.

Boomerang management Reverting from management's point of view to the worker's point of view.

Brainstorming Generating ideas without considering their drawbacks, limitations, or consequences (typically, a group activity).

Buddy system Training method in which an old hand shows a new worker the ropes; also known as the **Big Brother/ Big Sister training method**.

Budget An operational plan for the income and expenditure of money for a given period.

Cafeteria plans An employee can "select" the parts of a plan that they want.

"Can do" factors An applicant's or employee's job knowledge, skills, and abilities.

Cardiopulmonary resuscitation (CPR) A procedure employed after cardiac arrest (when the heart stops beating) in which massage, drugs, and mouth-to-mouth resuscitation are used to restore breathing.

Carrot-and-stick motivation The use of promised rewards plus punishment to motivate performance.

Chain of command Lines along which responsibility and authority are delegated from top to bottom of an organization.

Channels of communication The organizational lines (corresponding to the chain of command) along which messages are passed from one level to another.

Civil Rights Act of 1964, Title VII An act that makes it unlawful to discriminate against applicants or employees with respect to recruiting, hiring, firing, promotions, or other employment-related activities, on the basis of race, color, religion, gender, or national origin.

Coach A person who trains others.

Coaching Individual, corrective, on-the-job training that is focused on improving performance.

Coaching style Within the managerial grid, a supervisory style that uses a lot of directive and supportive behaviors with an employee.

Cohesive team A team that joins together well, has defined norms, unity, respect, and trust among its members.

Collective bargaining Process by which a labor contract is negotiated.

Communication zones See **Personal space**; **Public distance**; **Social distance**.

Communications The sending and receiving of messages.

Compensation philosophies The approach a company takes toward the compensation of associates.

Competition When there is high concern for one's own interest—two different individuals/groups become rivals.

Compromise Concern for both one's own and the other parties' ideas or position, finding ways of agreeing (give and take) on positions.

Conceptual skill The ability to see the whole picture and the relationship of each part of the whole.

Conditions and limitations Factors such as rules, policies, specific requirements, and limiting factors (e.g., time) that may apply when a problem is being defined.

Conflict Discord, a state of disharmony of open or prolonged fighting, strife, or friction.

Conflict management The application of strategies to settle opposing ideas/goals.

Consensus General agreement within a group.

Consumer price index Measures the cost of living by averaging several elements over time.

Contingency plan An alternative plan for use in case the original plan does not work out.

Contract An agreement between two parties that is fully understood and accepted by both.

Control Built-in method for measuring performance or product against standards.

Controlling Measuring and evaluating results to goals and standards previously agreed upon, such as performance and quality standards, and taking corrective action when necessary to stay on course.

Coordinating Meshing the work of individuals, work groups, and departments to produce a smoothly running operation.

Counseling Occurs when a counselor meets with a client in a private and confidential setting to explore a difficulty the client is having, distress he or she may be experiencing, or perhaps his or her dissatisfaction or loss of a sense of direction or purpose.

Culture The socially transmitted behavior patterns, art, beliefs, institutions, and all other products of human work or thought characteristic of a community or population.

Culture bound Believing that your culture and value system are better than all others.

Decision A conscious choice among alternative courses of action directed toward a specific purpose.

Decision making Using a logical process to identify causes and solutions to problems or to make decisions.

Decision-making leave with pay The final step in a positive discipline system in which the employee is given a day off with pay to decide if he or she really wants to do the job well or would prefer to resign the position.

Decisiveness The ability to reach a firm conclusion.

Defined benefit plan and defined contribution plans Both plans have eligibility requirements—must work for the company for a specified period and must be full-time employees.

Dehiring Avoiding termination by making an employee want to leave, often by withdrawing work or suggesting that the person look elsewhere for a job.

Delegating style Within the managerial grid, a supervisory style that is low on directive and supportive behaviors because responsibility is being turned over to an employee.

Delegation Giving a portion of one's responsibility and authority to a subordinate.

Demographics Characteristics of a given area in terms of data about the people who live there.

Demotivator An emotion, environmental factor, or incident that reduces a person's motivation to perform well.

Direct recruiting On-the-scene recruiting where job seekers are, such as at schools and colleges.

Directing Assigning tasks, giving instructions, training, and guiding and controlling performance.

Directing style Within the managerial grid, a supervisory style that uses a lot of directive and few supportive behaviors with an employee.

Discipline (1) A condition or state of orderly conduct and obedience to rules, regulations, and procedures. (2) Action to enforce orderly conduct and obedience to rules, regulations, and procedures.

Dissatisfier A factor in a job environment that produces dissatisfaction, usually reducing motivation.

Diversity Physical and cultural dimensions that separate and distinguish individuals and groups: age, gender, physical abilities and qualities, ethnicity, race, sexual preference.

Doing the right things right To be both a leader and a manager; to be both effective and efficient.

Drug-Free Workplace Act of 1988 A federal law that requires most federal contractors and anyone who receives federal grants to provide a drug-free workplace.

Due process An employee's right of self-defense in a disciplinary process.

Economic person theory The belief that people work for money alone.

Employee assistance program (EAP) A counseling program available to employees to provide confidential and professional counseling and referral.

Employee handbook A written document given to employees that tells them what they need to know about company policies and procedures.

Employee Polygraph Protection Act of 1988 A federal law that prohibits the use of lie detectors in the screening of job applicants.

Employee referral program A program under which employees suggest to others that they apply for a job in their company. If a person referred gets a job, the employee often receives recompense.

Employee self-appraisal A procedure by which employees evaluate their own performance, usually as part of a performance appraisal process.

Employee stock ownership plans (ESOP) Plans based on the performance of the company, plus employees may purchase company stock at a discount.

Employee turnover The rate of employee separations in a company—usually expressed as a percentage.

Employment agencies Organizations that try to place persons into jobs. **Private agencies:** privately owned agencies that normally charge a fee when an applicant is placed. **Temporary agencies:** agencies that place temporary employees into businesses and charge by the hour. **Government agencies:** employment agencies run by the government.

Employment requisition form A standard form used by departments to obtain approval to fill positions and to notify the recruiter that a position needs, or will need, to be filled.

Empowering To give employees additional responsibility and authority to do their jobs.

Environmental sexual harassment A type of sexual harassment in which comments or innuendos of a sexual nature, or physical contact, are considered a violation when they interfere with an employee's work performance or create an intimidating, hostile, or offensive working environment.

Equal employment opportunity (EEO) The legal requirement that all people be treated equally in all aspects of employment, regardless of race, creed, color, national origin, age, gender, or disability unrelated to the job.

Equal Employment Opportunity Commission (EEOC) A federal office responsible for enforcing the employment-related provisions of the Civil Rights Act of 1964 as well as other EEO laws.

Equal Pay Act of 1963 A law that requires equal pay and benefits for men and women working in jobs requiring substantially equal skills, effort, and responsibilities under similar working conditions.

Evaluating See **Controlling**.

Evaluation form A form on which employee performance during a given period is rated.

Exempt employees Employees, typically managerial personnel, who are not covered by the wage and hour laws and therefore do not earn overtime pay. To be considered an exempt employee, the following conditions must be met: The employee spends 50 percent or more of his time managing, supervises two or more employees, and is paid $250 or more per week.

External recruiting Looking for job applicants outside the operation.

Fact-finding The process of collecting all the facts about a certain situation.

Fair Labor Standards Act (FLSA) The law that covers wages and salaries that applies to employers with two or more employees.

Family and Medical Leave Act of 1993 An act that allows employees to take an unpaid leave of absence from work for up to 12 weeks per year for the birth or adoption of a child or a serious health condition of the employee or his or her spouse, child, or parent.

Feedback Giving information about the performance of an individual or group back to them during or after performing a task or job.

Fee-for-service Traditional plans offered by insurance companies that act as an intermediary between the patient and the healthcare provider—an example is Blue Cross, which has a plan that pays for 80 percent of most medical expenses.

First aid Emergency treatment given before regular medical services can be provided.

First-line supervisor A supervisor who manages hourly employees.

Forecasting Predicting what will happen in the future on the basis of data from the past and present.

Formal authority See **Authority, formal**.

Formal group Groups established by a company.

Formal leader The person in charge based on the organization chart.

Formally appointed team A team that has a formally appointed leader who may have more influence and decision-making authority than other team members.

Formative evaluation An ongoing form of evaluation that uses observation, interviews, and surveys to monitor training.

Generation X The group of Americans from age 18 through 40, born between the late 1960s and 1980.

Generation Y The group of Americans born in the 1980s and 1990s.

Grievance procedures A formal company procedure that employees can follow when they feel they have been treated unfairly by management.

Group decision making A process in which a group of people work together to come to a decision.

Halo effect The tendency to extend the perception of a single outstanding personality trait to a perception of the entire personality.

Harassment Intimidating, hostile, or offensive behavior toward someone, or the creation of an intimidating, hostile, or offensive environment for someone based on the person's national origin, race, color, religion, gender, disability, or age.

Hazard communication standard A regulation issued by the Occupational Safety and Health Administration that gives employees the right to know what hazardous chemicals they are working with, the risks or hazards, and what they can do to limit the risk.

Health insurance plans A voluntary benefit, meaning that employers are not obliged to offer all employees a health insurance plan.

Health Maintenance Organizations (HMOs) Offer full-service medical services to employees and their families; the most cost-effective and popular.

Hierarchy of needs A theory proposed by Abraham Maslow that places human needs in a hierarchy or pyramid. As one's needs at the bottom of the pyramid are met, higher-level needs are encountered on several levels up through the top of the pyramid.

Hourly workers Employees paid on an hourly basis who are covered by federal and state wage and hour laws and therefore guaranteed a minimum wage.

Human immunodeficiency virus (HIV) The virus that causes acquired immunodeficiency syndrome (AIDS).

Human relations theory A theory that states that satisfying the needs of workers is the key to productivity.

Human skill The ability to manage people through respect for them as individuals, sensitivity to their needs and feelings, self-awareness, and good person-to-person relationships.

Humanistic management A blend of scientific, human relations, and participative management practices adapted to the needs of the situation, the workers, and the supervisor's leadership style.

Hygiene factors Factors in the job environment that produce job satisfaction or dissatisfaction but do not motivate performance.

Immigration Reform and Control Act (IRCA) A federal law that requires employers to verify the identity and employment eligibility of all applicants and prohibits discrimination in hiring or firing due to a person's national origin or citizenship status.

Inclusion To include, to make a person feel welcome.

Informal groups Groups that form naturally in the workplace.

Informal leader The person who, by virtue of having the support of the employees, is in charge.

Informally appointed team A team that evolves on its own.

Internal labor market Giving internal candidates opportunities, for a brief period, to apply for a position ahead of external advertising.

Internal recruiting Searching for job applicants from within an operation.

Interpersonal communication The sending and receiving of messages between people.

Interviewing Conversation with the purpose of obtaining information, often used in screening job applicants.

JIT Job instruction training.

Job A specific group of tasks prescribed as a unit of work.

Job analysis Determination of the content of a given job by breaking it down into units (work sequences) and identifying the tasks that make up each unit.

Job description A written statement of the duties performed and responsibilities for a given position, and used to provide opportunity for achievement, recognition, learning, and growth.

Job evaluation The process of examining the responsibilities and difficulties of a series of jobs to determine which are worth the most and should therefore be paid more.

Job instruction For every detail of a given job in a given enterprise, instruction in what to do and how to do it.

Job instruction training (JIT) A four-step method of training people in what to do and how to do it on a given job in a given operation.

Job loading Adding more work to a job without increasing interest, challenge, or reward.

Job posting A policy of making employees aware of available positions within a company.

Job Service Center An office of the U.S. Employment Service.

Job setting The conditions under which a job is to be done, such as physical conditions and contact with others.

Job skills approach A method of assessing the various skills required to do a particular job.

Job specification A list of the qualifications needed to perform a given job.

Job title The name of a job, such as cook or housekeeper.

Just-cause termination Employee termination based on the commission of an offense that affected detrimentally the specific work done or an operation as a whole.

Labor contract The written conditions of employment that are negotiated between management and a union.

Labor market In a given area, the workers who are looking for jobs (the labor supply) and the jobs that are available (the demand for labor).

LBWA Leadership by walking around; spending a significant part of your day talking to your employees, your guests, your peers while listening, coaching, and trouble-shooting.

Leader A person in command who people follow voluntarily.

Leadership Direction and control of the work of others through the ability to elicit voluntary compliance.

Leadership style The pattern of interaction that a manager uses in directing subordinates.

Leading Guiding and interacting with employees regarding getting certain goals and plans accomplished; involves many skills, such as communicating, motivating, delegating, and instructing.

Learning The acquisition of knowledge or skill.

Level of performance Employee performance measured against a performance standard. **Optimistic level:** superior performance, near-perfection. **Realistic level:** competent performance. **Minimum level:** marginal performance, below which a worker should be terminated.

Line functions The personnel directly involved in producing goods and services.

Listening Paying complete attention to what people have to say, hearing them out, staying interested but neutral. See also **Active listening**.

Maintenance factors See **Hygiene factors**.

Management by example Managing people at work by setting a good example—by giving 100 percent of your time, effort, and enthusiasm to your own job.

Manager One who directs and controls an assigned segment of the work in an enterprise.

Managerial skills The three types of skills that a manager needs: technical, human, and conceptual. See the individual skills.

Mass communication Messages sent out to many people through such media as newspapers, magazines, books, radio, and television.

Material safety data sheet (MSDS) An information sheet put out by the manufacturer of a hazardous product that explains what a product is, why it is hazardous, and how it can be used safely.

Mellennials Those born between 1980 and 2000.

Mentor An experienced and proficient person who acts as a leader, role model, and teacher to those less experienced and less skilled.

Merit raise A raise given to an employee based on how well the employee has done his or her job.

Morale Group spirit with respect to getting a job done.

Motivation The why of behavior; the energizer that makes people behave as they do.

Motivator Anything that triggers a person's inner motivation to perform. In Herzberg's theory, motivators are factors within a job that provide satisfaction and that motivate a person to superior effort and performance.

Negative discipline Maintaining discipline through fear and punishment, with progressively severe penalties for rule violations.

Negligent hiring The failure of an employer to take reasonable and appropriate safeguards when hiring employees to make sure that they are not the type to harm guests or other workers.

Nonexempt employees Employees who are paid by the hour and are not exempt from federal and state wage and hour laws. Also called hourly employees.

Nonverbal communication Communication without words, as with signs, gestures, facial expressions, or body language.

Obstacle thinkers Those who focus on why a situation is impossible and retreat from it.

Occupational Safety and Health Administration (OSHA) A federal agency created to assure safe and healthful working conditions and to preserve the nation's human resources.

Open (or two-way) communication The free movement of messages back and forth between supervisor and worker and up the channels of communication as well as down.

Opportunity thinkers Those who concentrate on constructive ways to deal with a challenging situation.

Organization chart A diagram showing a company's levels of management and lines by which authority and responsibility are transmitted.

Organizational communication The sending of messages from the top of an organization down—usually the same message to everyone.

Organizing Putting together the money, personnel, equipment, materials, and methods for maximum efficiency to meet an enterprise's goals.

Orientation A new worker's introduction to a job.

Overgeneralization In interviewing and evaluation, translation of a single trait or piece of information about a person into an overall impression of that person.

Participative leadership A system that includes workers in making decisions that concern them.

Patterned interview A highly structured interview in which the interviewer uses a predetermined list of questions to ask each applicant.

Pay incentives Programs designed to reward employees for good performance.

Pension plan Plans that accrue pretax income that employees set aside, and in some cases, employers make a matching contribution.

Performance dimensions or categories The dimensions of job performance chosen to be evaluated, such as attendance and guest relations.

Performance evaluation, performance appraisal, performance review Periodic review and assessment of an employee's performance during a given period.

Performance standard Describes the what and how of a job, and explains what an employee is to do, how it is to be done, and to what extent.

Performance standard system A system of managing people using performance standards to describe job content, train personnel, and evaluate performance.

Personal space The area within close proximity of a person that "belongs" to the person and should not be invaded (the space varies according to culture).

Planning Looking ahead to chart the best course of future action. See also **Strategic planning**.

Points factor method Key jobs are examined by taking important factors into consideration.

Position Duties and responsibilities performed by an employee.

Positive discipline A punishment-free formula for disciplinary action that replaces penalties with reminders and features a decision-making leave with pay.

Positive reinforcement Providing positive consequences (praise, rewards) for desired behavior.

Power The capacity to influence the behavior of others.

Pregnancy Discrimination Act of 1978 An act that makes it unlawful to discriminate against a woman on the basis of pregnancy, childbirth, or related medical conditions.

Pretest Testing an experienced worker's job performance before training.

Primary dimensions of diversity Cultural and physical dimensions of individuals or groups that cannot be changed, such as age, gender, and race.

Problem solving Using a logical process to identify causes and solutions to problems or to make decisions.

Productivity How efficiently an operation converts an input (e.g., food) into an output (e.g., a meal).

Productivity standards A definition of the acceptable quantity of work that an employee is expected to do (e.g., the number of rooms that can be cleaned in 60 minutes).

Progressive discipline A multistage formula for disciplinary action.

Project teams Teams that are brought together for the completion of a project.

Projection Investing another person with one's own qualities.

Promoting from within A policy in which it is preferable to promote existing employees rather than filling the position with an outsider.

Public distance Often defined as from 7 to 25 feet away from a person—too far for giving directions or conversing.

"Quid pro quo" sexual harassment A type of sexual harassment: submission to or rejection of a sexual favor used as the basis for employment decisions regarding an employee.

Rating system A system, usually a scale, for evaluating actual performance in relation to expected performance or the performance of others.

"Reading people" The emotional awareness ability to read people by identifying their emotions.

Real authority See **Authority, real**.

Reasonable accommodation Any change or adjustment to a job or work environment that will enable someone with a disability to perform essential job functions.

Recruiting Actively looking for people to fill jobs. **Direct recruiting**: going where the job seekers are, such as colleges, to recruit. **Internal recruiting**: looking for people within a company to fill jobs. **External recruiting**: looking for people outside a company to fill jobs.

Representing Representing an organization to customers and other people outside an enterprise.

Resistance to change A reaction by workers to changes in their work environment that may be accompanied by feelings of anxiety, insecurity, or loss.

Responsibility The duties and activities assigned to a given job or person, along with an obligation to carry them out.

Retention The extent to which employees are retained by a company, thus reducing turnover.

Retraining Additional training given to trained workers for improving performance or dealing with something new.

Reverse delegation A situation in which you delegate a job to an employee and he or she tries to give it back to you.

Reward and punishment A method of motivating performance by giving rewards for good performance and punishing for poor performance.

Risk A degree of uncertainty about what will happen in the future.

Role model A person who serves as an example for the behavior of others.

Safety committee A committee that meets periodically to discuss safety matters and to perform other functions related to workplace safety, such as inspecting a facility and overseeing safety training.

Safety program A plan, consisting of elements such as safety rules and employee training, that attempts to keep a workplace safe.

Scheduling Determining how many people are needed when, and assigning days and hours of work accordingly.

Scientific management Standardization of work procedures, tools, and conditions of work.

Secondary dimensions of diversity Cultural and physical dimensions of individuals or groups that can be changed, such as occupation, education, and income.

Security program A plan to protect company assets and people by preventing theft and other unlawful acts.

Self-actualization The desire to fulfill one's own potential.

Sexual harassment Unwelcome advances, requests for sexual favors, and other verbal or physical conduct of a sexual nature when compliance with any of these acts is a condition of employment, or when comments or physical contact create an intimidating, hostile, or offensive working environment.

Single-use plan A plan developed for a single occasion or purpose.

Situational leadership Adaptation of leadership style to the needs of a situation.

Small-group communication Communication that takes place when two or more group members attempt to influence one another, as in a meeting.

Social distance Often defined as from 4 to 7 feet away from a person—suitable for communication between boss and subordinate.

Social man (person) theory The idea that fulfillment of social needs is more important than money in motivating people. See **Human relations theory**.

Social Security Gives financial support to retirees and their survivors if they have paid into the system for 10 years or more.

Span of control The number of employees that a manager supervises directly.

Staff functions Personnel who are not directly involved in producing goods and services but advise those who do, such as human resource and training directors.

Staffing Determining personnel needs and recruiting, evaluating, selecting, hiring, orienting, training, and scheduling employees.

Standing plan An established routine, formula, or set of procedures used in a recurring situation.

Stereotype A belief that a person will have characteristics generally attributed to members of a particular racial or social group simply because he or she is a member of that group.

Strategic planning Long-range planning to set organizational goals, objectives, and policies and to determine strategies, tactics, and programs for achieving them.

Strike A work stoppage due to a labor dispute.

Summative evaluation A form of evaluation that measures the results of training after a program has been completed.

Supervisor A person who manages employees who are making products or performing services.

Supporting style Within the managerial grid, a supervisory style marked by highly supportive behaviors with an employee.

Symbols Words, images, or gestures used to communicate messages.

Synergy The actions of two or more people to achieve outcomes that each is individually incapable of achieving.

Task In job analysis, a procedural step in a unit of work.

Teaching methods Ways in which teachers and trainers convey information to learners.

Team A special kind of group.

Team morale The extent to which a team has confidence, cheerfulness, and willingness to perform assigned tasks.

Team players Individuals that participate in a collective effort and cooperation to get the job done effectively.

Teamwork The cooperative actions that a team performs.

Technical skill The ability to perform the tasks of the people supervised.

Theory X The managerial assumption that people dislike and avoid work, prefer to be led, avoid responsibility, lack ambition, want security, and must be coerced, controlled, directed, and threatened with punishment to get them to do their work. A **Theory X manager** is one whose direction of people is based on these assumptions.

Theory Y The hypothesis that (1) work is as natural as play or rest; (2) people will work of their own accord toward objectives to which they feel committed, especially those that fulfill personal needs of self-respect, independence, achievement, recognition, status, and growth; and (3) arranging work to meet such needs will do away with the need for coercion and threat. A **Theory Y manager** is one who holds and practices this view of employee motivation.

Third-party sexual harassment A type of sexual harassment that involves a customer or client and an employee.

Timing Selecting that time when taking action will be most effective; making a decision at the moment it is most needed.

Total quality management (TQM) A process of total organizational involvement in improving all aspects of the quality of a product or service.

Training Teaching people how to do their jobs; job instruction.

Training objective A trainer's goal: a statement, in performance standard terms, of the behavior that shows when training is complete.

Training plan A detailed plan for carrying out employee training for a unit of work.

Transactional leadership Leadership that motivates workers by appealing to their self-interest.

Transformational leadership Leadership that motivates workers by appealing to their higher-order needs, such as providing workers with meaningful, interesting, and challenging jobs, and acting as a coach and mentor.

Truth in hiring Telling an applicant the entire story about a job, including its drawbacks.

Two-way communication In communication, when messages move freely back and forth from one person to another.

Unemployment insurance Gives financial support to employees who are laid off for reasons they cannot control.

Uniform discipline system A system of specific penalties for each violation of each company rule, to be applied uniformly throughout a company.

Union An organization that employees have designated to deal with their employer concerning conditions of employment, such as wages, benefits, and hours of work.

Union steward An employee designated by a union to represent and advise employees of their rights as well as to check on contract compliance. Also called a shop steward.

Unit of work Any one of several work sequences that together form the content of a given job.

Unity of command The organizational principle that each person should have only one boss.

Unpaid leave Under the Family Medical Leave Act of 1993 (FLMA) employees may take unpaid leave for up to 12 weeks under certain conditions.

Wage and salary survey A survey to assess what comparable companies are paying employees.

"Will do" factors An applicant's or employee's willingness, desire, and attitude toward performing a job.

Win/win problem solving A method of solving problems in which supervisor and worker discuss a problem together and arrive at a mutually acceptable solution.

Work climate The level of morale within a workplace.

Work rules Rules for employees that govern their behavior when working.

Work simplification The reduction of repetitive tasks to the fewest possible motions, requiring the least expenditure of time and energy.

Work supervisor A supervisor who takes part in the work task itself in addition to supervising.

Worker's compensation Insurance paid by the employer that gives medical care, income continuation, and rehabilitation expenses for people who sustain job-related injuries or sickness.

Index